SOVIET
SECRET PROJECTS

BOMBERS SINCE 1945

SOVIET SECRET PROJECTS

BOMBERS SINCE 1945

TONY BUTTLER & YEFIM GORDON

MIDLAND
An imprint of
Ian Allan Publishing

Soviet Secret Projects – Bombers since 1945

ISBN 1 85780 194 6

First published in 2004 by
Midland Publishing
4 Watling Drive, Hinckley, LE10 3EY, England
Tel: 01455 254 490 Fax: 01455 254 495

Midland Publishing is an imprint of
Ian Allan Publishing Ltd.

Worldwide distribution (except North America):
Midland Counties Publications
4 Watling Drive, Hinckley, LE10 3EY, England
Tel: 01455 233 747 Fax: 01455 233 737
E-mail: midlandbooks@compuserve.com
www.midlandcountiessuperstore.com

North America trade distribution by:
Specialty Press Publishers & Wholesalers Inc.
39966 Grand Avenue
North Branch, MN 55056, USA
Tel: 651 277 1400 Fax: 651 277 1203
Toll free telephone: 800 895 4585
www.specialtypress.com

Layout by Sue Bushell.

Printed by Ian Allan Printing Limited
Riverdene Business Park, Molesey Road
Hersham, Surrey, KT12 4RG, England.

Photograph on half-title page:
The Tsybin NM-1 model test aircraft.

Photograph on title page:
The prototype Ilyushin Il-54.

Contents

	Introduction and Acknowledgements	6
Chapter One	**Last of the Piston Bombers**	7
Chapter Two	**The First Jet Bombers**	16
Chapter Three	**The First Heavyweights**	32
Chapter Four	**The First Supersonic Bombers**	52
Chapter Five	**Composite Bombers and Reconnaissance Aircraft**	70
Chapter Six	**Nuclear Power and Flying Wings**	78
Chapter Seven	**Ground Attack Aircraft**	86
Chapter Eight	**Tactical Strike Aircraft**	102
Chapter Nine	**Maritime Patrol**	117
Chapter Ten	**Ultimate Performance**	132
Chapter Eleven	***Backfire* and *Blackjack***	144
	Soviet Secret Bomber Colour Chronology	161
	Glossary	167
Appendix One	**Soviet and Russian Bomber Project Summary**	168
	Bibliography and Source Notes	174
	Index	175

Introduction and Acknowledgements

Since 1990, following the break-up of the former Soviet Union, a great deal of new information has been made available describing the Warsaw Pact's post-war military aircraft. This includes plenty of drawings of unbuilt bomber projects and a number of Russian-language publications have now published a good selection of these. Quite a few English-language books and journals have also published drawings and material relating to selected service aircraft but it is believed that this is the first work to be produced in English that gives a full coverage of Soviet post-war bomber projects. There are still many gaps to be filled and, on occasion, different original sources have given conflicting information (particularly in regard to dates), but this is undoubtedly the most complete study to be produced so far.

In putting together these bomber projects in a logical sequence we have also been able to tell the story of Soviet jet bomber design in general. The expected heavy strategic types are here, but the term 'bomber' embraces other categories such as ground attack, strike, maritime patrol and anti-submarine, and space has been devoted to these as well. Several of the Soviet Union's strategic jet bomber and flying boat designs make up some of the most exotic and astonishing aircraft projects one could imagine. With few exceptions, Soviet military aircraft were never given names, but for identification purposes the Western powers eventually introduced a codename for known types and these are included for completeness.

Working though the text one was surprised to find just how many projects and programmes were cancelled before they entered production or even flew – one always associates the British aviation scene with cancellations but in fact this was a phenomenon that affected most countries. It has not been possible to dip into Soviet politics as much as one would have liked, but the rivalries between the design bureaux and the 'difficulties' between important personalities was just as strong in the Communist Soviet Union as in democracies like Britain and the USA, in fact probably more so. Nevertheless, the reader will appreciate that the argument for and against manned or unmanned strategic systems was just as powerful for the Warsaw Pact during the 1950s as it was for the Western nations.

For the English half of the authorship, *Soviet Secret Projects: Bombers since 1945* made a fascinating contrast to his earlier book *British Secret Projects: Jet Bombers since 1949*. In some areas British and Soviet areas of research and design came very close together, in others they were quite different, and it is worth mentioning the USSR's constant desire to find ever greater range for its bombers or, an associated factor, the great attention it gave to cruise missiles and their carrier aircraft. We hope the reader finds learning about these designs as fascinating as it was for us to research and write about them.

Acknowledgements
Unless stated, all of the illustrations in this book come from the Yefim Gordon collection. Special thanks must be given to Nigel Eastaway of the Russian Aviation Research Trust for his considerable help in finding extra information and filling gaps, to George Cox for allowing us to take photographs of the models in his collection (some of which were created by Alex Panchenko), and to Joe Cherrie and John Hall for kindly offering to make models that filled gaps in the illustration coverage. Sincere thanks also to Vladimir Rigmant (Tupolev archivist), Jens Baganz for his drawings, Peter Kostelnik for translations, Tophe Meunier, Thomas Mueller, and the authors of numerous articles published in *The Bulletin of the Russian Aviation Research Trust*. Finally we must thank the team at Midland for their support and Keith Woodcock for his front cover painting.

Tony Buttler, Bretforton
and Yefim Gordon, Moscow, October 2004

Chapter One

Last of the Piston Bombers

A production Tupolev Tu-4.

During the Second World War the Soviet Union concentrated on the development of fighters, attack aircraft and light/medium bombers for its airborne forces. Following the invasion by German forces in June 1941 the acquisition of these types was critical for the country's defence during the Great Patriotic War of 1941 to 1945. As a result the Petlyakov Pe-8 became the only four-engined heavy bomber produced in Russia during the war and by the end of the conflict this aircraft was itself bordering on obsolescence. Relatively little effort went into producing another heavy bomber and so, as a consequence of the low priority given to this category, the development of large bombers fell behind the standards of America and Great Britain. In the autumn of 1943 however, the first moves were made to correct this situation and in 1944 a resolution was passed by the GKO (State Committee for Defence) ordering the development of new four-engined high-speed high-altitude long-range heavy bombers that were intended to strike deep into enemy territory. This resulted in a competition sometimes termed the 'Flying Fortress Contest'.

High-Performance Bombers

Soviet design teams were formed into what were called OKBs, a term that actually described an Experimental Construction Bureau but which really embraced Design, the primary objective. Some OKBs had sufficient facilities to build prototypes of their new designs but none of them undertook any production, which instead was passed to a GAZ (State Aviation Factory) nominated and arranged for by MAP (the Ministry of Aviation Industry). From 1939 MAP also administered the OKBs themselves. The latter also hardly ever became involved in research, which was the task of centralised establishments such as the TsAGI (the Central State Aerodynamic and Hydrodynamic Institute) and TsIAM (Central Institute of Aviation Motors). As stated, the OKB's real purpose was simply to design aeroplanes. The following projects became involved in the new heavy bomber requirement.

Ilyushin Il-14

Although apparently included in the heavy bomber competition at some stage, the Il-14 was really a high-speed type intended to replace the bureau's wartime Il-4. Work began in spring 1944 and the resulting layout was quite unorthodox with a powerplant of four Mikulin AM-43 engines, with direct fuel injection, mounted in tandem pairs at either end of the centreplane (this particular engine did pass its state acceptance trials but did not enter production). In each nacelle the forward engine drove a tractor propeller, the rear unit a pusher propeller and the main gears of a tricycle undercarriage were fitted between the two inside the nacelle. Both main and nose legs had twin wheels and a fourth retractable landing gear strut was housed in the rear fuselage to prevent the rearmost propellers from striking the ground on rotation. Ilyushin approved the project on 12th July 1944.

Initially it was thought that the Il-14's high speed would remove the need for defensive guns but, after assessing the project during July, the Air Force's representatives requested the addition of both offensive and defensive weapons. A redesigned Il-14 retained the basic layout and its unusual powerplant but introduced a fixed forward-firing 23mm nose cannon and two movable 20mm cannon in dorsal and ventral mounts behind the wings; the extra guns meant that the crew was increased from three to four and Mikulin AM-43NV engines were now fitted. The normal take-off

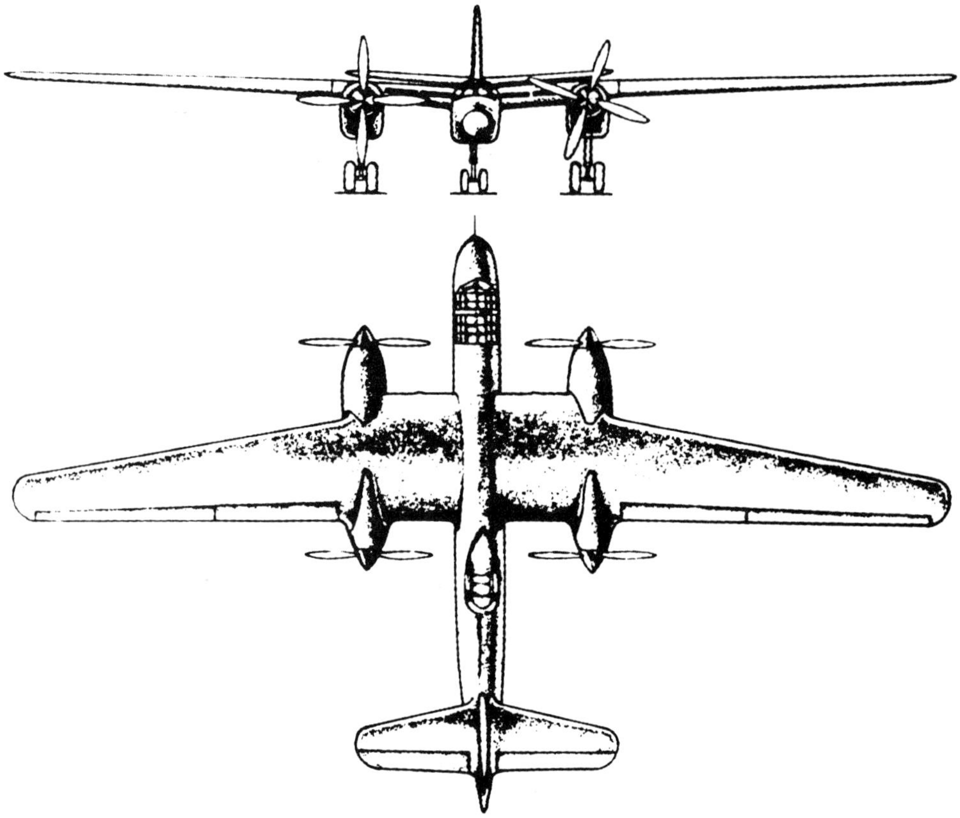

The Ilyushin Il-14 shown after the addition of offensive and defensive guns (mid-1945).

Model of one version of the DVB-302 (1944/45).

weight for the original unarmed bomber had been 44,092 lb (20,000kg); adding the guns increased this figure but it was still much lower than that given for the Myasishchev and Tupolev designs (around 74,956 lb [34,000kg]), which indicates how much smaller the Il-14 would have been.

The Il-14's service ceiling was estimated to be 41,010ft (12,500m) and range with a 4,409 lb (2,000kg) warload 1,554 miles (2,500km). Sufficient progress was made to see a full scale mock-up completed and work start on a prototype in 1945. It seems likely that this would have been fitted with 2,500hp (1,864kW) M-45 engines but the machine was never completed. In its unfinished state it was examined by a State commission who requested that the single-gun rearward defensive turrets should now be twin mounts. Eventually the project was abandoned and the Il-14 designation was later re-used for an airliner.

Myasishchev DVB-202 (VM-22) and DVB-302 (VM-23)

The preliminary designs of the Myasishchev DVB-202/VM-22 proposal (DVB = long-range, high-altitude bomber) had been presented to the NKAP (the People's Commissariat for Heavy Industry) by the end of December 1945. The initial VM-22 would have turbo-supercharged M-71FTK engines while an alternative VM-22a had ASh-72TKM units (and a different defensive gun arrangement with remotely-controlled turrets). A high wing position would allow space for a large bomb bay, the tapered wing had two spars and a trailing edge with Fowler-type flaps while the fuselage was to be built in five sections. Up to 24,250 lb (11,000kg) of fuel would be carried in both wing and fuselage tanks, the former mounted between the spars and the latter above the bomb bay. Nine crew were carried, a tricycle nosewheel undercarriage was employed and an alternative study examined a naval variant with ACh-31 diesel engines. Estimated service ceiling was 41,010ft (12,500m) and range with 11,023 lb (5,000kg) of bombs 3,729 miles (6,000km).

The DVB-202 was not taken up but during 1945 the OKB continued its studies in the form of the DVB-302/VM-23 which had a revised forward cabin and crew layout; the tail was also modified. This new design also featured remotely-controlled gun turrets or barbettes. Range with a 11,023 lb (5,000kg) warload was 2,486 miles (4,000km) and the ceiling was unchanged from the DVB-202.

Nyezval Bomber Projects

The task given to OKB-22 was to undertake a full and thorough update of the Pe-8 and the result, completed in 1944, described an aeroplane offering a substantial potential improvement in performance. However, that same year the OKB also began an all-new design based on, and possessing features and a performance similar to, the B-29 with the turbo-supercharged Shvetsov ASh-72TK as the preferred engine choice. Estimated time to 16,404ft (5,000m) was eleven minutes, service ceiling 36,089ft (11,000m) and range with an 17,637 lb (8,000kg) load 3,108 miles (5,000km). Between eight and ten crew would have been carried.

Tupolev 'Aircraft 64'

Work on this bomber first began in September 1943 and right from the start it was to have a pressurised cabin. The preliminary work saw the examination of several dozen alternative configurations before detail design began in May 1944 on the most favoured choice. The

Myasishchev DVB-202 (1944/45). The additional sideview shows the DVB-302.

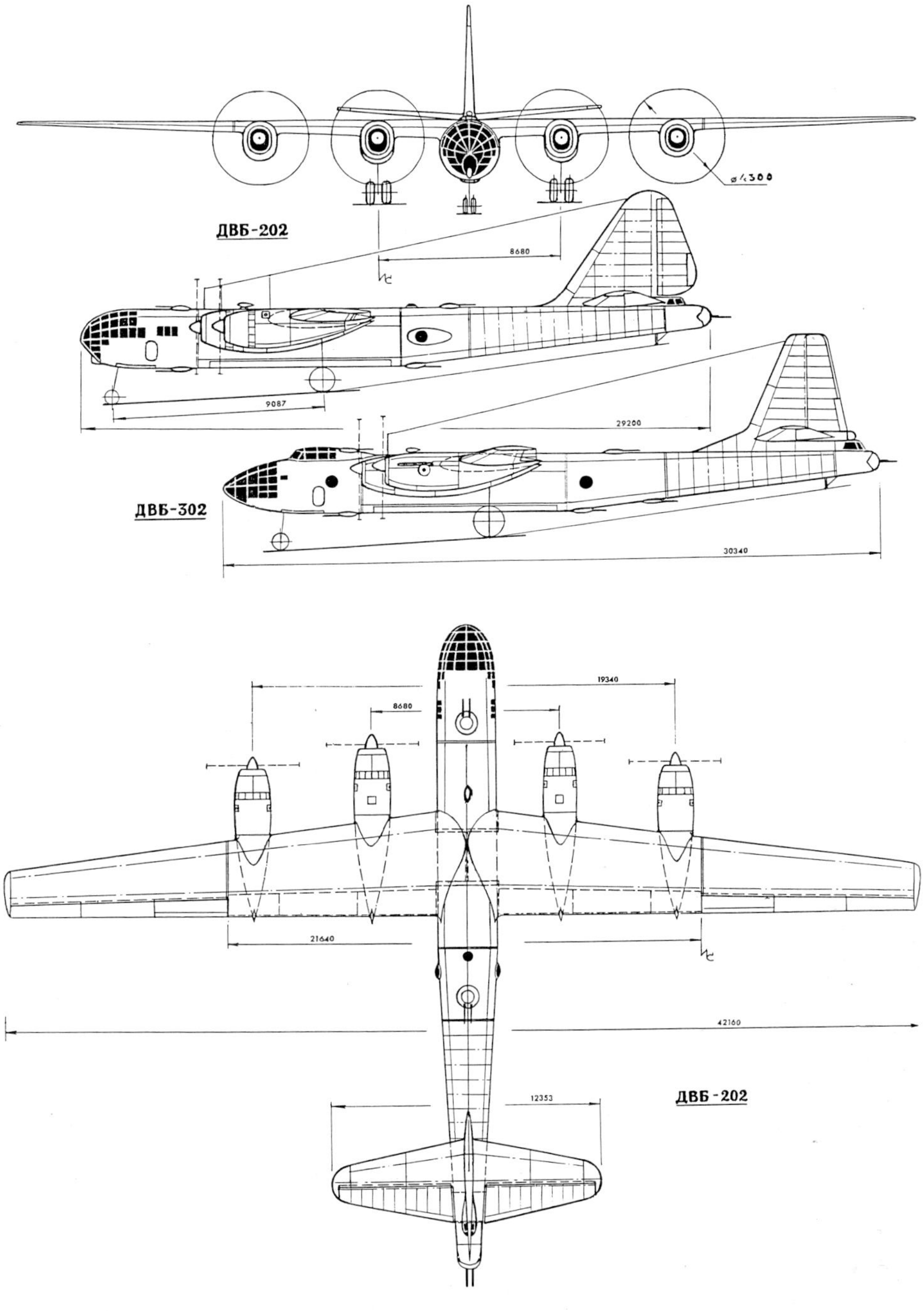

basic objectives were a maximum 39,683 lb (18,000kg) of bombs, including loads of two 11,023 lb (5,000kg) or 5,511 lb (2,500kg) bombs or a mix of smaller weapons, a speed of 311mph (500km/h) at 32,808ft (10,000m) and, depending on the bomb load, a range of up to 3,729 miles (6,000km). This project's length was 98ft 4in (29.975m), wing span 137ft 9½in (42m), normal take-off weight with 11,023 lb (5,000kg) of bombs 77,160 lb (35,000kg), maximum speed 373mph (600km/h) and cruise 249mph (400km/h), and ceiling 39,370ft (12,000m). There were several suitable engines – three liquid-cooled units comprising the 2,000hp (1,491kW) Mikulin AM-42TK or 2,300hp (1,715kW) AM-43TK-300B and the 1,900hp (1,417kW) Charomsky ACh-30BF diesel, the 1,900hp Shvetsov ASh-83FN radial or the massive liquid-cooled 2,500hp (1,864kW) Dobrynin/Skubachevsky M-250; performance calculations and project studies were completed for an installation of each engine type.

There were two bomb bays, forward and rearward of the wing centre section, and the aircraft had a tricycle undercarriage. Defensive armament would comprise four twin Nudel'man/Suranov NS-23 23mm cannon turrets plus twin NS-23s or a single Berezin B-20 20mm cannon in the end of the fuselage. The all-metal aircraft would be built using well-established manufacturing techniques and materials. The fuselage employed a monocoque structure with thick stressed skins and the wing, which had two spars, received several devices to help increase its lift for take-off and landing; the empennage would have twin fins. However, the OKB realised that an aircraft built to the specification within the required dimensions would actually be about twice as heavy as intended, a situation that would create numerous problems, not least with the structure.

By mid-1944 work on a modified and slightly smaller 'Aircraft 64' had begun which had a lower projected altitude than the earlier variant (ceiling 36,089ft [11,000m]). The VVS (Soviet Air Force) specification was finally issued in August which also requested that the '64' should be capable of day and night photo-reconnaissance and be suitable for adaptation as a transport with sufficient capacity to take certain types of vehicle, including T-60 light tanks, or seventy troops. The 'Aircraft 64' defence now comprised the four two-gun turrets with either 20mm or 23mm cannon plus one 45mm or 57mm cannon in a tail turret. The chief designer for all the variants of the '64' was D G Markov.

The evidence suggests that Myasishchev's efforts may have represented something of a cosmetic exercise because on 24th May 1945 Andrei Tupolev wrote to Aleksey Shakhurin to discuss the question of 'series production of the heavy bomber'. Shakhurin had been made the People's Commissar of the Aviation Industry, the head of NKAP, in January 1940. Tupolev noted that it would be necessary to use a factory that was a specialist in the field and he quoted GAZ-22 as the most suitable, both from the experience point of view and its available capacity. The factory was currently producing the Petlyakov Pe-2 twin-engined bomber and Tupolev stated that Pe-2 work should cease, that the factory should be fully prepared to build the Pe-8 in its final variant with ASh-82 engines and that its entire experimental production facilities should now be switched to heavy aircraft construction.

He added 'the comrade Nyezval's OKB and the OKB which previously developed the Pe-2 in Kazan (Myasishchev) are to be amalgamated with that of A N Tupolev'. They would then support the further production of the Pe-8 and modify it to use new alternative powerplants, the most suitable, in Tupolev's opinion, being

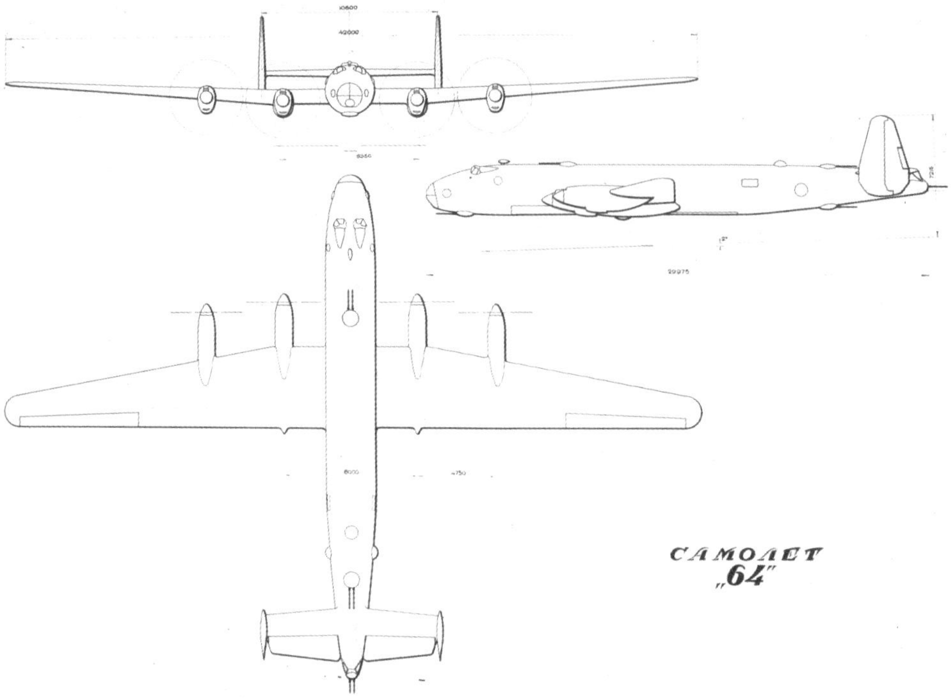

The Tupolev 'Aircraft 64' as first proposed with twin pilot canopies (1943/44).

Early wind tunnel models of the Tupolev 'Aircraft 64', one of which shows a single fin.

the AM-43 with TK-300 turbo-superchargers. In addition to improving the Pe-8's flight performance, 'it would also allow us to determine the suitability of the power unit for the newly developed '64' four-engined aircraft'.

The 'Aircraft 64's preliminary design was finished in August 1944 and a full-scale mock-up was completed in September; however, there were criticisms of the mock-up, in particular because it lacked a bomb-aiming radar that gave 360° coverage. In February 1945 an updated VVS specification was raised which added an extra crew number (making ten in all) who was to operate the new onboard radar and the mock-up was finally approved on 27th April. The definitive aircraft was now expected to be powered by four Mikulin AM-43s or AM-46s, fitted with TK-300 turbo-superchargers, and a maximum of 31,305 lb (14,200kg) of fuel would be carried. Maximum range with a 11,023 lb (5,000kg) load would be 3,108 miles (5,000km), with a 39,683 lb (18,000kg) load 4,040 miles (6,500km) and service ceiling 36,089ft (11,000m).

The airframe and power units (except for the turbo-superchargers) were not expected to present any problems but the modern equipment needed, such as radio and navigation aids and the technology for the remote control operation of the gun turrets, was a different matter – the component supply factories sim-

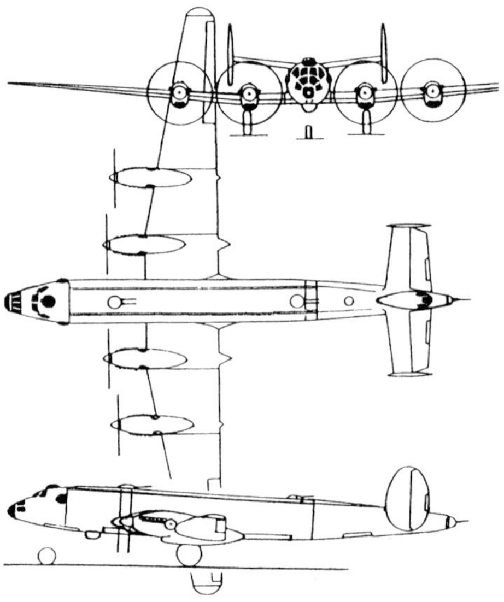

Tupolev 'Aircraft 64' in its final form which incorporated B-29 experience (1946).
Russian Aviation Research Trust

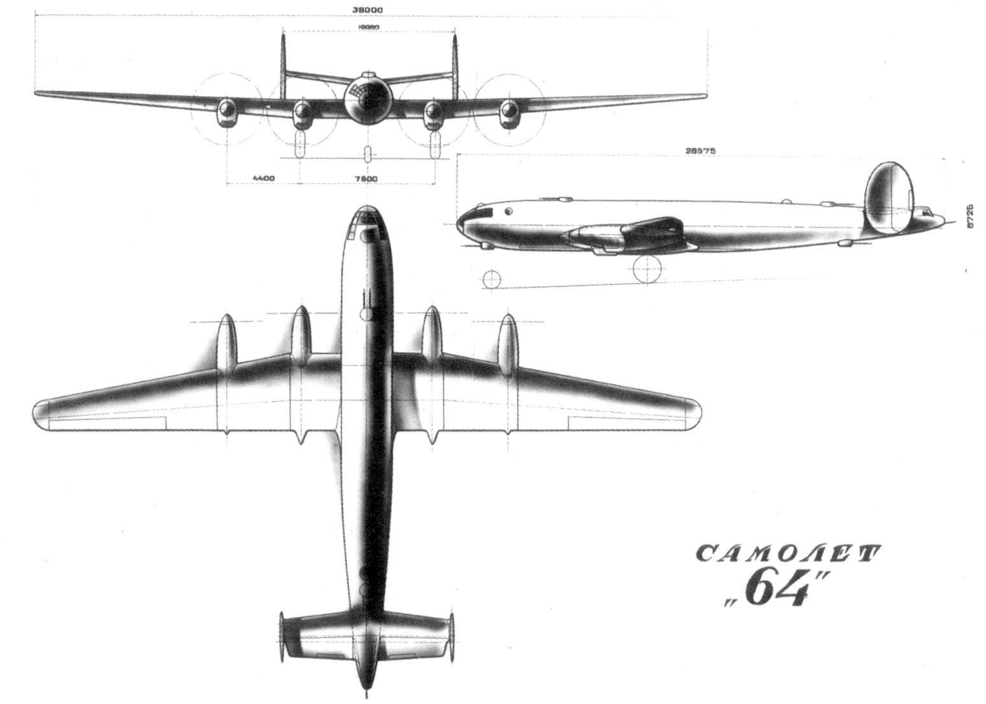

Tupolev 'Aircraft 64' from mid-1944.

ply could not provide it. Part of the problem was that the war had prevented the Soviets from acquiring some of the latest materials and equipment and, as a result, work on the '64' became rather bogged down. As described shortly, a Soviet copy of the American B-29 bomber eventually took the place of the planned in-house designs and the situation with 'Aircraft 64' helped to accelerate this move. Every OKB and Research Institute involved in the '64' would have required a great deal of time and effort to complete the project.

In service the 'Aircraft 64' was to receive the official Soviet designation Tu-10 (this was actually used for one of the final developments of the wartime Tu-2 light bomber) and an airliner variant called 'Aircraft 66' was also proposed. However, work on the '64' (and its competitors) did continue during 1946 until it was clear that the B-29 copy programme would proceed on schedule, at which point the NKAP officially terminated all of the projects. Nyezval did indeed become Tupolev's deputy and his OKB-22 and Myasishchev's OKB-482 were closed, a move was brought an abrupt end to the DVB-202, DVB-302 and Nyezval's bomber. The Il-14 was also abandoned in 1946. The GKO order authorising the go-ahead for the B-4 (the initial designation of the B-29 copy) was made on 6th June 1945, by which time drawings for the '64's construction were being issued and the OKB's experimental department was preparing to build the prototype.

In the end the 'Aircraft 64' was held as a back-up programme for the B-29 well into 1947. In fact a third development was considered through to the end of 1946 which was basically a rework that introduced much of the knowledge gained from close inspections of the B-29. In this form service ceiling was calculated to be 36,089ft (11,000m) and maximum range with a 8,818 lb (4,000kg) bomb load 4,040 miles (6,500km). All of the 'Aircraft 64' work was finally cancelled by a MAP order dated 16th April 1947, at the point when the first production B-29/Tu-4 was complete, but the '64's wing was later re-assessed for fitting onto the follow-on 'Aircraft 85' described shortly. There remained a substantial amount of effort to be put into piston bomber design in the Soviet Union but, for the time being, this effort was concentrated on the B-29 copy.

Tupolev B-29 Copy

It was the B-29 which eventually filled the Soviet Union's gap in heavy bomber knowledge, principally because some examples of America's most advanced bomber had literally dropped into the Soviet Union's hands. Before the end of the war four of them had had to land in Soviet territory after receiving battle damage during operations over Japanese territory. The first of these made a forced landing at the Soviet Navy's airbase at Tsentral'-naya-ooglavaya, not far from Vladivostok, on 20th July 1944 and its arrival was something of a miracle for the Soviet Union had frequently and persistently asked for deliveries of B-29s. These aircraft were legally interned in the USSR (a fifth example was eventually returned) but their loss caused little worry in American circles since no-one expected the Soviets to be capable of completing a full copy of their best bomber.

Tupolev Tu-4

During 1945 three of the B-29s were restored to flight condition and a thorough assessment of the type was completed. Tupolev was given the job of copying the B-29 and the OKB's entire resources were swiftly concentrated upon the task; the highest priority was also given to the project. This was a vital moment in Soviet history – the Great Patriotic War had been won but the future 'Cold War' that would develop with the USSR's former allies in the West was already seen as a distinct possibility. In early 1945 the lack of heavy bomber development and the technical capability needed to produce such aircraft was seen by Soviet leader Josef Stalin, who always took great interest in his country's military equipment, as a massive weakness. The USSR now needed such an aeroplane to deliver the new nuclear weapons that were considered to be vital to the nation's security.

Typically, several people in high office who Stalin considered were to blame for the problem were punished (including Shakhurin who was arrested and replaced by M V Khrunichev) when, in truth, the real culprit was the war. A suggestion to re-start Pe-8 production did not help and, indeed, proved to be the catalyst for Stalin's anger. The lack of progress with the Tupolev Aircraft '64' was also a key aspect and it had become imperative to find an alternative aeroplane.

Just how the idea of copying the American B-29 became fact is not yet clear. On 22nd June 1945 the NKAP issued an order that the production of the B-4 (B-29 copy) should be carried out at Factory No 22 in Kazan and Tupolev would be responsible for the drawings. Of the three restored B-29s, one was allocated to the Ramenskoye Flight Research Institute (the equivalent to the UK's RAE Farnborough which today is more often called Zhukovsky), a second was dismantled to make a full examination of the structure and the third was held in reserve as a reference. The task was huge and is described in detail in *Red Star Volume 7: Tupolev Tu-4*; however, the choice of engine proved to be straightforward. The Shvetsov ASh-73TK was similar to the B-29's Wright Cyclone and was therefore

The sole example of the Tupolev Tu-80.

a logical replacement which removed any need to copy the American engine.

The first B-4 made its maiden flight on 19th May 1947 and during the Tushino Air Display, held in August, the first three B-4s made a fly past at low level. For some time, western observers believed that these aeroplanes were in fact the three former USAAF airframes restored to flight condition, and not new aircraft. In truth their completion had been an extraordinary achievement for the Soviets which, in an incredibly short time, had now furnished its Air Force with a top-class bomber. By October the type had been redesignated Tu-4 and it was to be built in substantial numbers, production ending in 1952. The decision to copy the B-29 proved to be a wise and correct move for the Soviet Union to take because it enabled its aircraft industry to gain experience in new technology and equipment and it also put the Soviet Air Force back on a level with its Western opponents. The Tu-4 stayed in frontline service until the early 1960s.

Final Piston Developments

Before the development of jet engines had reached a stage which had made piston-powered bombers obsolete, there were several attempts to produce a piston bomber that was superior to the Tu-4. The biggest weakness of the B-29 Copy was its lack of range – it could not reach the USA without in-flight refuelling – and Tupolev's first step to rectify this was the 'Aircraft 80'. Prior to the arrival of this aeroplane, Tupolev assessed many designs within the 'Project 471', '473' and '474' series of designations and these covered different powerplants, dimensions and weights.

Tupolev 'Aircraft 80' or Tu-80

This represented a thorough redesign of the Tu-4/B-29 which dealt with several weaknesses and gave a boost to its weaponry. The lift/drag ratio was also improved by reducing the cross-section of the engine nacelles and cutting the drag from the defensive turrets – one (the forward dorsal) was made retractable for cruise flight while two more were recessed into the fuselage; the radar was also given a new streamlined chin radome. There were many other changes, particularly to the wing, and the bomb bays were lengthened.

Design work began in March 1948 and the official directive covering the development of a single prototype was issued on 12th June. Construction began in November and the Tu-4's production facilities were used extensively in the bomber's manufacture. In July 1949 the Tu-80 was essentially complete except for its equipment and the only example to be built flew on 1st December. Its performance, however, was behind the required standards of the day and, with the follow-on and more advanced Tu-85 under way, the decision was taken not to submit the Tu-80 for state acceptance trials. The aircraft also suffered from a problem that required the carriage of 1,984 lb (900kg) of ballast to keep the CofG in an acceptable position, but it continued flying until 1951. It ended its career on the range as a target for gunfire and bombs from newer aeroplanes.

Tupolev 'Aircraft 85' or Tu-85

The need to be able to deliver a nuclear weapon to North America prompted some more studies for a successor to the Tu-4, powered either by piston or jet engines. The initial designs were given numbers in the 480-series, the first being 'Project 485' on which work began in the summer of 1948. Variants were considered with four or six engines, either ASh-2TK or M-251TK (VD-3TK) power units, and they were to be armed with guided bombs or cruise missiles up to 15,432 lb (7,000kg) in weight. Ten or twelve crew would be carried, a 'Kobal't' radar was fitted, maximum range was 3,729 miles (6,000km) and ceiling 49,213ft (15,000m).

The work progressed onto 'Project 489', a derivative of the '485' initiated in mid-1948. Comparative studies were made, jointly between the Tupolev OKB and TsAGI, of six and eight-engine designs for a long-range strategic bomber, the objective being to find the optimum size, wing area and powerplant arrangement. The alternatives looked at an aircraft with piston engines only plus turbo-superchargers (ASh-2TK or M-251TK), M-224 diesel engines combined with RD-45 turbojets, various piston engines in unison with

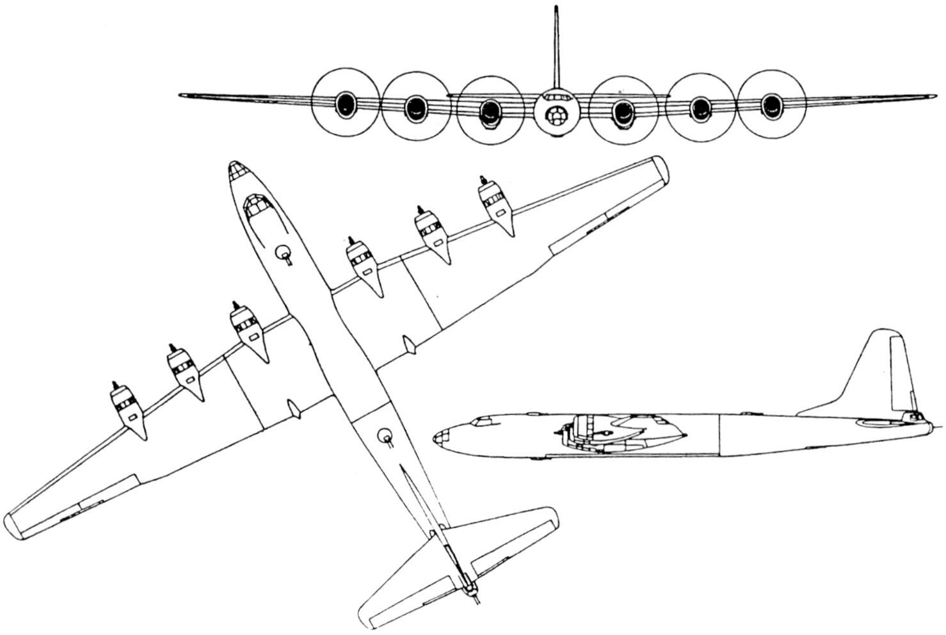

Tupolev 'Project 485' (1948).

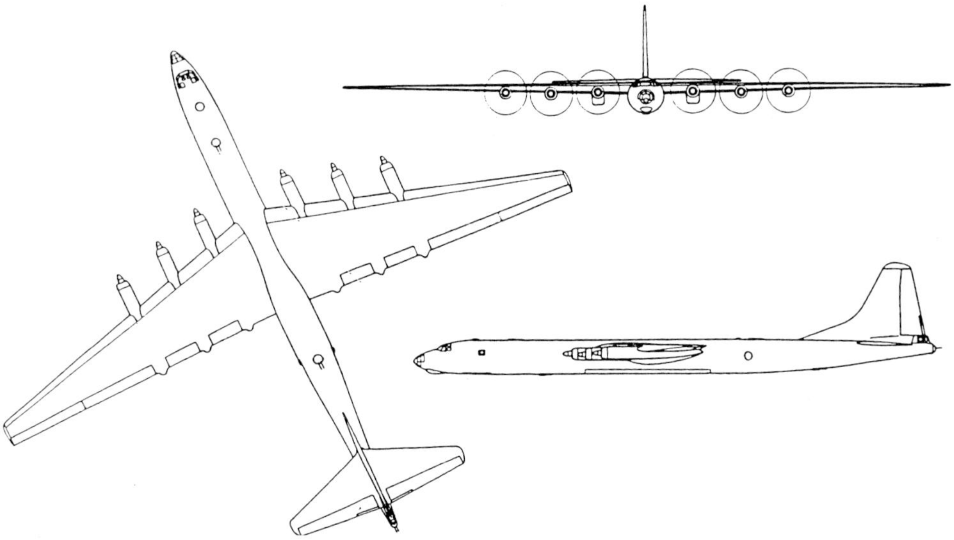

Tupolev 'Project 489' (1948).

Tupolev Tu-85.

RD-45 jets or a turboprop only arrangement using VK-2 engines. The defensive armament, which became standard for all of these pre-projects and the resulting 'Aircraft 85', was four remote-controlled turrets on the upper or lower fuselage and another in the tail, each fitted with two 23mm cannon. Maximum range was 7,458 miles (12,000km) and service ceiling 36,089ft (11,000m). The '489' never progressed beyond paper studies.

Yet another set of designs were embraced by 'Project 487' which covered long-range and extra-long-range alternatives, the latter featuring outer wing panels of greater span. The '487' design team was led by B M Kondorsky and its work made use of 'Aircraft 80' experience. With four ASh-2TK engines the range of the extra-long-range bomber with a 11,023 lb (5,000kg) load was 8,080 miles (13,000km) and service ceiling 36,089ft (11,000m). One version with four M-35s offered a range of 11,063 miles (17,800km) while four 3,700ehp (2,759kW) VK-2 turboprops gave an estimated 7,147 miles (11,500km). At the end of 1948 work on the long-range '487' was halted, leaving only the extra-long-range variant that was eventually to gel into a real aeroplane, the 'Aircraft 85'.

On 16th September 1949 the Council of Ministers issued its directive calling for the construction of the 'Aircraft 85', an aircraft that was to have a range of between 6,837 and 8,080 miles (11,000 and 13,000km); in fact work on the manufacture of the first of two prototypes had actually begun in late July. Both Tupolev's Tu-4 and Tu-80 did not possess the aerodynamics or powerplant that would ever give them the capability to reach America, regardless of how much tweaking might be done to improve their performance. Consequently the result was a wholly new aeroplane and its detail design was completed by the close of 1949. Initially it was intended that the 'Aircraft 85' should have four Shvetsov ASh-2K engines with five-blade propellers and these may have been fitted when the aircraft was taxied for the first time on 14th November 1950; however, when it made its maiden flight on 9th January 1951 the 'Aircraft 85', or Tu-85 as it became known, was fitted with VD-4Ks (the nacelles could accept both). On 12th and 13th September 1951 the aircraft completed a long distance test flight that covered 7,469 miles (12,018km).

The second prototype '85/2' featured a slightly reduced wing area (by 48.5ft^2 [4.5m^2]) and flew on 28th June 1951 but by now the Korean War had revealed that the jet fighter offered a substantial superiority over the piston bomber – American B-29s had taken heavy losses against the Soviet-built MiG-15 fighter. As a result work on the development of piston-engined bombers was brought to a close both in the Soviet Union and America, and this included the Tu-85 despite plans for production at three factories and that the aircraft had performed to specification (the production plans had been laid down on 23rd March 1951). Current piston designs were to be replaced by turbojet and turboprop alternatives and the official directive that ended the Tu-85's flight testing and production planning was received in July 1951.

However, much of the 'Aircraft 85' knowledge and equipment found its way into 'Aircraft 95' (Chapter 3) but Tupolev also studied ways of fitting Kuznetsov TV-2F or TV-10 turboprops onto the Tu-85's airframe. Despite an estimated range of over 9,944 miles (16,000km) and a top speed reaching 460mph (740km/h), the overall results were unsatisfactory and the project was not proceeded with. There was also the Tu-85 powered by four VD-4K piston engines and two VK-1 jets.

The Tu-85 is considered to have had, from an aerodynamic point of view, one of the best unswept wings ever designed for a piston-engined aircraft and during its development two aspects were examined for the first time – the load distributions along the span of a flexible wing during manoeuvring or in turbulence and the definition of the real loads applied during rough landings. This work was carried out under the initiative of A M Cheryomukhin and resulted in the Tu-85 being the first native Soviet aeroplane to take into account the influence of the wing's deformation on the redistribution of the calculated loads. In order to lessen the bending moment in flight it was suggested that the order of the emptying of the fuel tanks should be changed and so the fuel in the outer wing tanks was now used last of all.

Tupolev 'Aircraft 94'

Another study made in the early 1950s looked at installing new engines, namely the 5,153hp (3,843kW) Kuznetsov TV-2 turboprop, into the Tu-4 airframe. This programme was raised under a Council of Ministers order dated 22nd August 1950 and the result was 'Aircraft 94'.

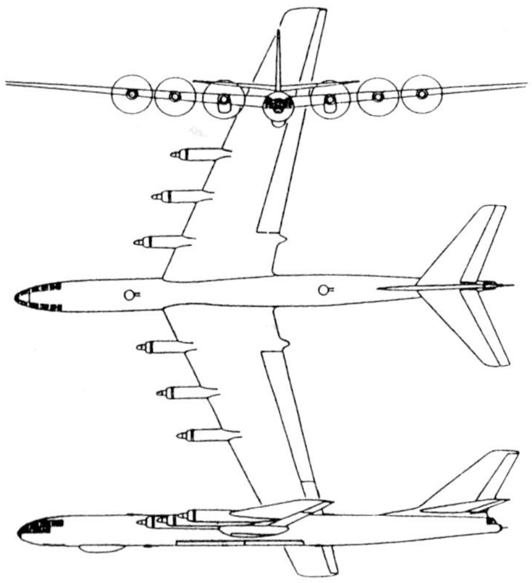

One of the projects studied as part of the Tupolev OKB's research into a heavy bomber, which in the West was known as the 'Tu-200'.

However, the estimated performance improvement over the Tu-4 was not sufficient to make the job worthwhile and the project was put on hold until Tupolev returned to it with suggestions to fit 4,000hp (2,983kW) Kuznetsov NK-4 turboprops. By then, however, all-new designs of much higher performance were making their appearance and these ensured that any further developments of the Tu-4 airframe would be pointless. (Note: In due course China acquired a quantity of Tu-4s to form a strategic bombing force of its own and eventually some examples were re-fitted with home-built Ivchenko AI-20 turboprop engines. A selection of these were further modified and flown as Advanced Early Warning [AEW] aircraft while others served as drones.)

Tupolev Heavy Bomber

Yet another investigation into intercontinental multi-engine bombers which embraced the 'Project 485' described above plus 'Projects 471', '473' and '474', all of them developments of the Tu-4. The work resulted in designs for six- and eight-engine bombers offering ranges of up to 12,430 miles (20,000km). 'Project 473' had a span of 183ft 9in (56.0m), a maximum take-off weight of 209,436 lb (95,000kg) and was fitted with six Shvetsov ASh-73TK; some of the designs were given moderate sweepback on all of their flying surfaces. When reports of the possible existence of these large bombers was received in the West, the provisional designation 'Tu-200' was allocated to them, but this number was never used by the Soviet Union.

Myasishchev Long-Range Bomber

In 1946 Myasishchev draw a family of studies for six-engine bombers. The first had six pusher piston engines and five twin 23mm turrets and the second tractor turboprops which allowed the addition of two more turrets, one each in the rear portion of the underwing nacelle also used to house the main undercarriage. The piston aircraft could have alternative 1,850hp (1,380kW) AM-39TK engines to replace the ASh-73TK, which gave a maximum speed of 388mph (625km/h), and their respective ceilings were 32,808ft (10,000m) and 41,010ft (12,500m) and range with a 8,818 lb (4,000kg) load 7,458 miles (12,000km) and 8,080 miles (13,000km). A crew of fourteen would be carried and 'Kobal't' radar fitted. The turboprop aircraft was expected to offer a top speed approaching 466mph (750km/h) while the third design had jet engines but, so far, only a drawing has been found without data.

Ilyushin Il-26

During 1947 and 1948, in response to a new specification, the Ilyushin OKB undertook preliminary studies for a heavy long-range strategic bomber. Designation Il-26 embraced a full series of designs centred on one basic layout but with different types of engine and either four or six power units, with a consequent variation in final dimensions. The choices included 4,500hp (3,356kW) Shvetsov ASh-2TK pistons, 6,000hp (4,474kW) Yakovlev M-501 diesels and 5,000hp (3,729kW) Klimov VK-2 turboprops, and wind tunnel research clearly showed that the VK-2 offered the best performance in regard to range and speed.

Overall, the Il-26 had a straight mid-fuselage mounted wing and a single fin and the bomb bay could take one 22,046 lb (10,000kg) or three 6,614 lb (3,000kg) bombs. Twin 23mm radar-controlled turrets were fitted in two dorsal and two ventral positions and in the fuselage tail and, in all, twelve crew would have been carried. To obtain the maximum possible fuel load (and therefore range), and reduce the weight of the airframe, the undercarriage was to have been fitted with some auxiliary support units that were to be jettisoned after take-off (the carriage of a very large volume of fuel meant that the difference between the Il-14's take-off and landing weight would have been considerable). The bomber's ceiling was estimated to be 32,808ft (10,000m) and range with a 11,023 lb (5,000kg) load 7,185 miles (11,560km). Work on the Il-14 was halted after Ilyushin was ordered to concentrate on his Il-28 medium bomber described in Chapter 2.

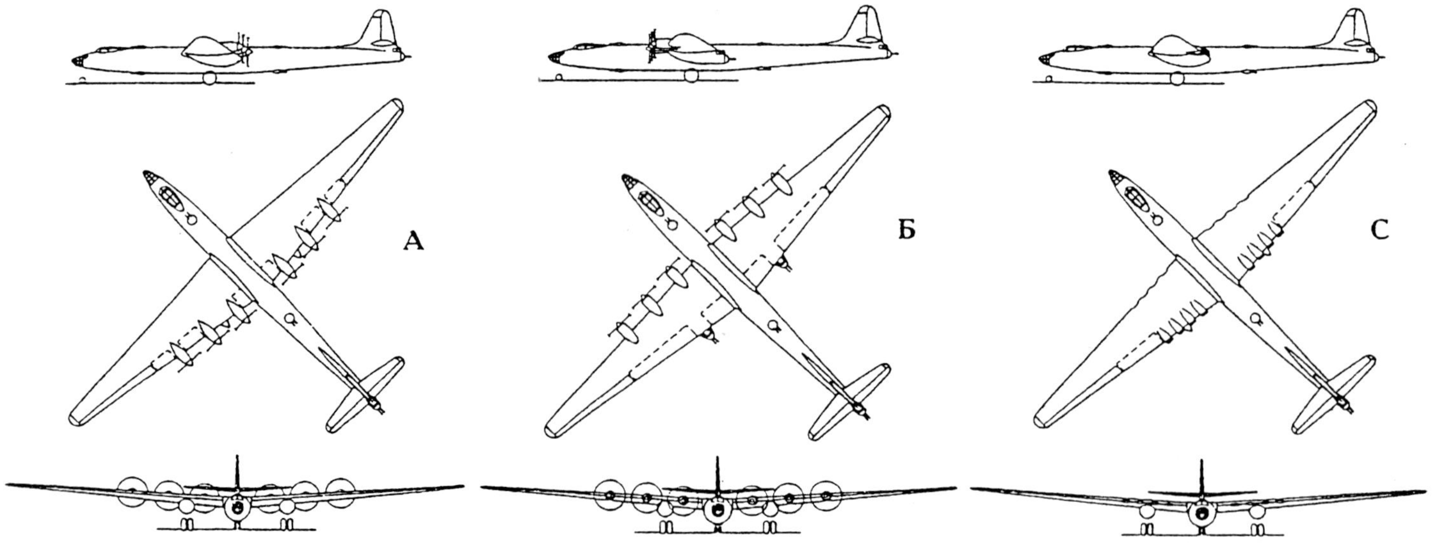

The three forms of Myasishchev's long-range bomber, A with piston, B turboprop and C turbojet engines (1946). Russian Aviation Research Trust

Model of the piston-powered Myasishchev long-range bomber.

This wind tunnel model shows one form of Ilyushin's heavy Il-26 bomber.

Piston-Engined Bombers – Estimated Data

Project	*Span* ft in (m)	*Length* ft in (m)	*Gross Wing Area* ft² (m²)	*Max Weight* lb (kg)	*Engine* hp (kW)	*Max Speed / Height* mph (km/h) / ft (m)	*Armament*
Ilyushin Il-14	?	?	901 (83.8)	49,383 (22,400)	4 x AM-43NV 2,460 (1,834)	373 (600) at S/L, 472 (760) at 30,512 (9,300)	1 x 23mm + 2 x 20mm cannon, 5,511lb (2,500kg) bombs
Myasishchev DVB-202	137 9.5 (42)	95 9.5 (29.2)	1,613 (150.0)	97,002 (44,000)	4 x ASh-72TKM 2,100 (1,566)	404 (650) at 32,808 (10,000)	5 x 2 x 20mm or 4 x 2 x 20mm & 1 x 23mm cannon, 48,280lb (21,900kg) bombs
Myasishchev DVB-302	137 9.5 (42)	95 9.5 (29.2)	1,398 (130.0)	99,206 (45,000)	4 x AM-46TK 2,300 (1,715)	423 (680) at 32,808 (10,000)	5 x 2 & 1 x 1 x 20mm cannon, 35,273lb (16,000kg) bombs.
Nyezval Bomber	?	?	?	110,229 (50,000)	4 x ASh-72TK 2,250 (1,678) max	376 (605) at 31,168 (9,500)	4 x 12.7mm machine guns, 4 x 20mm & 1 x 23mm cannon, 26,455lb (12,000kg) bombs
Tupolev 'Aircraft 64' (mid-1944)	124 8 (38)	93 9 (28.575)	1,400 (130.2)	100,309 (45,500)	4 x AM-43TK-300B 2,300 (1,715)	323 (520) at S/L, 379 (610) at 31,168 (9,500)	5 x 2 x 23mm or 4 x 2 x 23mm & 1 x 1 x 20mm cannon, 47,840lb (21,700kg) bombs
Tupolev 'Aircraft 64' (1946 design)	137 10 (42.0)	98 4 (29.98)	1,616 (150.3)	110,229 (50,000)	4 x AM-46TK-3PB 2,300 (1,715)	404 (650) at height	5 x 2 x 23mm or 20mm, 52,249lb (23,700kg) bombs.
Tupolev Tu-4 (flown – early aircraft)	141 3 (43.05)	99 0 (30.2)	1,739 (161.7)	140,212 (63,600) maximum	4 x ASh-73TK 2,400 (1,790)	270 (435) at S/L, 347 (558) at 33,629 (10,250)	4 x 2 & 1 x 3 x 20mm cannon, 26,301lb (11,930kg) bombs
Tupolev Tu-80 (flown)	143 0 (43.58)	112 7 (34.32)	1,861 (173.1)	133,598 (60,596)	4 x ASh-73KFN 2,720 (2,028) at t/o	339 (545) at 32,808 (10,000)	5 x 2 x 23mm cannon, c26,301lb (11,930kg) bombs
Tupolev '485' (typical)	183 8 (55.98)	121 5 (37.0)	?	209,436 (95,000)	6 x ASh-73TKF 2,400 (1,790)	435 (700) at height	5 x 2 x 20mm or 23mm cannon, 44,092lb (20,000kg) bombs
Tupolev '489' (typical)	154 10 (47.2)	115 6 (35.2)	2,256 (209.8)	200,353 (90,880)	See text	404 (650) with 4,000hp (2,983kW) units	5 x 2 x 23mm cannon, 44,092lb (20,000kg) bombs
Tupolev '487' Very Long-Range Bomber	183 9 (56.0)	115 6 (35.2)	2,581 (240) max	216,446 (98,180)	4 x ASh-2TK 4,000 (2,983)	379 (610) at height	5 x 2 x 23mm cannon, max bomb load unknown
Tupolev Tu-85 1st Prototype (flown)	183 7 (55.96)	128 11 (39.31)	2,942 (273.6)	236,389 (107,226)	4 x VD-4K 4,300 (3,207)	397 (638) at 32,808 (10,000)	5 x 2 x 23mm cannon, up to 39,683lb (18,000kg) bombs
Tupolev 'Aircraft 94'	141 3 (43.05)	99 0 (30.2)	1,739 (161.7)	139,555 (63,302)	4 x TV-2 5,163 (3,850)	420 (676) at 19,685 (6,000)	4 x 2 & 1 x 3 x 20mm cannon, 26,301lb (11,930kg) bombs
Myasishchev Long-Range Bomber	?	?	?	246,914 (112,000)	6 x ASh-73TK 2,400 (1,790)	398 (640)	5 or 7 x 2 x 23mm cannon, 44,092lb (20,000kg) bombs
Ilyushin Il-26 (typical)	?	?	5,108 (475)	418,871 (190,000)	6 x VK-2 5,000 (3,729)	339 (545) at S/L, 348 (560) at 30,512 (9,300)	5 x 2 x 23mm cannon, max 26,455lb (12,000kg) bombs

Chapter Two

The First Jet Bombers

Tupolev's 'Aircraft 77' (Tu-12) only just missed out on being the first Soviet jet bomber to fly; it became airborne on 26th July 1947.

Throughout the world the advent of the jet engine brought a revolution in aircraft design and the Soviet Union proved to be no exception. The first all-Soviet jet engine was actually drawn by A M Lyulka in 1937 at the Kharkov Aviation Institute in Ukraine but the German invasion in June 1941 put this work on hold. However, by the end of February 1942 Soviet Air Force Command was insisting that work on jet development should resume and in August Lyulka's team of engineers returned to its research. When it became clear that Germany had jet aircraft in front-line service and both Britain and America were making good progress with jet engines and aircraft, the USSR was faced with the problem that it still had no native turbojet or jet-powered aircraft available. The Soviet leaders had to take steps to catch up and in February 1944 the GKO established the NII-1 Scientific Research Institute which brought together various groups of engineers who had all been working on 'jet' engines in some form – turbojets, ramjets, pulsejets or rockets; Lyulka was put in charge of the Institute's turbojet department.

In early 1945 the design work was completed on the best Soviet turbojet so far, the S-18 which featured an eight-stage axial compressor; the 'flyable' version was designated TR-1 and gave a maximum thrust of 2,860 lb (12.7kN). The TR-1 was flown from 28th May 1947 onwards in the prototype Sukhoi Su-11 twin-jet fighter, and from 24th July it also powered the Ilyushin Il-22 bomber. Both types were demonstrated at the Tushino Air Show in August and these were the first jet aircraft to be powered by a Soviet-designed engine. In 1947 Lyulka's team also completed the TR-3 with a seven-stage axial compressor which was eventually rated at 10,140 lb (45.1kN) thrust. Two examples were installed in the prototype Ilyushin Il-30 bomber but, unfortunately, this aircraft was never flown.

The end of the war also enabled the Soviets to get their hands on Germany's most advanced jet engines; indeed numerous German jet aircraft and engines plus complete factories specialising in such types fell into the hands of the Soviet Army. The engines were thoroughly evaluated and eventually the Jumo 004 and BMW 003A were respectively copied as the RD-10 and RD-20. Several upgraded versions were also developed and much of the work was undertaken by German specialists who were formed into their own OKBs. The last engines to have German participation in their development were the '022' turboprop of 1950 (later designated TV-022 and TV-2), which entered series production as the NK-2M, and the NK-12 turboprop used in the Tupolev Tu-95 (Chapter 3). An improved TV-2F was used by the Tupolev Tu-91 (Chapter 7).

In 1946 a USSR Council of Ministers Decree ordered that the development of home-grown Soviet engines should be intensified in parallel with improvements to various types of captured German jets. Well established OKBs such as Klimov and Mikulin, which until then had been engaged purely on piston engine design, switched their attention to turbojets and, as a result, some new and very powerful piston engines such as the VK-110 and VK-150 were abandoned. (This practice occurred in other countries; for example the complex 3,500hp [2,610kW] Rolls-Royce

Myasishchev RB-17 (DSB-17) (1945/46).

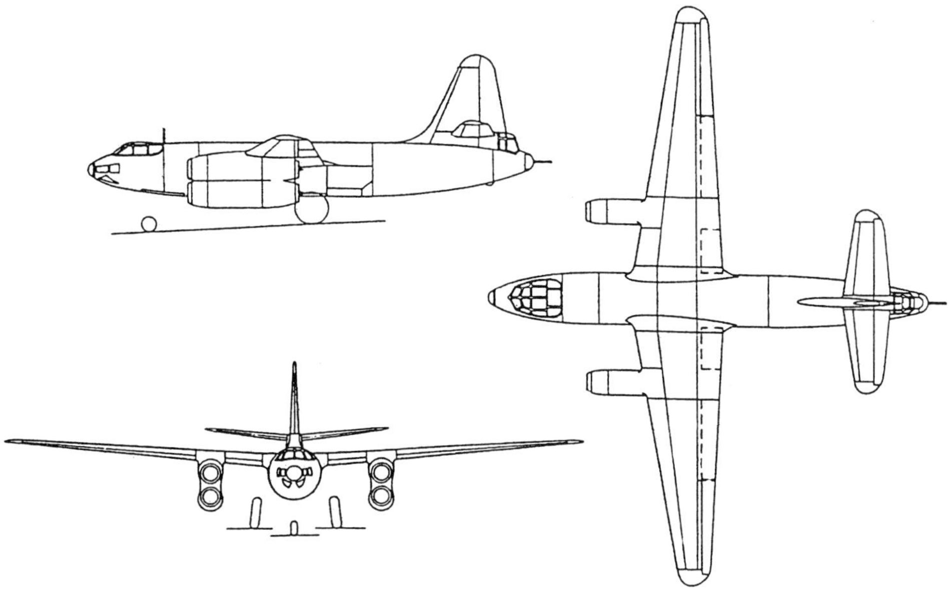

Eagle, the ultimate development in British piston engines, was a victim of the new jet technology.) In 1946 Alexander Mikulin began working on his first turbojet (the TKRD-1 or AM-TKRD-1) which was bench tested in 1947 and intended to go into Baade and Ilyushin bomber designs.

By April 1946 it had become clear that the quality of German-designed engines fell below world standards and so their application was generally considered as a temporary measure that would prepare the USSR's aircraft industry and Air Force for the superior powerplants expected to be available in the near future. The decision was also made to try to obtain the most advanced engines then available in Great Britain who, at the time, was the world leader in turbojet development. Since the end of the war relations between Moscow and London had cooled but the two countries had remained allies. To the Soviet's surprise they were allowed to buy, in some numbers, both the Rolls-Royce Derwent 5 and Nene and these were delivered between January and November 1947.

Many of the British-made units were put into prototype aircraft, most of which were fighters but some of the bombers described shortly were also included. Both British engines were copied and put into service as the RD-45 (Nene) and RD-500 (Derwent), each type clearing its 100-hour State tests during 1948. Production of the more important RD-45 ceased in 1950 but eventually an entire family of engines was developed from it. The first in the series was the VK-1, created by a USSR Council of Ministers Decree of 9th April 1946 and well before the first Nenes and Derwents arrived in the Soviet Union, and its development was assigned to OKB-117 headed by V Ya Klimov. Klimov chose the Nene as his pattern for the new engine but, at the time, the OKB had very limited information available, simply articles and descriptions published by Western magazines plus some photographs including a longitudinal cross section. The specification requested a thrust of 5,950 lb (26.4kN), that is, nearly 1,000 lb (4.4kN) more than a Nene.

Development of the VK-1 began in the middle of 1946. While retaining the general layout and size of the Nene (VK-1 was slightly larger), most of the Soviet engine's parts and sub-assemblies were new; the arrival of the British engines did prompt some modifications but the VK-1's original components were retained. The first VK-1 was ready for factory testing in October 1947 (that is, earlier than the RD-45) and the type entered production in late 1949. This Soviet home-grown unit was built in far greater numbers that the British copies, the key reasons being its higher thrust and better compatibility with current Soviet technology. It was produced in four versions for four different production aircraft; the 'A' went into the Ilyushin Il-28 bomber and the 'B' into the Tupolev Tu-14, while the 'V' and 'S' served in Mikoyan's MiG-15*bis* and MiG-17 fighters. G M Beriev also used two VK-1s in his experimental R-1 jet flying boat (Chapter 9).

Many published sources have roundly criticised the sale of British jet engines to the Soviet Union in 1946/47. However, recent research has shown that in the VK-1 the Soviets had actually 'copied' the British Nene long before they had their hands on an example. In the author's opinion, where the Soviets really benefited from acquiring British-made engines was the possession of new materials that were resistant to very high temperatures, primarily the nickel-base Nimonic alloy. This allowed their own designs to function reliably and for longer periods between overhauls. Most of the Soviet's early jet engine designs had used the more complex axial compressor but, for the short term, the virtues of the simpler centrifugal type were soon recognised. The development of the VK-1 and the purchase of Nenes and Derwents filled this important gap.

Like most of the rest of the aviation world, the first Soviet jet aircraft designs, drawn during and just after the war, were predominantly fighters since the new type of powerplant lent itself to high speeds and high rates of climb. The engine's high fuel consumption initially prohibited its rapid adoption for any practical bomber design because it automatically necessitated the carriage of a much larger volume of fuel than on an equivalent piston-powered type. However, it was not too long before jet bomber designs were receiving attention from Soviet design bureaux.

Because the jet consumed so much fuel, a large number of experimental studies had to be made to find the optimum airframe arrangement that could be powered by such engines. An effective bomber still needed to combine an adequate warload with a satisfactory speed and range while also carrying a sufficiently strong defensive armament. Like Great Britain, the first Soviet essays into jet bomber design that were to lead to substantial production orders fell within the medium bomber category, roughly equivalent to the English Electric Canberra. However, there were some experimental types which preceded these.

First Attempts

Myasishchev RB-17

By the end of the war five OKBs were working on jet-powered bomber aircraft. Andrei Tupolev was trying to match his Tu-2 piston bomber with jet engines (below), two unfinished examples of the German wartime Arado Ar 234 were in the hands of I V Chetvyerikov with the intention of making them airworthy, and both Ilyushin and Sukhoi were working on new four-engined designs (the Il-22 and Su-10 described shortly). The fifth Bureau was Myasishchev who initially approached its introduction into jet aircraft in

Model of the Myasishchev RB-17. George Cox

two directions, the first being to restore and simplify the highly successful German Messerschmitt Me 262 jet fighter for flight and possible production. Myasishchev envisaged making the fighter much lighter and possibly introducing more powerful engines but Government officials brought a halt to the project when they opted for the new home-grown MiG-9 and Yak-15 projects.

The alternative was an all-new fast day bomber project initially called the R-1 but quickly redesignated RB-17. In the beginning Myasishchev looked closely at fitting turbojets into the Petlyakov Pe-2I light bomber, a streamlined high-performance development of the Pe-2 dating from the final years of the war, and the initial studies were very encouraging. However, as the work progressed it became clear that fitting jet powerplants would require far more detail design throughout the whole aircraft than first thought, plus extra equipment and numerous new subassemblies. It was realised that an all-new design would be an easier task and this feeling was backed up by the problems Tupolev experienced in fitting jet engines to the Tu-2 (the resulting Tu-12 was essentially an all-new type). The turbojet Pe-2I did not progress beyond the technical proposal stage but much of the knowledge gained from Myasishchev's Me 262 and Pe-2I work found its way into the RB-17.

Much study and testing was required before Myasishchev selected what it considered to be the most suitable way of installing four German Jumo 004 engines, the optimum choice being two underwing pods with two engines in each mounted one above the other. It was expected that a new more powerful engine offering 4,409 lb (19.6kN) of thrust would be available relatively quickly to replace the 004s and the vertical pairing of the current engines was expected to assist in the relatively simple installation of a single more powerful unit under each wing (which 'new' engine this statement refers to is unknown). Stainless steel panels would cover those portions of the wing that would be exposed to hot engine gases because an aluminium covering would deteriorate in such temperatures.

The preliminary RB-17 proposals were submitted to the NKAP on 28th November 1945 and, despite fierce criticism from NKAP and the Air Force NII regarding the project's aerodynamics and weight, the construction of a mock-up was begun. TsAGI had contributed much to the wing design, Vladimir Myasishchev had confidence in his bomber and when the mock-up was officially reviewed there was relatively little unfavourable comment. Cruising at 423mph (680km/h), the RB-17 was expected to be capable of bombing a target 1,865 miles (3,000km) away when carrying 2,205 lb (1,000kg) of bombs; estimated ceiling was 37,730ft (11,500m). Two 20mm or 23mm cannon would be carried for defence, one in a tail turret worked by the wireless operator and the other fixed in the lower nose just off centre to starboard, and a tricycle undercarriage was fitted with narrow-track main wheels. The pilot would do the navigating and to help him various navigation aids were to be carried.

The RB-17 was considered a practicable project but fate intervened long before it was ready to fly. Firstly, in December 1945 an NKAP order was made cancelling the bomber, although afterwards Myasishchev continued to work on it. Then in early 1946, shortly after Shakhurin had been replaced by Khrunichev as the chief of the NKAP (see Chapter 1), the Soviet aircraft OKBs were reorganised and the Myasishchev Bureau was

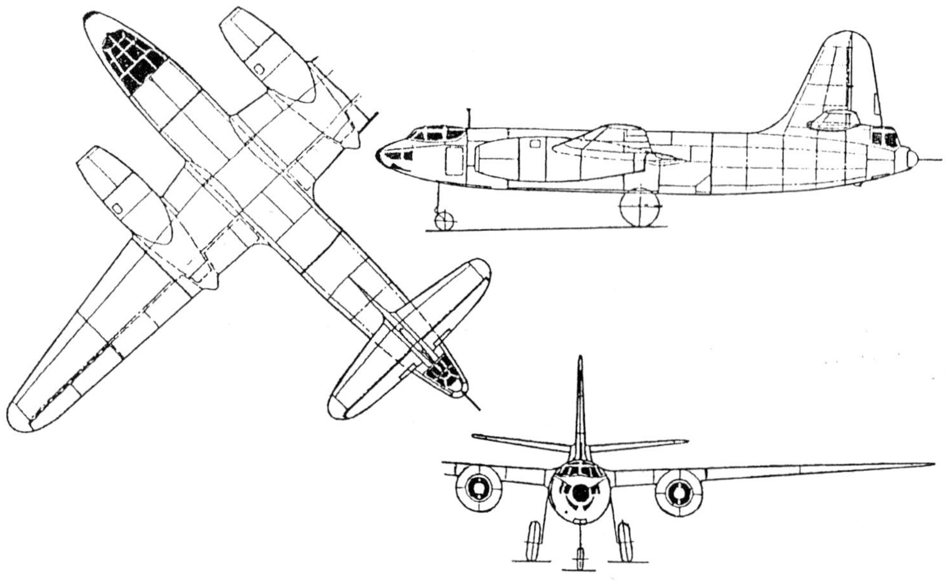

Myasishchev DSB-17-D5 (1945/46). This is the only known drawing showing the RB-17 re-engined with a single more powerful jet under each wing. The dimensions are unchanged.

Artist's impression of the RB-17 which shows clearly the positions of the nose and tail defensive guns.

closed down. Following this, both the RB-17 and the Chetvyerikov Ar234 programmes were officially terminated by a Ministry resolution. It did not help that Myasishchev was well ahead of the more traditional bomber OKBs in designing the Soviet's first practical jet bomber and some of his competitors, who had more influence with the Government than he had, were on hand to help get his OKB-482 shut down.

There had still been time, however, to evaluate three versions of the RB-17 fitted with two British Derwent V engines (which gave an estimated top speed of 510mph [820km/h] at 16,404ft [5,000m] and a ceiling of 41,995ft [12,800m]), the DSB-17 bomber, RB-17 reconnaissance aircraft and IDD-17 long-range air defence fighter (DSB = long-range mid-wing bomber). The RB-17 itself was also known as the VM-24.

Ilyushin Il-22

The Il-22 resulted from a VVS requirement issued during 1946 for a jet bomber and it was to become the very first entirely Soviet-designed jet-powered bomber to fly. Ilyushin's response to the requirement was a straight wing layout which, after much investigation, carried its four TR-1 engines in individual nacelles. The nacelles were not large enough to take the main wheels and consequently the landing gear was housed entirely in the fuselage, which made for a narrow track undercarriage. After the Myasishchev OKB was closed in 1946 several key design staff were transferred to Ilyushin and they became involved in the Il-22's design. For example, the wing structure was rather like that used on the wartime Myasishchev DVB-102 piston bomber while this narrow main undercarriage resembled the arrangement on the RB-17.

The Il-22 made its maiden flight on 24th July 1947 but only one example was built. Its performance was found to be inadequate, principally because the engines had had to be derated due to development problems, and consequently the aircraft was not submitted for state testing. The test pilots also disliked the flightdeck glazing with its heavy framework, the former giving reflections and a distorted view while the frames created blind spots. However, the Il-22 did fly over Tushino during the Aviation Day display of 3rd August 1947 and the US Air Force allocated Type Number 10 to it (the ASCC codename system was not introduced until 1954).

Ilyushin Il-24

In response to an official order, Ilyushin eventually redesigned the Il-22 with two 7,275 lb (32.3kN) Mikulin AM-TKRD-01 jets, designating the project Il-24. However, the need to get the Il-22 up and running prevented any real progress being made with the follow-on project until November 1946. To begin the Il-24 was essentially just an Il-22 airframe with new engines, a new twin-cannon tail barbette and other minor changes and on 11th March 1947 a Council of Ministers directive was passed authorising a prototype to be built which would commence its state acceptance trials during the following September. The design was completed in the spring and Sergey Ilyushin gave his approval on 16th May. The estimated time for the aircraft to reach

Ilyushin's Il-22 was the first Soviet designed jet bomber to fly, making its maiden flight on 24th July 1947.

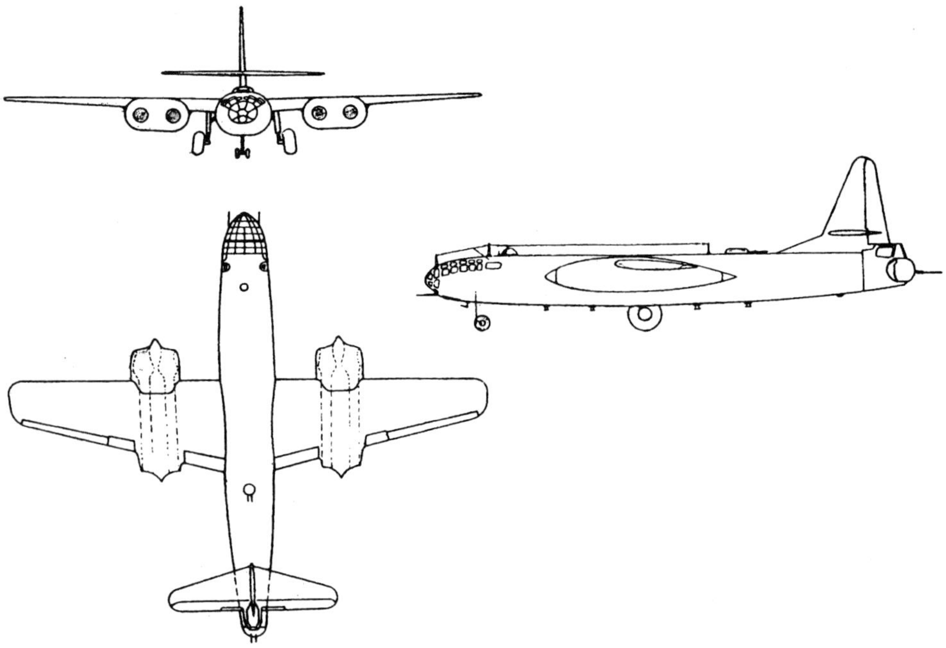

The version of the Ilyushin Il-24 fitted with four Rolls-Royce Nenes (1946/47).

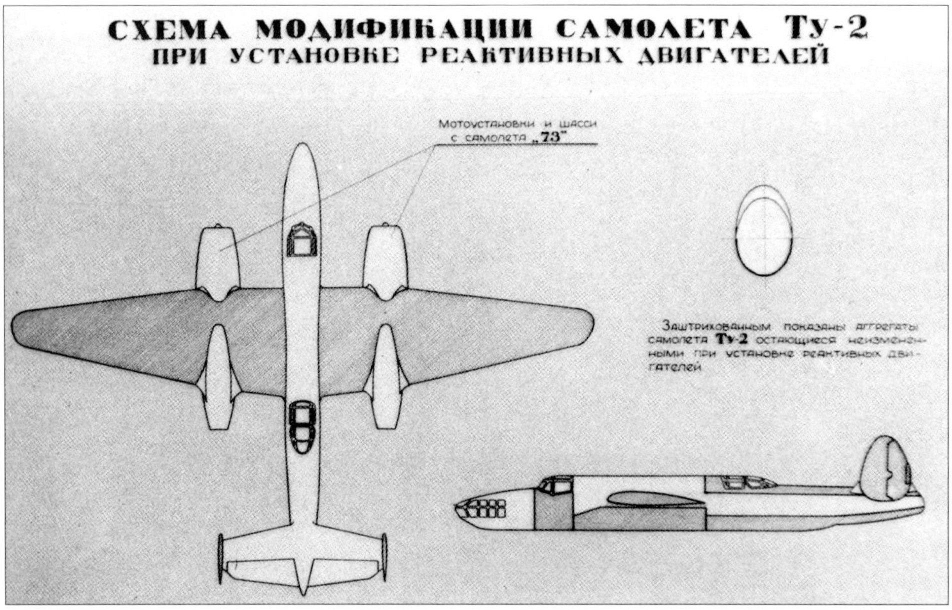

One of the first Tupolev 'Aircraft 77' proposals for a jet bomber based on the piston Tu-2.

16,404ft (5,000m) was 5.6 minutes (8.6 minutes for the Il-22), service ceiling 47,900ft (14,600m) (Il-22 36,417ft [11,100m]) and range with a 4,409 lb (2,000kg) load 1,864 miles (3,000km).

However, MAP decided that the engines allocated to the Il-24 should in fact go to OKB-1, where Semyon Alekseyev was working on the '140' bomber (Chapter 3), so an alternative arrangement with four RD-45 (Nene) engines mounted in two twin underwing nacelles had to be prepared; two RD-45s alone would not supply enough power for such a large aeroplane. Despite redesigning the Il-24 to take four engines, the fact could not be avoided that the aircraft was rather heavy and, at 61,728 lb (28,000kg) for the RD-45 variant, could not be operated from unpaved strips when few hard runways were, as yet, available in the Soviet Union. As a result another Council of Ministers directive was issued on 21st June 1948 to cancel the Il-24 programme, at which point the prototype was about 65% complete. In addition, the design bureau had by now also started work on the Il-28 bomber with two Nenes. Both Il-22 and Il-24 designations were later reused for variants of the Il-18 airliner.

Tupolev 'Aircraft 77'

This was an update of an older piston type, the highly successful wartime Tu-2S powered by ASh-82FN engines. The Tu-2 itself was used by the Soviets as a test-bed on which jet engines could have their performance assessed in flight, the units being housed in a pod mounted beneath the forward fuselage, and ex-German, ex-British and the Soviet's own native engines were all tested in this way. The first official request for Tupolev to develop a jet bomber was received on 9th April 1946 and design work began on two projects, 'Aircraft 72' and '73'. 'Aircraft 72' was a modified version of the Tu-8 piston bomber prototype but it did not get very far because by early 1947 it was clear that 'Aircraft 73' was the better design (both are described shortly). During 1947 the bureau became heavily involved on 'Aircraft 73' because the prototype had to be flying by the middle of the year in readiness for it to participate in the traditional annual air parade at Tushino. Eventually, however, the original layout had to be reworked to incorporate some new requirements, which prevented the type from being ready in time. As an alternative the VVS and Tupolev decided to replace, as quickly as possible, the piston engines of several production Tu-2s with Nene-1 jets.

The result was a simpler design, 'Aircraft 77', which was officially designated Tu-10 and later Tu-12. It was based on the Tu-2 but had a longer nose, tricycle landing gear and a slightly modified wing. In April Andrei Tupolev proposed 'Aircraft 77' to the Government, supporting the pure technical and organisational arguments on this work by the desire of his OKB to develop a turbojet-powered 'transitional bomber' based on the well established Tu-2. At the same time, without waiting for an official Decree, the bureau started to prepare the preliminary project and to build a full-size mock-up. The latter was completed in May 1947 and officially evaluated between the 24th to 28th, after which the design bureau was informed of any problems. By this time the preliminary project data was essentially ready but the required changes delayed its completion. Nevertheless the construction of the first prototype began in May and the Decree covering 'Aircraft 77' was finally issued on the 31st with the first example to be modified from a production Tu-2 by the experimental factory and five more to be reconfigured from Tu-2s at Factory No 23.

Flight testing began on 26th July 1947 and in August the aircraft was demonstrated at the Tushino air show. After a series of modifications had been made, the type's official state testing was completed in February 1948. Compared to the Tu-2, the Tu-12 revealed a much higher top speed and an improved rate of climb and service ceiling, but its take-off and landing performance was not so good. In order to get the same range as the Tu-2, the jet bomber had to carry a much greater volume of fuel which gave a higher take-off weight; thus, the significant increase of airspeed was paid for by a big drop in take-off and landing characteristics. In addition the lack of pressurised cabins with heating and ventilation considerably reduced the effectiveness of the Tu-12's operations at high altitude (all of Tupolev's subsequent jet bombers had this feature). Finally, because of the increased air pressure, in flight it was found that it was practically impossible to use the dorsal and belly defensive machine-gun mounts while the 12.7mm machine-guns themselves were clearly insufficient; it was suggested that they should be changed for a powered turret with cannons of at least 20mm calibre.

During state testing the Tu-12 took part in (for the first time in the USSR) aerial 'combats' between jet bombers and jet fighters (MiG-9s and Yak-23s). These experiments gave valuable results for improving the gun armament of both fighters and bombers (for all of the types currently under development) as well as evaluating the specific tactics needed in combat between jet aircraft of different classes. This attempt to match the Tu-2 airframe with jets actually produced a new and quite different aircraft from its ancestor and the '77' was indeed looked upon as a transitional type to give experience in the operation and servicing of jet bombers. The small batch of Tu-12s, with home-built RD-45 engines, was completed by Factory 23 and used for training VVS crews, although one example was modified into the Tu-12LL flying test-bed to assess ramjet engines, a role in which it was employed until the beginning of the 1950s. The experience gained by the Tupolev design bureau in the development, manufacture and testing of the Tu-12 allowed the OKB to develop more sophisticated successors – the Tu-12 was an important landmark.

Medium Bombers

Following the initial series of jet bomber designs and prototypes, the development of a medium bomber for service operation became a clear objective. The players were Sukhoi, Ilyushin and Tupolev.

Tupolev's 'Aircraft 77' (Tu-12) only just missed out on being the first Soviet jet bomber to fly; it became airborne on 26th July 1947.

The earliest proposed layout for the Sukhoi Su-10 with six fuselage-mounted RD-10 engines (5.46).

Sukhoi Su-10

On 26th February 1946 a SovMin resolution was issued authorising the development of a bomber powered by four Jumo 004 (RD-10) engines. At the time the Sukhoi OKB was working on several fighters (including the forthcoming Su-9 jet fighter) and the design of tactical jet bombers was an area in which it was unfamiliar, but an NKAP order made on 27th March requested that Sukhoi should build this new aeroplane. It was to have a maximum speed at sea level of 497mph (800km/h) and at 26,247ft (8,000m) 528mph (850km/h), a maximum take-off weight of 30,864 lb (14,000kg) and, with full fuel and 2,205 lb (1,000kg) of bombs, possess a range of 932 miles (1,500km). With an additional

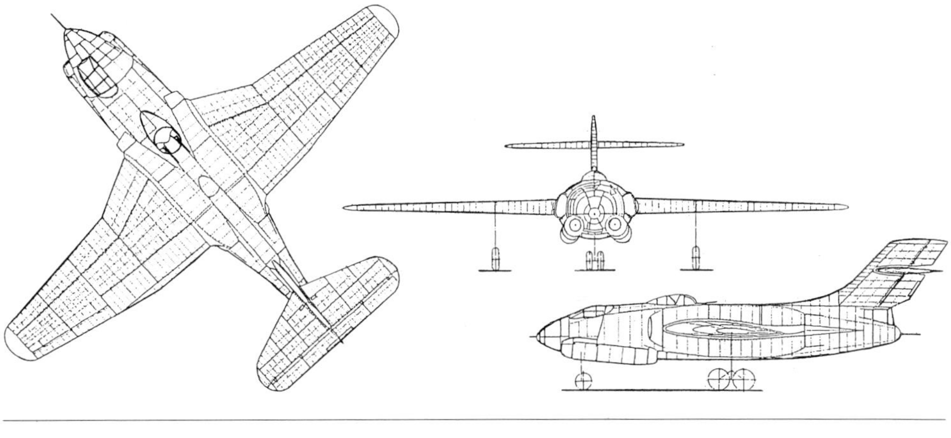

Artist's impression of the Sukhoi Su-10 in initial form.

The next version of the RD-10-powered Su-10 had the engines mounted in a curious arrangement in groups of three under the wing.

1,102 lb (500kg) bombs carried externally, the maximum warload was to be 4,409 lb (2,000kg); defensive armament was four 20mm cannon and the aircraft had to have a ceiling of 36,089ft (11,000m). There were to be two prototypes and the first had to be ready to fly by 1st February 1947.

A preliminary project was initiated on 24th April and Sukhoi's earliest studies revealed that six engines would be needed to perform the required tasks. By May 1946 this research had led to a project, of rather ugly appearance, which had all six power units housed in the fuselage and on 6th May work was started on a full-scale mock-up. This aircraft, called the Su-10, would have a two-spar mid-set tapered wing with a t/c ratio of 12% and an almost symmetrical aerofoil section formed by using ribs and stringers. The wing would have Fowler-type flaps and outboard ailerons and house two 2,728gal (600lit) fuel tanks plus the two-wheel main undercarriage which retracted inwards. The streamlined oval cross-section fuselage, made of bulkheads, longerons and stringers, had four of the engines stacked in pairs at the wing roots, the fuselage ahead of these being scalloped to ensure that the air flowing into them was undisturbed. The intake canal was split above and below the main wingspar and then combined again on reaching the RD-10 compressors; the other pair of RD-10s were housed individually beneath the flightdeck.

A large weapon bay was placed in the fuselage centre and four crew would be carried – pilot, navigator, wireless operator/gunner and tail gunner. The defensive armament actually comprised a single-gun nose turret and twin dorsal and tail turrets, the wireless operator being responsible for the nose gun and also the dorsal position, for access to which he had to pass through a tunnel in No 1 fuel tank. There were three fuselage tanks and four RATOG rocket motors were fitted to keep the take-off run a minimum. The six-engine Su-10's maximum fuel load eventually reached 17,108 lb (7,760kg), estimated time to 32,808ft (10,000m) was 20.4 minutes, service ceiling 38,386ft (11,700m) and range with 2,205 lb (1,000kg) bombs and maximum fuel 1,243 miles (2,000km).

The preliminary project was completed on 19th June, which was the point where Sukhoi officially proposed his six engine powerplant, while at the same time the SovMin approved its '1946 Plans for Experimental Aircraft Construction' which included the new bomber. Su-10 project design was initiated on 1st July, the first wind tunnel model was delivered to TsAGI on 19th July and in mid-August work began on prototype drawings. The mock-up was officially inspected on 7th October and, following some modifications to the design, the manufacture of parts for the first prototype went ahead.

The critical aerodynamic research into high-speed flight in general, which was needed to help produce an ideal aerodynamic shape for the Su-10, was still to be carried out in the USSR. However, Sukhoi's Aerodynamics department, under the control of I Ye Baslavsky, worked out its own aerodynamic calculations for establishing the range of turbojet-powered aeroplanes at various weights and at different heights and power settings. As the Su-10 research and TsAGI wind tunnel testing continued, Sukhoi moved on to some alternative engine arrangements, principally six RD-10s or four TR-1s now mounted on the wing. The tunnel testing eventually brought an end to the six fuselage-mounted RD-10 layout and established that a maximum critical Mach number could be attained by having three engines per side mounted in such a way that their thrust lines coincided with the wing chord line; the air intakes would also have to be placed ahead of the wing leading edges.

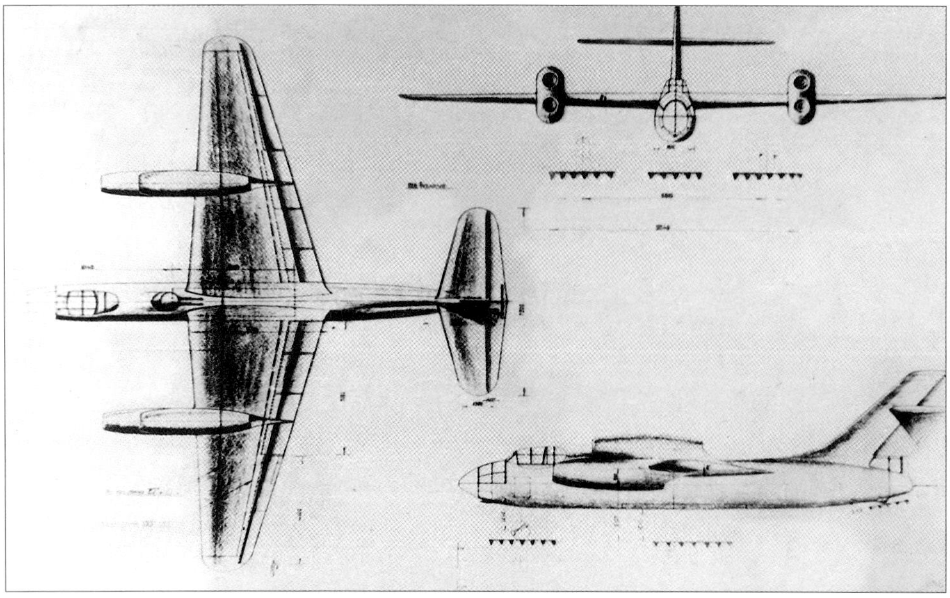

The Su-10's layout as built. The prototype was never completed.

Artist's impression of the Su-10 in its final form.

Sukhoi Su-10 static test airframe.

The Su-10 full-scale mock-up was officially accepted on 2nd December 1946 but two days later Shishkin, the deputy-head of MAP (the Ministry of Aircraft Industry), gave permission for Sukhoi to proceed with the development of a new Su-10 variant with four TR-1s. By the end of the year the construction of a static test prototype had begun and preparations for the flight prototype were under way. The preliminary project for the new design was approved on 7th January 1947 but a week later an Su-10 rear mainspar ground test specimen, in fact the section which carried the aileron root bracket, failed at 100% load. On 11th March a SovMin resolution was issued which authorised the construction of the Su-10 with four TR-1s and this was followed by a MAP order on 12th April. Work now moved forward at a steady rate although on 19th August Pavel Sukhoi did send a request to the director of TsAGI asking for a new wing to be designed which would be adapted to take four RD-500 (Derwent) engines.

The Su-10's powerplant now comprised four engines placed well forward on the wing and housed in vertically staggered pairs with the lower unit forward; this was intended to cut down both the size of the nacelle and its cross-sectional area. In addition the engine thrust lines were arranged to have a slight vertical angle which also helped to reduce drag. Initially it was thought that four RD-10 units would be suitable but soon afterwards these were replaced by the more powerful TR-1 which offered an improved performance. The larger engines added 2,205 lb (1,000kg) to the weight but this was offset by the extra thrust and better aerodynamics. The aircraft itself would be all-metal with a fuselage of semi-monocoque construction using D16T alloy skins. The four crew were retained and a similar defensive turret arrangement to the original six-engine Su-10 was used, although the forward gun was now fixed beneath the nose and the dorsal turret, located between the weapon bay and a fuel tank, was remotely controlled. The tail turret could be elevated to 30°, depressed by 45° and traversed to each side by 30°.

The two-spar wing had stressed skin panels, the trailing edge had flaps which were split into four sections and the ailerons on the outer wing were split into two. The Su-10 would carry a substantial bomb load, all internally in a large bomb bay in the fuselage centre section, which could comprise either a single 6,614 lb (3,000kg) or 3,307 lb

(1,500kg) bomb or eight 1,102 lb (500kg), twelve 551 lb (250kg) or twenty 220 lb (100kg) bombs; the aircraft would also carry an AFA-33 camera. The take-off was still to be RATOG-assisted using four solid-fuel motors attached to the side of the fuselage; these rocket packs had already been tested and proved on Sukhoi's Su-9 jet fighter and when expired they were to be jettisoned by the Su-10's pilot. Maximum fuel load was 16,534 lb (7,500kg), time to 16,404ft (5,000m) 6.7 minutes and 32,808ft (10,000m) 20.4 minutes, service ceiling (with four more powerful 3,307 lb [14.7kN] TR-1As) 39,370ft (12,000m) and maximum range (with 2,204 lb [1,000kg] bombs and full fuel) 1,243 miles (2,000km).

By mid-December 1947 the first Su-10 had reached the final stages of assembly. Its construction, however, had been delayed by a shortage of workspace at Factory No 134 where it was to be built and it was only after this facility had completed building the Sukhoi Su-12 prototype (a reconnaissance and observation aircraft powered by two piston engines) that it was able to begin putting together the Su-10. In addition the bomber was to carry a good deal of new equipment, all of which required testing and modifications, and this brought further delays from the equipment sub-contractors; in fact during 1947 there had been delays in supplying the engines, defensive guns and autopilot. On 12th March 1948 studies to fit four 4,156 lb (18.5kN) thrust TR-2 engines were completed but it was decided that prototype flight development should proceed with the TR-1s as fitted; however, the aircraft was expected to undergo its state acceptance trials four months after the TR-2s had been delivered.

This chapter shows that a high proportion of the Soviet Union's earliest jet bomber designs were actually built and achieved flight status; the Su-10 however, did not quite get that far. On 4th June 1948 the SovMin issued a new resolution detailing a reduction to the aircraft industry research budget for 1948 and eight days later it issued a second document that confirmed a revised MAP production plan for 1948; this also terminated the Su-10 programme. As a result the Sukhoi Su-10 prototype was never completed or flown and was eventually scrapped.

Ilyushin Il-28

This bomber, to be powered by two British Nenes, was first conceived during 1947 as a private venture and was intended to be a scaled down Il-22 offering a higher speed and the capacity to operate from tactical airfields. A key change from the earlier aircraft was an improved rear defensive armament that comprised a single lightweight tail turret which was capable of covering much of the rear hemisphere. The Ilyushin OKB's private research into defensive weaponry eventually resulted in the very efficient electro-hydraulically powered remote-controlled Il-K6 ball turret which initially mounted NS-23 cannon but later housed two Nudelman/Richter NR-23 23mm cannon. This new turret was a great help in keeping down the aircraft's weight; in addition only three crew would be carried and the pilot would have a fighter style cockpit.

Sergey Ilyushin wrote to Nikolay A Bulganin, the Soviet Minister of Aircraft Industry, on 31st October 1947 to outline this new proposal,

Two views of the prototype Ilyushin Il-28 tactical bomber.

Side view of the Tupolev 'Aircraft 72' jet bomber (1947).

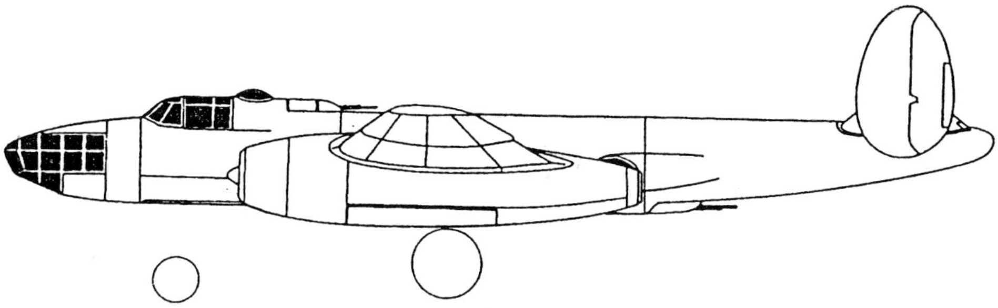

This model of the Tupolev '72' has a V-tail.

and he backed it by offering a prospective first flight in July 1948. However no official go-ahead was forthcoming (despite the Tupolev OKB receiving an order to build a similar tactical jet bomber) but Ilyushin approved his project's development on 12th January 1948 and made the decision to built a prototype at his own risk. His belief in the design was very strong but the Soviet Air Force also badly needed a new tactical bomber. Finally, on 12th June the Ministry of Aircraft Industry included the new type in its experimental aircraft construction programme and the Soviet Union's Council of Ministers issued a directive to cover it. The Il-28 made its first flight on 8th July 1948 powered by two of the original Nenes supplied from Britain. Two Soviet RD-45Fs were installed in the second prototype and this machine undertook and passed the Il-28's state testing between February and April 1949.

Tupolev's design was called the Tu-14 (below) and it was a more complex aeroplane to build but did offer a longer range. A winner had to be selected, each type had many supporters and the debate eventually reached a very high Ministerial level; one strong opponent against the Il-28 was the head of the NII VVS. Finally, on 14th May 1949 Stalin chaired a special commission held specifically to discuss the problem and, after comparing the two sets of performance data and absorbing the comments of his military advisers, he selected the Il-28. To back this up however, the aircraft's speed had to be increased to 559mph (900km/h) by fitting it with Klimov VK-1 engines. Preparations were now put in hand for production to get under way at three different factories.

On 8th August 1949 the first VK-1-powered Il-28 made its maiden flight. Fitting the Soviet engine had required certain modifications to the nacelles but Ilyushin also added some streamlining to the airframe. The first Il-28 bombers entered service in the spring of 1950 and the Il-28T torpedo bomber (for the Naval Air Arm) and Il-28R reconnaissance aircraft also flew that year. The Il-28 was to stay in production for a long time and, having initially intended to build around 3,000 examples, the final total completed in the Soviet Union reached 6,000. It was also built in China and served with many other air forces around the world; the Western recognition codename given to the aircraft was *Beagle*.

Tupolev 'Aircraft 72' and 'Aircraft 73'

As noted, the first official request for Tupolev to develop a jet bomber was sent on 9th April 1946 and in the following autumn design work began on two projects, the 'Aircraft 72' and '73'. However, the first design to be allocated the 'Aircraft 72' designation was an early 1946 high-altitude medium bomber development of the wartime Tu-2 powered by two ASh-2TK engines. An order was made clearing the way for a prototype on 16th April 1947 with state testing planned for September 1948, but the programme was cancelled later in 1947 because of the bureau's work on the Tu-4 and the growing introduction of jet-powered aircraft.

The jet-powered 'Aircraft 72' short-range bomber was a modified version of the Tu-8 prototype ('Aircraft 69' powered by ASh-82FN piston engines) which introduced two 5,000 lb (22.2kN) Nenes in underslung nacelles, a new nosewheel landing gear and different crew and armament positions. The aircraft would have had four crew, a service ceiling of 39,370ft (12,000m) and a range with 2,205 lb (1,000kg) of bombs 1,243 miles (2,000km). However, this project only reached the preliminary design stage because at the beginning of 1947 'Aircraft 73', whose design had run in parallel with the '72', was found to be the more promising aircraft. For a brief period 'Aircraft 72' received the official designation Tu-18.

In January 1947 Tupolev began work on a new short-range front-line jet bomber to attack targets behind enemy lines which in-house was called 'Aircraft 73' but officially Tu-20; this was essentially a new aeroplane that had nothing in common with its piston-engined ancestors. The early calculations revealed that an aircraft fitted with two Nenes could achieve 528mph (850km/h) at altitude and in June the performance specification was settled. Initially the aircraft had a wing and engine nacelle arrangement which resembled the British Gloster Meteor jet fighter while the equipment followed practice used on 'Aircraft 69'. The nose-mounted forward-firing armament comprised one NS-37 37mm and two NS-23 cannon while the rear and lower hemispheres were protected by two single B-20 cannon mounts. Four crew were carried, the service ceiling was estimated to be 41,175ft (12,550m) and the standard range 622 miles (1,000km).

The mock-up received its preliminary examination in February 1947 and work began on the first drawings. However, by this time the design was being revised to increase its size and alter the defensive guns and on 10th March the VVS rejected the initial idea for a short-range aircraft and suggested the development of a larger class of aeroplane with the same powerplant but twice the bomb load and range. The next day the Council of Ministers issued a Decree to this effect which also noted that a prototype was to be ready for its state testing in December 1947. However, the English-built Nenes were slightly down on the specified thrust 4,410 lb (19.6kN) rather than 5,000 lb (22.2kN) and it became clear that a '73' fitted with two Nene 1s could not achieve the specification.

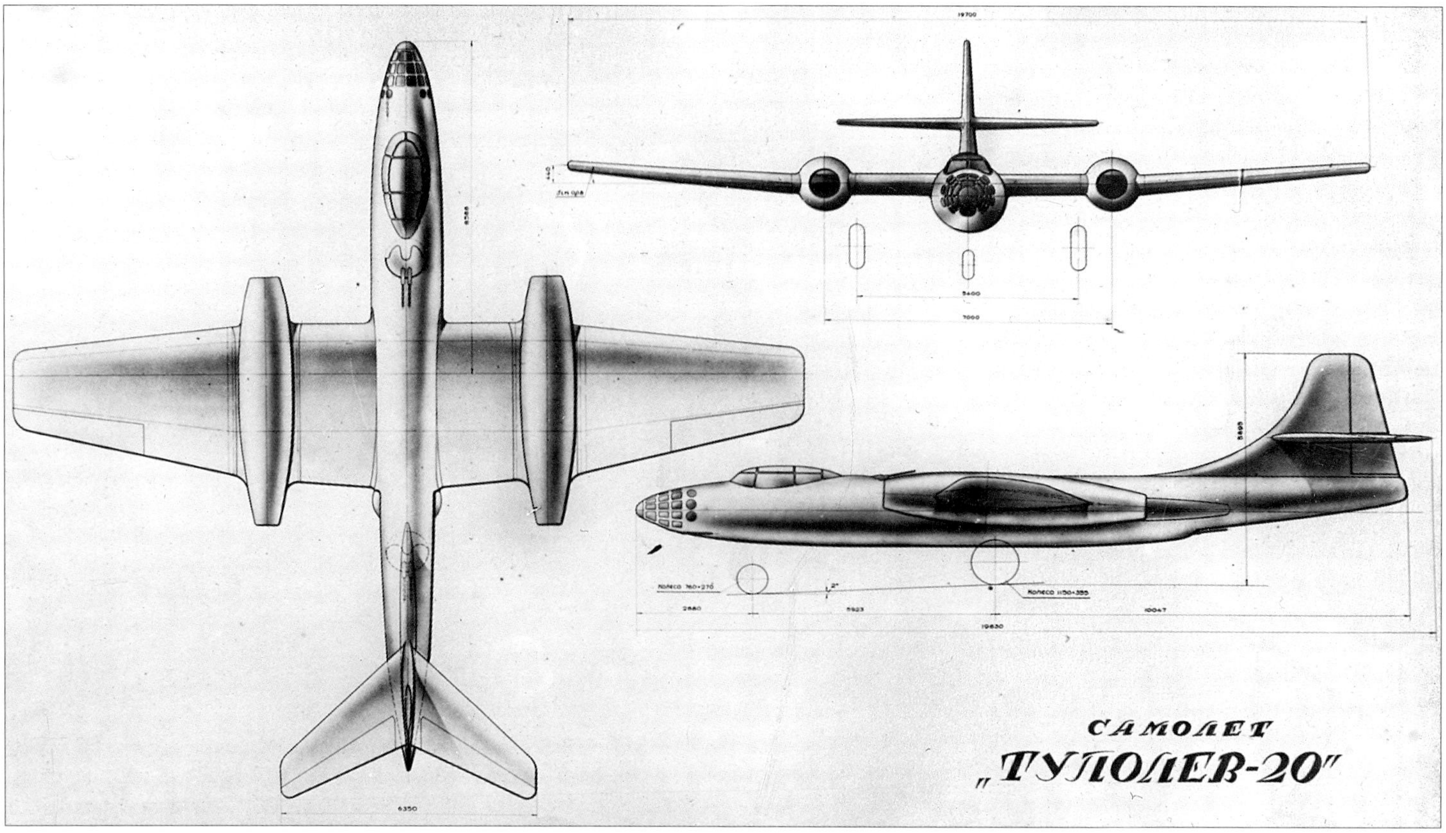

Tupolev 'Aircraft 73' as first proposed (1.47). It was initially given the official designation Tu-20.

At Andrei Tupolev's suggestion a third engine, a British Derwent, was installed in the rear fuselage, a move which produced the world's first trijet. The extra engine would only be used for take-off and to help avoid enemy fighters and in fact 'Aircraft 73' was completed with a retractable fairing for the Derwent so that it could be closed off during cruise flight. D S Markov was responsible for the team working on the aircraft. The Decree approving the additional engine was issued on 31st May and design work was completed in July. Construction of the prototype began in August (when it received the official designation Tu-14), it was rolled out from the experimental workshop in October and on 29th of that month made its maiden flight. The gun armament was now intended to be two fixed 23mm in the nose and two twin 23mm turrets facing rearwards. Before the factory's own trials were completed, the decision was made on 9th April 1948 to build an experimental production batch of ten aeroplanes fitted with two RD-45s and one RD-500. The '73' itself was found to be a bit short on the required range but met most of the rest of the specification, but a lack of radar and certain navigation equipment was a weakness.

However, a further Decree, released on 14th May 1949 after the results of the state testing on both the Il-28 and 'Aircraft 73' had been collected, cancelled the production run. The production jigs and assembly line, which by this time had been put in place at Factory No 23, were later moved to Factory No 39 for building 'Aircraft 81' below. The '73' prototype, however, was kept flying and served in several research programmes, particularly as a flying laboratory for rocket testing.

Tupolev's 'Aircraft 73' prototype.

Tupolev 'Aircraft 78' and 'Aircraft 79'

One more three-engined prototype was built, 'Aircraft 78' which flew on 7th May 1948 with a longer fuselage and modified empennage and was intended to serve as a reconnaissance type. A Decree dated 22nd August 1947 had told Tupolev to build 'Aircraft 73' in two versions, the Tu-14 medium bomber and the Tu-16 reconnaissance aircraft; the OKB

The Tupolev 'Aircraft 77' (Tu-30) bomber (1946).

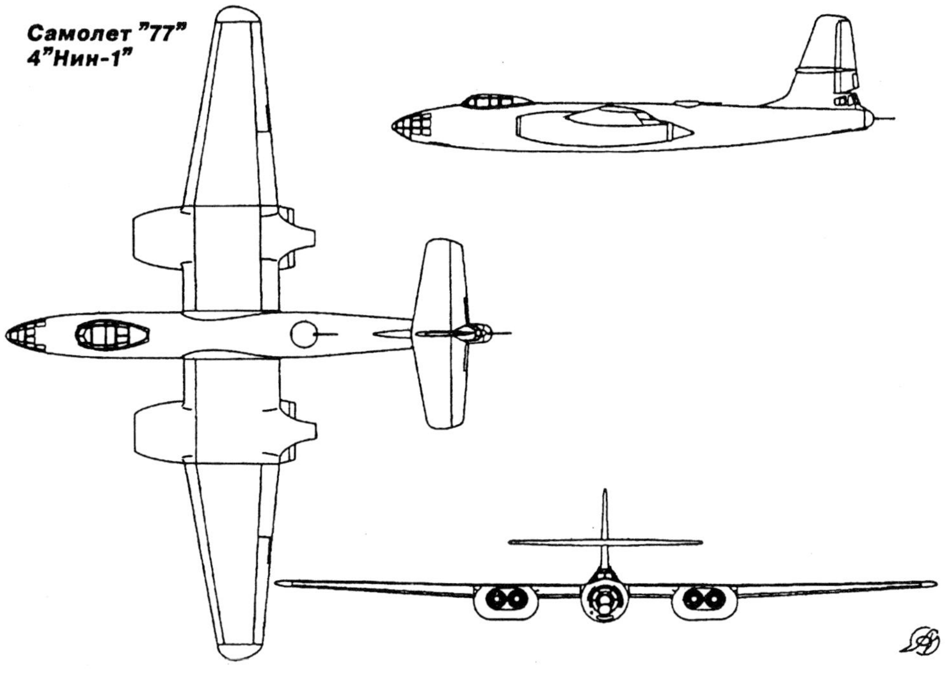

called the latter 'Aircraft 78' (this aircraft was initially designated 'Aircraft 79' and Tu-30, which must have made the paperwork quite confusing). Like the '73', 'Aircraft 78' was cancelled in favour of the Il-28, after the OKB had spent two years working on these aeroplanes. The 14th May 1949 Decree stated that all work on the '73' and '78' had to stop but 'Aircraft 78' was also kept in service for research duties.

Another reconnaissance development of 'Aircraft 73' was called 'Aircraft 79' and this was expected to use one RD-500 and two VK-1 engines. A go-ahead was given on 23rd December 1948 and the main external differences to the '78' was the absence of the air intake shutter on the rear RD-500 and new nacelles for the VK-1s. The prototype was to be produced from an available '73c' production airframe but again the Decree issued on 14th May 1949, which terminated the '73' and '78', also brought an end to 'Aircraft 79'. However, another modified 'Aircraft 73' with just two VK-1s, called 'Aircraft 81', was to prove more successful.

During its development 'Aircraft 73' became a rival to the Il-28 and both the '73' and '78' prototypes undertook their manufacturer's flight tests at the same time and venue as the Il-28 (the Flight Research Institute in Zhukovsky, south of Moscow). When Andrei Tupolev saw the Il-28's neat tail turret he realised that the heavy defensive armament fitted on his bombers, based on that used on the Tu-2, had helped to make them too large and too heavy; the extra crew that the guns required here was a big factor. At a special commission held by Stalin on 14th May 1949 to discuss the winning tactical bomber design, Tupolev was asked to modify the Tu-14 ('Aircraft 73') with two VK-1s in preparation for service with the Soviet Naval Air Arm.

Tupolev 'Aircraft 77'

In 1946, at the same time as proposals were made for a 'Aircraft 74' piston-powered reconnaissance aircraft and its 'Aircraft 76' bomber derivative, Tupolev also suggested to the VVS a project for a medium jet bomber with four Nene 1s, which became 'Aircraft 77' (and officially Tu-30). In its appearance this '77', which was not the 'Aircraft 77' jet prototype described above, was fairly similar to the North American B-45 Tornado that flew in March 1947. The desire to have the aircraft perform at high altitude meant that a wing was selected with a relatively large area while the engines were housed in pairs in underwing pods. The defensive weaponry comprised three single 23mm cannon, one fixed in the nose, a second in a dorsal turret aft of the wing and a third in the tail of the fuselage. The OKB proposed that the bomber would be ready for its state testing in June 1948 but it was not taken up by the Air Force and Ministry. 'Aircraft 77's predicted ceiling was 37,730ft (11,500m) and range with a 2,204 lb (1,000kg) load 1,865 miles (3,000km).

Tupolev 'Aircraft 81'/Tu-14

After the decision made by Stalin on 14th May 1949 Tupolev turned 'Aircraft 73' into the 'Aircraft 81' medium bomber with wing-mounted VK-1s. In fact the development of this aircraft had been approved by a Council of Ministers Decree on 23rd December 1948

Wind tunnel model of Tupolev's 'Aircraft 77'/Tu-30.

A production Tupolev Tu-14T.

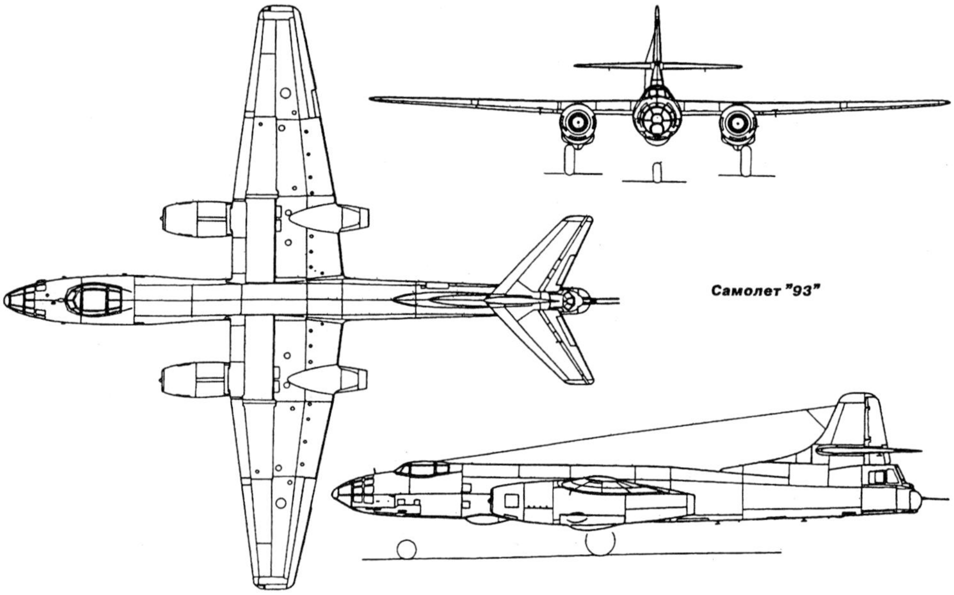

Tupolev 'Aircraft 93' (1951).
Russian Aviation Research Trust

and work on the preliminary design and a mock-up began in January 1949; the latter was inspected on 25th March. The main changes saw the '73's Nenes replaced by VK-1s and the RD-500 removed altogether, which was to prove a better arrangement than the '73's three power units. However, after the project had been reviewed and rejected by the VVS the OKB looked at reducing the crew and modifying the defensive gun positions along the lines of the Il-28. This was intended to counter the main reason why 'Aircraft 81' had been turned down, the bureau having continued its work in spite of the uncompromising tone made by the May 1949 Decree towards 'Aircraft 73' and '78'.

New and revised requirements were issued on 20th August with the gun armament reduced from three to two twin 23mm mounts and the crew from five to three. By the end of September the new preliminary design was ready, the tail cannon having increased the aircraft's length; the aircraft's flight characteristics were expected to be unchanged. The original Tu-81 had been officially designated Tu-18 but this redesign was renumbered Tu-14 and the first prototype flew on 13th October having been completed using sub-assemblies from production 'Aircraft 73s' built by Factory 23. State testing was completed on 27th May 1950 and series production at Factory 39 was given the go-ahead during 1949.

By July 1950 the first production aircraft had been rolled out at Irkutsk but this was modified into a Tu-14T torpedo bomber while the next example became the 'Aircraft 89' reconnaissance variant; altogether, five production aircraft were completed from cancelled 'Aircraft 73' airframes. In 1952 the type entered service with the Naval Air Force as the Tu-14T and around 100 were completed. 'Aircraft 89', officially dubbed Tu-16, first flew on 23rd March 1951 but did not begin its State trials because the Il-28R reconnaissance variant of Ilyushin's front-line bomber was the type chosen to go into production.

Tupolev 'Aircraft 93'

This was the last proposal to come out of the Tu-14 family and was a projected version of the Tu-14T which the OKB worked on during 1951 and 1952. The original layout and structure were retained and any changes were of a minor nature – the centre fuselage was made fatter to accommodate more fuel plus a longer bay for the aircraft's mine or torpedo load, which brought about the removal of the centre cabin, and the flaps were modified. Two 6,615 lb (29.4kN) VK-5 or 9,260 lb (41.2kN) VK-7 formed the main powerplant but two SU-1500 rocket motors were available for take-off assistance. The bomb and torpedo load, and the defensive armament, were basically unchanged but work on the '93' stopped after the completion of the preliminary design. Normal take-off weight was expected to be 45,194 lb (20,500kg), maximum speed at 16,404ft (5,000m) VK-5 561mph (902km/h) and VK-7 584mph (940km/h), service ceiling 39,370ft (12,000m) and range 2,175 miles (3,500km).

Advanced Developments

Most of the Soviet Union's first-generation jet bombers were eventually developed into more advanced versions.

Tupolev 'Aircraft 82'

This was the first native Soviet swept-wing bomber and came into being after the nation's first swept wing jet fighters had entered flight test in 1947 and showed a significant increase in flying speed. The Tupolev OKB was the first Soviet Bureau to look into the problem of fitting swept wings on bombers and in February 1948 it began working on a prototype fitted either with two 5,000 lb (22.2kN) RD-45F or 5,950 lb (26.4kN) VK-1 jets, which were expected to push the bomber's top speed up to around Mach 0.9 or

Tupolev 'Aircraft 82' prototype.

0.95. The project was created on Tupolev's own initiative and received the designation 'Aircraft 82'. To begin with the design represented a considerable modification of 'Aircraft 73' with the introduction of a 35° sweep wing, two RD-45Fs, a crew reduced to three and a different defensive armament; its span was 57ft 5in (17.5m), length 55ft 9in (17.0m) and normal take-off weight 28,660 lb (13,000kg). On 12th June the Council of Ministers raised the project to official status in a Decree which called the aircraft Tu-22.

However, during June the bureau completed a revised and slightly larger design which introduced new equipment and guns and also improved the flight characteristics. The normal take-off weight was now expected to be 29,762 lb (13,500kg), top speed 631mph (1,015km/h) at 9,843ft (3,000m) with VK-1s and 589mph (948km/h) at 16,404ft (5,000m) with RD-45Fs and respective ceilings 43,963ft (13,400m) and 39,862ft (12,150m). Construction of the first prototype began during the following month, the mock-up was inspected in mid-August and on 15th February 1949 the aircraft was completed. It made its maiden flight (with RD-45Fs) on 24th March and was found to possess satisfactory flight characteristics which proved the concept of swept wings on big aeroplanes. Only the prototype was built, which lacked guns although production aeroplanes were expected to have three cannon including two in a tail turret. The type was not selected for production because the VVS was quite happy with the straight-wing Il-28.

Tupolev 'Aircraft 83'

Tupolev's main combat variant of its 'Aircraft 82' project was intended to be the larger 'Aircraft 83' which had a longer fuselage stretched to 65ft 4in (19.925m) to accommodate an additional twin 23mm dorsal turret (and another crewman to operate it), plus a longer bomb bay and a modified front cabin. It was to be built as the second '82' prototype but was cancelled in 1949. Using 'Aircraft 83' as a basis, Tupolev also proposed the 'Aircraft 83P' heavy long-range interceptor to defend the northern regions of the USSR. This would have carried a powerful cannon armament but the Soviet Air Defence Command (the PVO) did not rate it because they felt that an aeroplane with a take-off weight of between 33,069 lb and 39,683 lb (15,000kg and 18,000kg) would not have much value. Seven years later the PVO was in dire need of such a type and this resulted in the OKB receiving orders for the Tu-128 heavy interceptor.

Ilyushin Il-30

Work on a new tactical bomber designated the Il-30, an ill-fated machine which was destined never to leave the ground, began on 21st June 1948 before the prototype Il-28 had flown. It was to be powered by two Lyul'ka TR-3 turbojets and had to meet a specification which included a normal bomb load of 4,409 lb (2,000kg), a range of 2,175 miles (3,500km) and a maximum speed of at least 622mph (1,000km). As distinct from the Il-28, which had VK-1 engines with centrifugal compressors, the new bomber had units with axial compressors housed in slimmer nacelles. The Il-30 full-scale mock-up was inspected by the so-called mock-up review commission at the beginning of March 1949 and the conclusion was favourable, although by then the prototype's construction was about 85% complete.

The new aircraft, while possessing basically similar dimensions to the Il-28, differed from it in several important respects. The most critical was the introduction of a swept-back wing, considered essential to meet the specified top speed, and the 35° sweep angle at quarter-chord was common to many Soviet jet aircraft designs of this period. The swept wings would benefit the aircraft's aerodynamics at high subsonic speeds and offer a higher performance, but they also brought certain drawbacks when compared to straight wings. In particular they possessed somewhat inferior lifting properties and attempts to make up for this by increasing the aircraft's angle of attack brought the onset of wingtip stall at an earlier stage than on straight wings. A remedy was found both by decreasing the wing taper and installing boundary layer fences on the upper wing surface. The rather moderate 12% thickness-to-chord ratio dictated by the decreased taper created some problems associated with obtaining the required stiffness and strength of the wing/fuselage joints. It also reduced the volume of fuel that might be housed in the wings and so the specified maximum range could only be obtained by using wingtip drop tanks.

In regard to the engine position, Ilyushin studied several configurations and finally decided to use the same arrangement as the

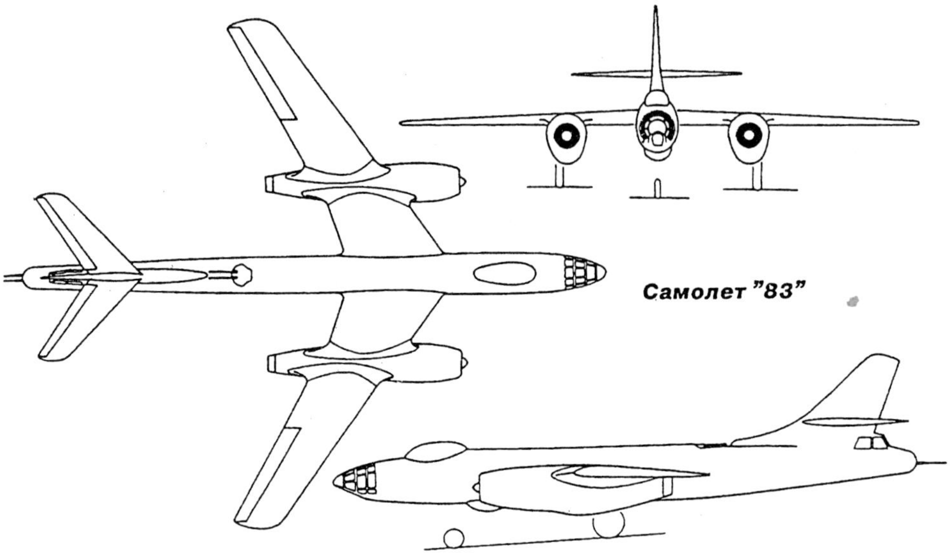

Tupolev 'Aircraft 83' (1948/49).

The Ilyushin Il-30 prototype.

Il-28. The engines were to be housed in slim nacelles attached directly to the underside of the wings which, like those on the Il-28, were 'area-ruled', their waist being narrowest where the wing section was at its thickest. The Il-30's mid-set wings, coupled with the wing's anhedral, reduced the clearance between the ground and the underwing-mounted engine nacelles and made it difficult to use the same arrangement to stow the main undercarriage as on the Il-28. Instead, the designers had to put the wheels in the fuselage making the Il-30 the first Soviet aircraft of any considerable size to incorporate a bicycle undercarriage. Compared to the Il-28 the new aircraft featured a more potent defensive armament of six 23mm Nudelman/Rikhter NR-23 cannon, two forward-firing fixed in the forward fuselage plus, for rear defence, two twin-turrets in dorsal and ventral positions just behind the cockpit. The swept-wing Il-30 was much heavier than the Il-28 and this was partly accounted for by the additional turrets.

The prototype was completed in the summer of 1949 and at the end of August was transported from the OKB's home to the flight test facility at Zhukovsky. Prior to flight testing the aircraft made several taxying runs, the purpose being to check the performance of the unorthodox undercarriage, but at slow speeds the aircraft tended to veer off the straight; however, as the speed increased the aircraft became more stable in its movement along the runway. The Il-30's first flight was imminent, yet it was destined never to take place because doubts arose as to the strength and aerodynamic properties of the swept wings – a contributive factor was an incident involving the Tupolev Tu-82. During a low-altitude training flight this aircraft passed through a area of strong upward air currents of varying intensity which caused the failure of one of the engine mounts. As a result, TsAGI insisted that the Il-30 should undergo further theoretical and experimental studies to evaluate the strength and aeroelastic properties of its thin swept wing.

During 1950 some wind tunnel experiments were made using an Il-30 scale model and these, together with new calculations, showed that the wing was strong enough. However, the delay caused by these investigations was fatal because the Ilyushin OKB was now instructed to concentrate on solving the tasks associated with introducing several versions of the Il-28 into service. On 20th August 1950 the government issued an order terminating all work on the Il-30 and the prototype stood idle on the OKB's premises for several years, until it was finally scrapped at the beginning of the 1960s. Interestingly, when the existence of this aircraft became known to the West, press reports about it contained much guesswork and erroneous information. For example, it was asserted that the Il-30 had a ventral gun barbette; it was also believed that the aircraft had been test-flown and 'became the first Soviet bomber to attain the then magical figure of 1,000km/h (621mph)'.

Ilyushin Il-28S

Ilyushin, however, kept on looking at ways of improving the Il-28 and during 1949 and 1950 put together the Il-28S project, a version of the original straight wing aircraft fitted with 35° swept wings and more powerful engines (S = *strelovidnoye* or swept wing). The powerplant was to be two 6,834 lb (30.4kN) Klimov VK-5 jets, an engine which began its bench tests in 1949 and, thanks to a new compressor, was expected to offer a lower fuel consumption; in the event, despite the development of a reheated VK-5F version which passed its State bench tests in 1953, this engine was to remain a prototype. Results also indicated that the Il-28S possessed no significant advantages over the standard Il-28 whilst the new wing would bring added complications to the production line; consequently the project was abandoned. The Il-28S would have been very similar in appearance to Tupolev's 'Aircraft 82'.

Ilyushin Il-38

The Il-30 had a sequel in the shape of a project designated Il-38, which was a twin-engined bomber powered by two TR-3 engines. It retained the Il-30's basic shape and carried the same bomb load but differed from its predecessor in having smaller dimensions, a lower weight and a revised defensive armament borrowed from the Il-28 (comprising two forward-firing NR-23 cannon supplemented by the Il-K6 tail turret with two NR-23 cannon). There were three crew and the bicycle undercarriage gave way to a tricycle arrangement which differed from the Il-28, being more along the lines of that used on the Il-46 (Chapter 3) except that the twin wheels on each main leg were moved further apart to either side of the nacelle; this arrangement resulted in the characteristic low sit of the aircraft on the ground. The project drew favourable comment from the Air Force but the military wished to see an improved airfield performance through the use of more powerful engines.

Ilyushin Il-42

In compliance with these wishes and a government directive dated 10th July 1950, the Il-38 project was reworked by the Ilyushin OKB to accept 11,023 lb (49.0kN) TR-3A (AL-5) engines; the new design was called Il-42 (a designation later reused for an attack aircraft). With a normal bomb load of 4,409 lb (2,000kg) it was to possess a range of 1,492 miles (2,400km), the same as the Il-28, and its airfield performance was to make it suitable to operate from the same airfields as the Il-28. The directive stipulated that the aircraft should be submitted for state acceptance trials in December 1951. More than 50% of the manufacturing drawings had been issued and the construction of the prototype was under way when the customer came up with a new, radically revised specification set down in a government directive dated 24th March 1951. This called for a two-fold increase in range and a maximum bomb load 50% higher than the previous figure of 8,818 lb (4,000kg). Ilyushin was required to start manufacturer's tests in a year's time, with submission for state acceptance trials (with the stipulated performance) in July 1952. This modification became the Il-46 described in Chapter 3.

Early Jet Bombers – Data / Estimated Data

Project	*Span ft in (m)*	*Length ft in (m)*	*Gross Wing Area ft^2 (m^2)*	*Max Weight lb (kg)*	*Powerplant Thrust lb (kN)*	*Max Speed / Height mph (km/h) / ft (m)*	*Armament*
Myasishchev RB-17	68 3 (20.8)	52 8 (16.06) (excl tail gun)	516 (48.0)	42,354 (19,212)	4 x Jumo 004 (RD-10) 1,985 (8.8)	451 (725) at S/L, 500 (805) at 16,404 (5,000)	2 x 20mm or 23mm cannon, 6,614lb (3,000kg) bombs
Ilyushin Il-22 (flown)	75 8 (23.06)	69 1 (21.05)	802 (74.5)	60,185 (27,300)	4 x TR-1 2,865 (12.7)	408 (656) at S/L, 446 (718) at 22,966 (7,000)	2 x 23mm + 2 x 20mm cannon, 6,614lb (3,000kg) bombs
Ilyushin Il-24	?	?	802 (74.5)	61,067 (27,700)	2 x AM-TKRD-01 7,275 (32.3)	518 (834) at S/L, 559 (900) at 22,966 (7,000)	2 x 23mm + 4 x 20mm cannon, 6,614lb (3,000kg) bombs
Tupolev Tu-77 (flown)	61 10.5 (18.86)	54 0 (16.45)	525 (48.8)	32,407 (14,700) (normal)	2 x Nene or RD-45 5,000 (22.2)	483 (777) at S/L, 487 (783) at 13,123 (4,000)	1 x 23mm cannon, 2 x 12.7mm guns, 6,614lb (3,000kg) bombs
Sukhoi Su-10 (at May 1946)	67 7 (20.6)	60 0 (18.3)	753 (70.0)	47,840 (21,700)	6 x RD-10 1,985 (8.8)	503 (810) at S/L, 528 (850) at 19,685 (6,000)	2 x 2 + 1 x 1 20mm cannon, 4,409lb (2,000kg) bombs
Sukhoi Su-10 (as built)	67 7 (20.6)	62 6.5 (19.06)	767 (71.3)	46,803 (21,230)	4 x TR-1 3,090 (13.7)	503 (810) at S/L, 528 (850) at 19,685 (6,000)	2 x 2 + 1 x 1 20mm cannon, 8,818lb (4,000kg) bombs
Ilyushin Il-28 (flown)	70 4.5 (21.45)	57 3 (17.45)	653.8 (60.8)	44,092 (20,000) (overload)	2 x VK-1 5,950 (26.4)	466 (750) at S/L, 524 (843) at 19,685 (6,000)	1 x 1 (fixed) + 1 x 2 23mm cannon, 6,614lb (3,000kg) bombs/torpedoes
Tupolev 'Aircraft 72'	72 4.5 (22.06)	52 10 (16.11)	658.7 (61.26)	35,494 (16,100) (normal)	2 x Nene 5,005 (22.2)	466 (725-750)	2 x 2 + 1 x 1 20mm cannon, up to 8,818lb (4,000kg) bombs
Tupolev 'Aircraft 73' (first study, 1.47)	61 0 (18.6)	64 5 (19.63)	595.7 (55.4)	33,069 (15,000) max	2 x Nene 5,005 (22.2)	530 (852) at 16,404 (5,000)	1 x 37mm + 1 x 2 x 23mm (fixed) and 2 x 1 x 20mm cannon, up to 3,307lb (1,500kg) bombs
Tupolev 'Aircraft 73' (flown)	71 2 (21.7)	66 8 (20.32)	724.5 (67.38)	46,296 (21,000)	2 x Nene 5,000 (22.2) & 1 x Derwent 3,500 (15.6)	522 (840) at S/L, 542 (872) at 16,404 (5,000)	2 x 2 + 2 (fixed) 23mm cannon, 6,614lb (3,000kg) bombs
Tupolev 'Aircraft 77' (project 1946)	c86 11 (26.5)	c64 0 (19.5)	c1,182.8 (110.0)	?	2 x Nene 5,005 (22.2)	497 (800) at 19,685 (6,000)	3 x 1 x 23mm cannon, c6,614lb (3,000kg) bombs
Tupolev 'Aircraft 81' (flown)	71 2 (21.7)	70 2 (21.4)	724.5 (67.38)	46,296 (21,000) (normal)	2 x VK-1 5,950 (26.4)	535 (861) at 16,404 (5,000)	2 x 2 x 23mm cannon, up to 6,614lb (3,000kg) bombs
Tupolev Tu-14T (flown)	71 2 (21.7)	72 0 (21.95)	724.3 (67.36)	46,142 (20,930) (normal)	2 x VK-1 5,950 (26.4)	527 (848) at 19,685 (6,000)	2 x 2 x 23mm cannon, 6,614lb (3,000kg) bombs/torpedoes
Tupolev 'Aircraft 82' (flown)	58 5 (17.81)	57 7 (17.57)	496.8 (46.2)	32,890 (14,919) (normal)	2 x RD-45F 5,950 (26.4)	577 (931) at 13,123 (4,000)	None fitted but 2 x 1 + 1 x 1 23mm cannon planned + bombs
Ilyushin Il-30	54 2 (16.5)	59 1 (18.0)	1,076 (100.1)	82,787 (37,552)	2 x TR-3 10,140 (45.1)	559 (900) at S/L, 622 (1,000) at 16,404 (5,000)*	3 x 2 x 23mm cannon, up to 8,818lb (4,000kg) bombs

*one source gives 653 (1,050)

Chapter Three

The First Heavyweights

The prototype Tupolev 'Aircraft 88'.

On 29th August 1949 the Soviet Union detonated its first atomic device, a move which changed the balance of world power in that America no longer had a monopoly in the possession of nuclear weapons. This added more fuel to the scenario that had become the Cold War and, indeed, probably sealed the events of the next forty years which would see two 'Super Powers', America and the USSR, in a stand-off with both capable of destroying the world at the flick of a switch. The term Cold War was actually coined in about 1947 and essentially defined all of the conditions for the confrontation – political, economical, military and ideological. However, to most people the Cold War will be represented by, and remembered as, two massive groups (the Warsaw Pact and the Western Allies) facing each other with enough nuclear weapons to saturate the world many times over.

The Soviet Union had begun the development of its own nuclear weapons in 1943. However, the design of the first device to be exploded was based on quite detailed descriptions and diagrams of America's first bomb, the documents having been obtained via the USSR's intelligence services. Work on the Soviet's own original design resumed in the spring of 1948 and this weapon was first tested on 24th September 1951 (the second Soviet nuclear test). Soviet nuclear stores were designed to fit inside the bomb bay of a Tupolev Tu-4 and the first explosion using an atomic weapon dropped from an aircraft was made on 18th October 1951 at the Semipalatinsk range.

Having lagged behind the Americans in the acquisition of atomic bombs, the Soviets took the lead in the development of the thermonuclear or hydrogen bomb and exploded the world's first hydrogen charge on 12th August 1953. Moreover, on 22nd November 1955 the Soviet's scientists gained a second 'first' when they detonated a hydrogen bomb released from an aeroplane. The successful development of hydrogen bombs allowed the world's designers to produce more powerful but smaller weapons (including tactical), and thus smaller aircraft to carry them, and the first such store to be developed in Russia was article '244N' or 'Tatiyana'. It was 18ft 0½in (5.5m) long and 2ft 9½in (0.85m) in diameter and could be carried by light bombers and even fighters.

The Soviets also became very skilled in designing stand-off flying bombs such as the AS-4 air-to-surface missile (codenamed *Kitchen* by the Western powers after its existence was revealed during the 1961 Soviet Aviation Day flypast) which could carry either nuclear or large conventional warheads at very high supersonic speeds. Finally, on 30th October 1961 a Tupolev Tu-95 bomber released a 50-megaton bomb over Novaya Zemlya which was detonated at an altitude of 2.5 miles (4km); it remains to this day the most powerful nuclear device exploded anywhere in the world.

Model of the Baade EF 131. George Cox

Model of the Baade EF 132. George Cox

The growing obsolescence of piston-powered bombers, brought about by the swiftly improving capabilities and performance of each new jet fighter design, meant that the next stage in Soviet bomber development must bring forth the introduction of some much larger jet-powered types and these would need to be capable of delivering nuclear weapons to America. However, the first 'heavy' jet bomber to fly in the Soviet Union, the Baade Type 150 which in reality falls inside the 'medium' bomber category, was actually a product of a team of German engineers.

Baade OKB-1 Designs

EF 131 and EF 132

The Baade OKB was formed on 22nd October 1946, primarily to make use of German Junkers company engineers who had been taken to Russia. Before the end of the Second World War the Red Army overran the factories and facilities of Junkers Flugzeug und Motorenwerke and in doing so captured the prototype Ju 287 forward-swept wing jet bomber, which had made its maiden flight on the 16th August 1944, plus many of the engineers who had worked on it. The second Ju 287, which was yet to fly, was to be powered by six engines and such was the interest in the aircraft that work on it was continued.

In the event the second machine was not completed but parts of it were used to build the EF 131 research aircraft at Dessau (EF = *Entwicklungsflugzeug*/research aircraft). The EF 131 was then dismantled and taken with many of its German engineers and pilots to Moscow where it was fully re-assembled and flown from Stakhanovo airfield (Zhukovsky) on or close to the 23rd May 1947. Two prototypes of the EF 131 were actually planned powered by six RD-10 engines but, as better engines and aircraft began to appear, the EF 131 was surpassed by the OKB's follow-on Type 140 project. The first prototype completed only a small amount of flying and on 21st June 1948 an order was made to cease all work on it.

The development of another Junkers design, the EF 132 long-range jet bomber with swept back wings and six Jumo 012 jets mounted in the wing roots three per side, had begun by the time the Soviet Army arrived. Dipl-Ing Brunolf Baade, a former chief designer at Junkers, and who with Hans Wocke was to become the leader of the Baade OKB, started the EF 132 project in 1945 and this was to become the second ex-German bomber project to be continued under

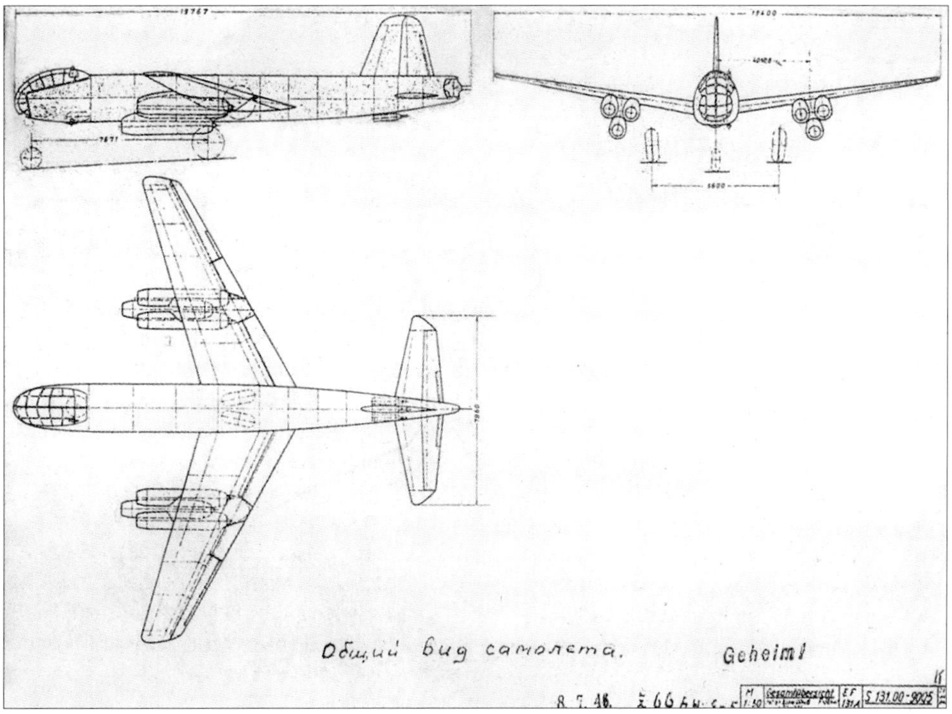

The Baade EF 131 (drawing dated 8.7.46).

The Baade EF 132 (1945).

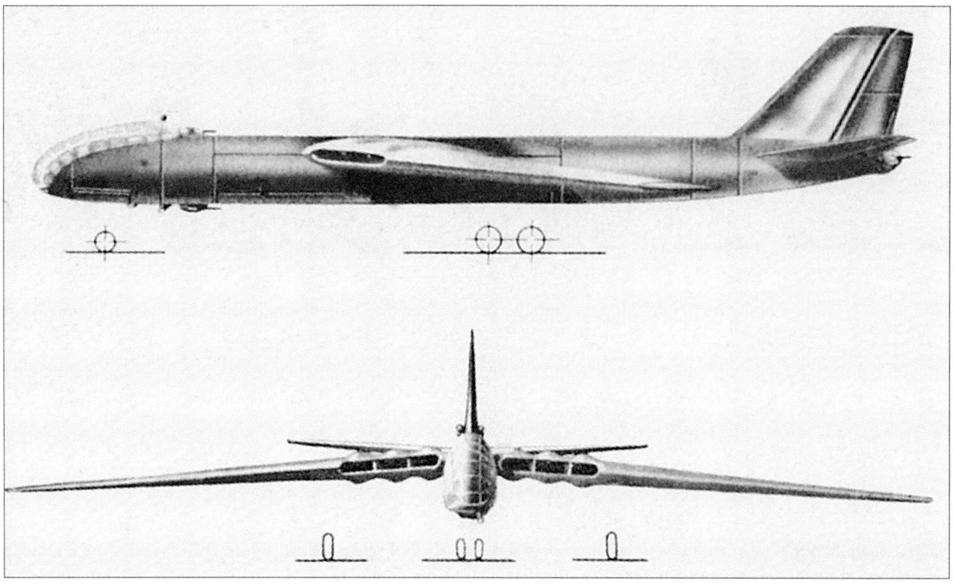

When the sole Baade 140/R first flew with Mikulin axial engines it carried tip tanks. These were removed when centrifugal Nene or VK-1 engines were fitted, as shown here.

This view of the 140/R shows the forward-swept wings to good effect.

Soviet control. An official order stated that its technical design was to be completed in 1946, the construction of two prototypes had to begin in 1947 and these were to be ready in 1948. Versions with forward and backward wing sweep were considered and then in 1947, after the OKB had been established in the USSR, the design was revised with six Mikulin AM-TKRD-1 axial jets replacing the original powerplant of Jumo 012s.

The EF 132 was to have three twin 20mm defensive turrets, one dorsal and one ventral just behind the cockpit well forward of the wings and another at the end of the fuselage, and five crew. Its normal bomb load was to be 8,818 lb (4,000kg) and the estimated range and service ceiling with Jumo engines had been 2,424 miles (3,900km) and 37,402ft (11,400m) respectively, and with Mikulin powerplants 2,486 miles (4,000km) and 43,635ft (13,300m). During 1947 work got under way on the construction of a full-size mock-up and by mid-1948 this had been completed and a number of sub-assemblies for the prototypes had been manufactured. However, a SovMin order terminated the EF 132 programme on 12th June 1948 because the aircraft had fallen behind schedule and would not be ready to start its test programme on time.

Baade Type 140

Such was the confidence held by the German OKB in its forward-sweep wing arrangement that the Type 140 was initially pursued as a private venture. The one weakness that did need to be addressed was the old powerplant which used early German jet technology and by 1947 the opportunity had arisen to fit a single Mikulin TKRD-1 jet engine under each wing. This was a new power unit which passed its State tests in 1948 and was also known as the AM-01. With these in place the aircraft's construction made good headway and was helped by including some major pieces originally intended for the second EF 131 prototype (the new engines were the principal difference from the EF 131). Originally called the EF 140 (until the German EF prefix was dropped), the aircraft made its maiden flight on 20th September 1948. The

A poor quality picture but rare because it shows the Baade 140R moving, in fact deploying its drag chute on landing.

Artist's impression of the Baade RB-2 bomber project (1948).

OKB had intended that it should be designed purely as a reconnaissance aeroplane (the 140R) but a SovMin decree placed in August 1948 declared that it should also be developed as a bomber (140B/R) and the second prototype was to be configured as such.

In October Baade was replaced as the head of OKB-1 by Semyon Alekseyev after the closure of the latter's own OKB in August. Up to now German morale had been low (the staff were still classified as prisoners) but both Baade and Alekseyev worked well together and the situation improved. Flight testing revealed that the 140 suffered from flutter but there were also difficulties in controlling the engine thrust, the TKRD-1 being prone to spontaneous changes in rpm which did not help the aircraft's flight characteristics. Later the thrust was increased to 8,333 lb (37.0kN) but Mikulin's engines never showed the desired reliability. They were eventually replaced by imported Nenes and then VK-1s but in July 1950, when the second machine had begun ground testing, the Type 140 was cancelled.

Prior to this, in early June 1949, OKB-1 had been ordered to convert the second aircraft into the 140R reconnaissance type, a move which halted work on the new Type 150 bomber (below) for about two months. There was another crash programme in September 1949 to convert the aircraft to the 140B/R bomber/reconnaissance configuration (bringing a further stoppage to the 150 until October), which was to have carried 3,306 lb (1,500kg) of bombs over 1,865 miles (3,000km) at a height of 39,370ft (12,000m), but the forward-sweep wings created structural problems which were never fully overcome while the aircraft's German background was also never going to be popular with Soviet officialdom. In many respects the EF series served merely as an insurance against the failure of the Soviet Union's home-grown bomber designs.

Baade RB-2

In late spring 1948 the bureau completed a new design for a short-range bomber called the RB-2. The aircraft's shoulder-mounted wings were swept back 38.33°, thickness/chord ratio was 9% to 11% and one 8,820 lb (39.2kN) Mikulin AM-TKRD-02 or one 11,025 lb (49.0kN) Lyulka TR-3 jet engine was mounted in a nacelle under each wing. There was a crew of five and the RB-2 had a bicycle landing gear with outrigger struts beneath the wingtips. A single 6,614 lb (3,000kg), two 3,307 lb (1,500kg), twelve 1,102 lb (500kg) or eighteen 551 lb (250kg) bombs could be taken on board, there was a single fixed forward-firing 23mm cannon plus two twin 23mm turrets in the dorsal and tail positions and a substantial quantity of avionics was to be carried. The RB-2's estimated service ceiling was 44,619ft (13,600m), time to 32,808ft (10,000m) 14.4 minutes and maximum range with a 3,307 lb (1,500kg) bomb load 3,294 miles (5,300km).

In July 1948 the project was submitted to the Minister of Aircraft Industry (Mikhail V Khrunichev) and NII VVS, and then on 17th August it was sent to Nikolai A Bulganin, the Minister of Defence, for review. A week later it was assessed by a MAP special panel, who suggested that TsAGI should be advised in regard to a more detailed analysis of the wing and tail units, and on 24th September the NII VVS gave the go-ahead for a design programme to proceed, although at the same time making several critical comments. The crew and tail gunner cabin mock-ups were approved on 5th October but, shortly afterwards, the RB-2 design was thoroughly revised to become the aircraft that was to be built as the Type 150.

Baade Type 150

As the RB-2 or Type 150, OKB-1's last bomber project presented a more conventional sweptback arrangement. Despite Baade and Wocke's confidence in their forward-sweep ideas, it was realised that making that layout work would be difficult and so, even before leaving Germany, they had sketched some alternative preliminary studies for a design with sweptback wings; in 1948 this work became an official OKB-1 project. Alekseyev took the place of P N Obrubov as chief designer and the aircraft, originally set at a weight of 55,115 lb (25,000kg) grew quickly in

Three views of the OKB-1 Type 150 bomber.

size until it was covered by a brief VVS specification which requested a take-off weight between 83,774 lb (38,000kg) and 103,616 lb (47,000kg). This put the bomber's dimensions between the Il-28 and Tupolev's new 'Aircraft 88' (Tu-16).

There was close collaboration with TsAGI on the bomber's design which still had two engines placed in nacelles mounted on underwing pylons, at this time a relatively new feature; in fact, the 150 was to be the first Soviet aircraft to feature swept pylons that kept the engine nacelles apart from the wings. The 150 also had a T-tail, instead of the RB-2's position near the base of the fin, but retained the bicycle undercarriage with wingtip outriggers. A full-scale mock-up was completed on 20th March 1949 and accepted officially on 11th April. At one stage the 150 was intended to be powered by the Mikulin AMRD-04 but in June 1949 MAP confirmed that two Lyulka TR-3A (AL-5) engines were to be used. During 1949 the 150's design work was slowed down considerably by interruptions from the 140 programme (above) but wind tunnel testing was completed in October of that year.

Just the one prototype was ordered and this flew on 14th May 1951 having been completed within the May 1951 deadline. It performed well for almost a year until, on 9th May 1952, the undercarriage, engines and fuselage were damaged when the 150 stalled during an approach to land. All of the 150's new innovations, including the swept wing (itself another new departure for big bomber design), had proved successful but there were some aerodynamic problems which were never cured. The aircraft passed all of its specified requirements but, despite efforts to the contrary, the decision was taken not to repair it because Tupolev's 'Aircraft 88' was showing far more promise. In late 1953 Baade and his team moved to Dresden to form the VEB company where they used the Type 150's general layout as a basis for the BB-152 airliner. The Type 150 itself was dismantled and parts of it were used as training aids at the Moscow Aviation Institute.

Before the first real Soviet heavy bomber arrived, the Tupolev Tu-16 described shortly, two other medium bomber types were produced.

Ilyushin Il-46

A SovMin resolution dated 24th March 1951 ordered Ilyushin to produce a high-level medium-range bomber prototype which could bomb from any height between 9,843ft (3,000m) and its ceiling, operate individually or with other aircraft and also perform in an arena which contained enemy fighters in numbers. Following the OKB's experience to date with swept wings, the biggest decision centred on whether to use them again or stick with straight alternatives; sweepback afforded a higher maximum speed but traditional straight wings gave better lift properties and promised a longer range. Faced with a tight schedule, Ilyushin eventually decided to produce two prototypes of his aircraft, called the Il-46, in parallel and the first of these would have a straight wing that made use of Il-28 structure and aerodynamics. His decision was guided, among other things, by apprehensions that the swept wings might become a source of unexpected problems and cause delay, while straight wings would enable speeds of 575mph (925km/h) to be reached while offering a much greater increase in range coupled with minimal increase to the aircraft's size and weight. The Il-46 resembled a scaled-up Il-28, but was much larger and heavier. The swept-wing variant was designated Il-46S (S = *strelovidnoye* or swept wings) and had 35° of sweep on the 25% chord-line. Work on the two designs proceeded concurrently and the Il-46S was expected to be capable of reaching 621mph (1,000km/h) and possess a range of 3,011 miles (4,845km) and ceiling 41,667ft (12,700m). The problem of providing suffi-

Side view of the sole prototype of the Ilyushin Il-46.

Planview of the swept wing Ilyushin Il-46S compared to the original straight wing Il-46.

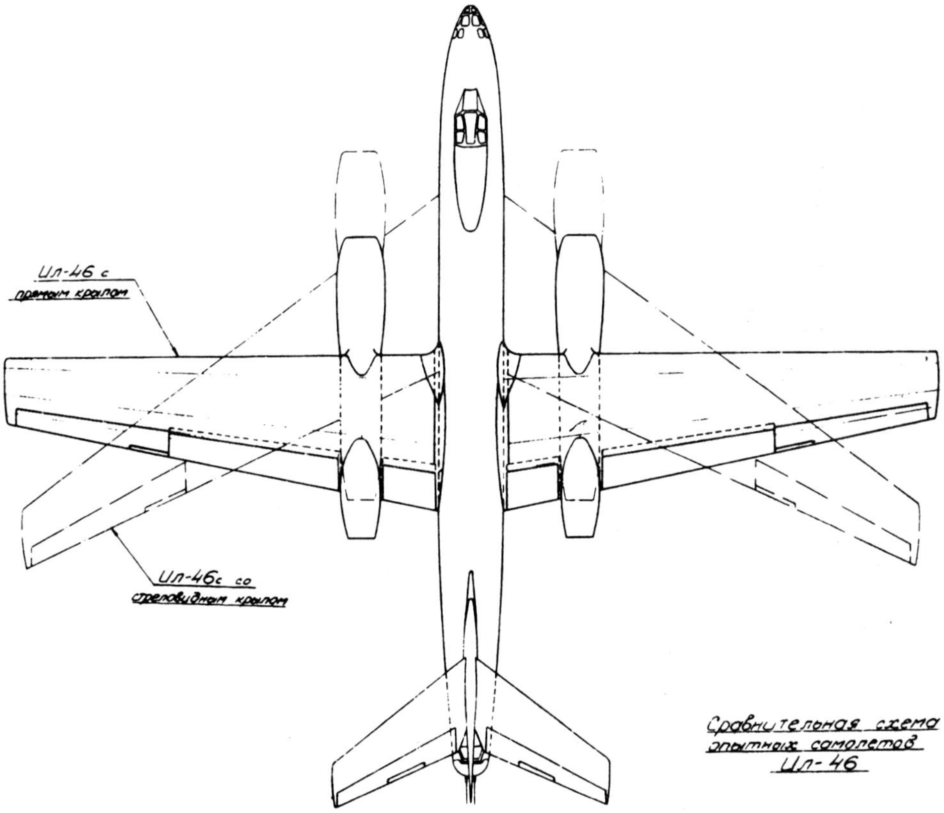

cient wing strength and rigidity was solved by increasing the root section to 16% thickness/chord ratio, which also made more space for fuel. Extra wing area was provided to offset the reduced lift characteristics of the swept wing itself but this did push up the take-off weight.

Another major decision to affect the layout was the choice of the optimum defensive armament. The customer (Air Force Command) was of the opinion that the arrangement used previously on the Il-28, featuring only a pair of forward-firing fixed guns and a tail turret, was inadequate in this case; Air Force specialists wanted the new bomber to possess an all-round protection in the front and rear hemispheres. Ilyushin contested these views, pointing out that this approach would result in an unjustifiable increase in aircraft weight and dimensions. In the end, Ilyushin's point of view prevailed and the Il-46 did feature a weapon arrangement similar to the Il-28. This determined the crew complement which comprised three persons – pilot, navigator and tail gunner. Interestingly, Ilyushin's competitor, Tupolev, chose a different approach and provided its rival bomber design, the '88' (Tu-16), with an impressive array of three turrets in addition to a fixed forward-firing cannon. Accordingly, its crew comprised six or seven people.

Both aircraft were to be powered by a pair of Lyulka TR-3A turbojets, an improved version of the TR-3 produced in 1948 which initially gave a maximum thrust of 10,174 lb (45.2kN); in 1952 this engine's rating was increased to 11,243 lb (50.0kN). At the beginning of the 1950s the TR-3A was manufactured in small numbers as the AL-5 but it was only ever installed in prototype aircraft – the Il-46, Mikoyan's I-350 fighter, the Yakovlev Yak-1000 experimental fighter (which never flew) and the Lavochkin La-190 fighter. In the Il-46 the engines were placed in the forward part of the nacelle well ahead of the wing leading edge so that enough room was left for the main undercarriage legs. In addition a 4,409 lb (19.6kN) thrust PSR-2000-15 rocket motor was placed on each side of the fuselage to supplement the available take-off power.

The Il-46 employed a novel undercarriage arrangement to allow it to operate from the same type of surfaces as the Il-28 – namely the main wheels were paired but given separate legs which retracted in opposite directions, the inner leg rearwards, the outer forwards, with both pivoting 90° to allow them to fit inside the engine nacelles. This design was similar to that used on the Il-38 (Chapter 2).

On 29th and 30th December 1951 the prototype Il-46 was transported from Moscow to the OKB's test facility in Zhukovsky. It made its maiden flight on 3rd March 1952 and the test pilot's first impressions were quite favourable. State acceptance trials followed between mid-August and mid-October and the results demonstrated that the bomber's performance met specification. The prototype was then fitted with improved AL-5F turbojets providing 12,676 lb (56.3kN) of afterburning thrust for take-off which improved the aircraft's performance. However, despite this, and the positive state acceptance results (the Il-46 achieved a range of 3,011 miles [4,845km] with a bomb load of 11,023 lb [5,000kg] dropped at the half-way point), the

Model of the Ilyushin Il-46S.

Artist's impression of the swept wing Il-46S.

type was not put into production. It lost out to its rival, Tupolev's prototype 'Aircraft 88', the future Tu-16, which was also submitted for State trials at that time.

Ironically, this was a reversal of the situation with the Il-28 and Tu-14 in which the Tupolev bomber lost out to Ilyushin's design but, in this case, the military opted for the '88' despite the fact that the aircraft had not yet passed its state acceptance trials. The decisive factors were the '88's swept-back wings, enabling the bomber to attain a maximum speed of 1,000km/h (621mph), and more powerful defensive armament – clearly Ilyushin had failed to convince the military of the soundness of his ideas for defensive armament.

The Tupolev Tu-16 was an altogether superior aircraft and, as a result, the Government ordered all work on the Il-46 to be terminated, a move which also put an end to the Il-46S prototype. The advanced development project for the Il-46S had been endorsed by Ilyushin in early December 1951, with a slightly increased wing area to compensate for the inferior lifting properties of the swept-back wing. Thicker wing roots, designed to ensure adequate wing stiffness, housed more fuel and the joints between the fuselage and the wing roots were moved forwards. This led to some changes in the fuselage structure, making it possible to increase the length of the bomb bay to 25ft 7in (7.8m). Another change was the more forward position of the engine nacelles because, to retain the same position of the main undercarriage units relative to the fuselage, the engine nacelles needed to be extended further aft.

Tupolev 'Aircraft 86'

In mid-1948 the Tupolev OKB began to evaluate some preliminary projects for high-speed medium and long-range bombers. Initially these studies fell within the confines of 'Project 486' and were basically developments of the 'Aircraft 73' (Chapter 2) but with powerplants offering a better thrust-to-weight ratio. The intention was to replace the '73's three engines, which gave a total static thrust of 12,346 lb (54.9kN), with two AM-TRD-02 (AM-02) engines offering a combined thrust of 21,076 lb (93.7kN); a cannon would take the place of the third engine. However, preliminary estimates showed that the excess power supplied by the new engines would push the airspeed into the critical Mach Number region and so it would be unwise to retain 'Aircraft 73's straight wing.

The solution was a swept wing, which had already been applied to 'Aircraft 82' (Chapter 2), but the higher fuel consumption of the more powerful engines would also require possibly as much as 26,455 lb (12,000kg) of internal fuel and so it became clear that the new engine arrangement would actually bring big changes to the aircraft's entire layout. The result was a new 'Project 486', with 34.5° swept wings and a crew of five that offered a range approaching 2,486 miles (4,000km) when carrying 2,205 lb (1,000kg) of bombs. Its defensive armament comprised a single NR-23 23mm and six G-20 20mm cannon. This work formed the basis of a new long-range bomber and at the end of 1948 the bureau began working on projects powered by the two 10,538 lb (46.8kN) AM-02s or, instead, two 9,921 lb (44.1kN) TR-3s. This step marked the raising of the '486' up to the stature of a full project, designated 'Aircraft 86' by the OKB.

Full development and wind tunnel testing (at TsAGI) got going in January 1949 and by 19th March the mock-up with AM-02s was 50% complete and the first drawings were available for the construction of a prototype; on that day a preliminary official evaluation was made of the mock-up. In April the VVS issued its technical and tactical requirements for the '86' project, which in many respects was similar to 'Project 486' although the wing sweep had reached 36° and there were significant differences to the fuselage, equipment and armament layout.

The biggest changes were more internal fuel and a larger bomb bay now configured for a normal load of 4,409 lb (2,000kg) and a maximum 13,228 lb (6,000kg). The aircraft was bigger, it had a fatter fuselage and a second pilot was introduced together with a new 'stepped' forward fuselage without the fighter-style canopy. A PSBN radar was housed in a nose blister and an additional nose cannon was carried while the rest of defensive weaponry was upgraded to 23mm. Service ceiling would be 43,963ft (13,400m) and range with a 4,409 lb (2,000kg) load 2,486 miles (4,000km). Also drawn were the '86R' reconnaissance and '86T' torpedo carrier

Two views of the Tupolev 'Project 486' model after swept wings had been introduced (1948).

The first swept wing version of the Tupolev 'Aircraft 86' (1948/49). Russian Aviation Research Trust

variants, the first with yet more fuel while the second had a modified weapon bay.

In due course 'Aircraft 86' had to be further redesigned with a longer fuselage and a greater span and fuel capacity, which included for the first time tanks inside the fuselage. A PSBN-M radar substituted the earlier set and the service ceiling would now be between 41,010ft and 42,651ft (12,500m and 13,000m) but the range figures were unchanged. In the meantime the OKB continued its research into long-range bombers and this eventually led to 'Aircraft 88' (below) which became the preferred choice. In due course 'Aircraft 86' was cancelled during its detail design phase but much of the engineering solutions, such as the fuselage layout and the positions of the equipment, guns and aircrew, found their way into the more advanced '88'.

In April 1949 Tupolev began working on 'Project 491' which was essentially a modified 'Aircraft 86' with a 45° swept wing. A pre-

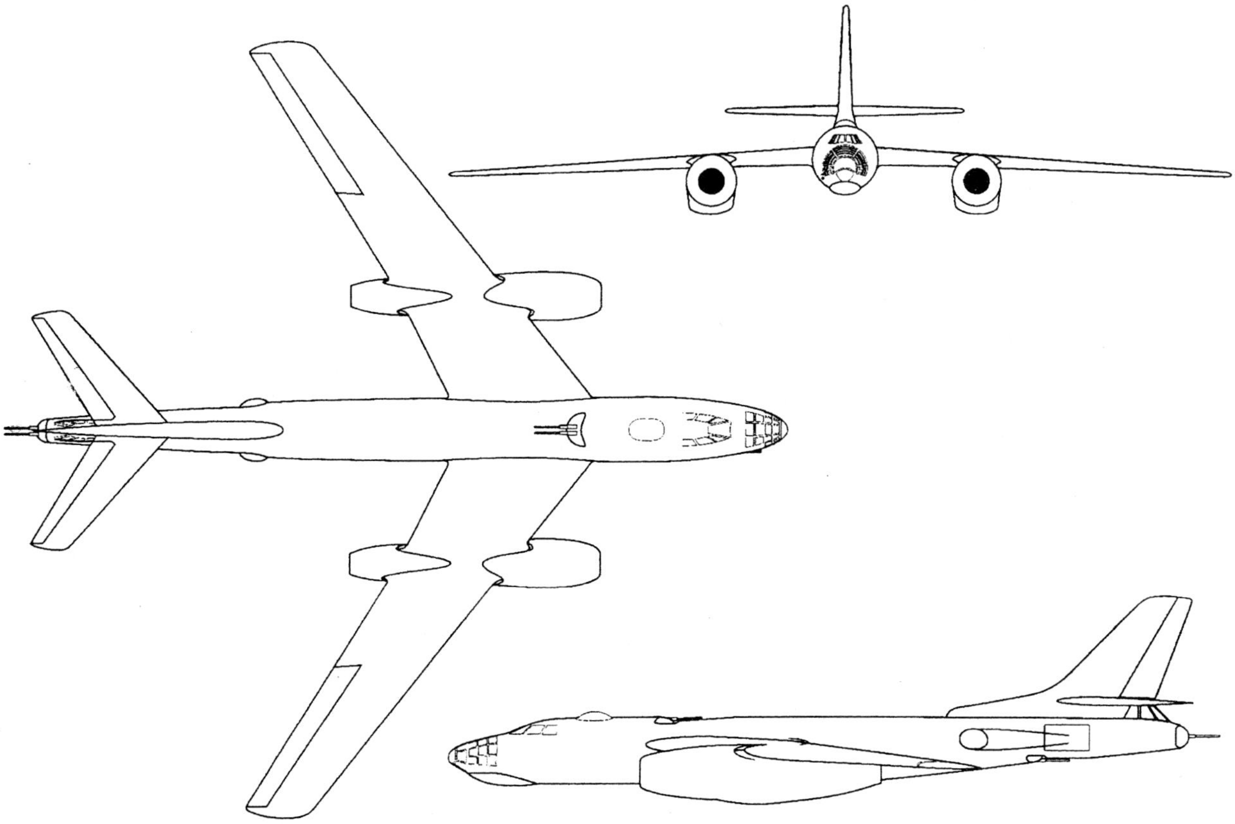

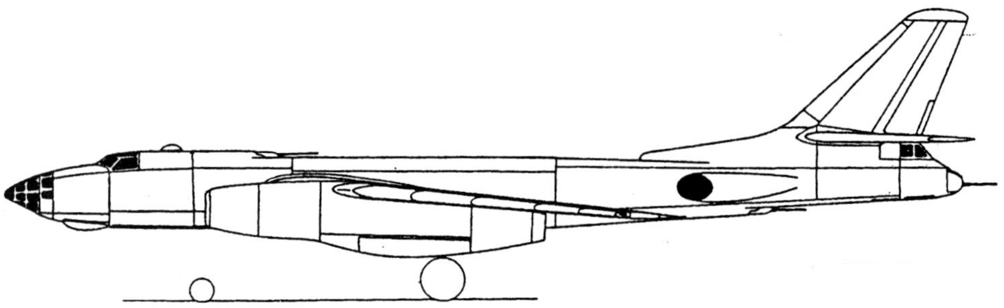

Side view showing the later arrangement of the 'Aircraft 86' with a modified fuselage (c1949).

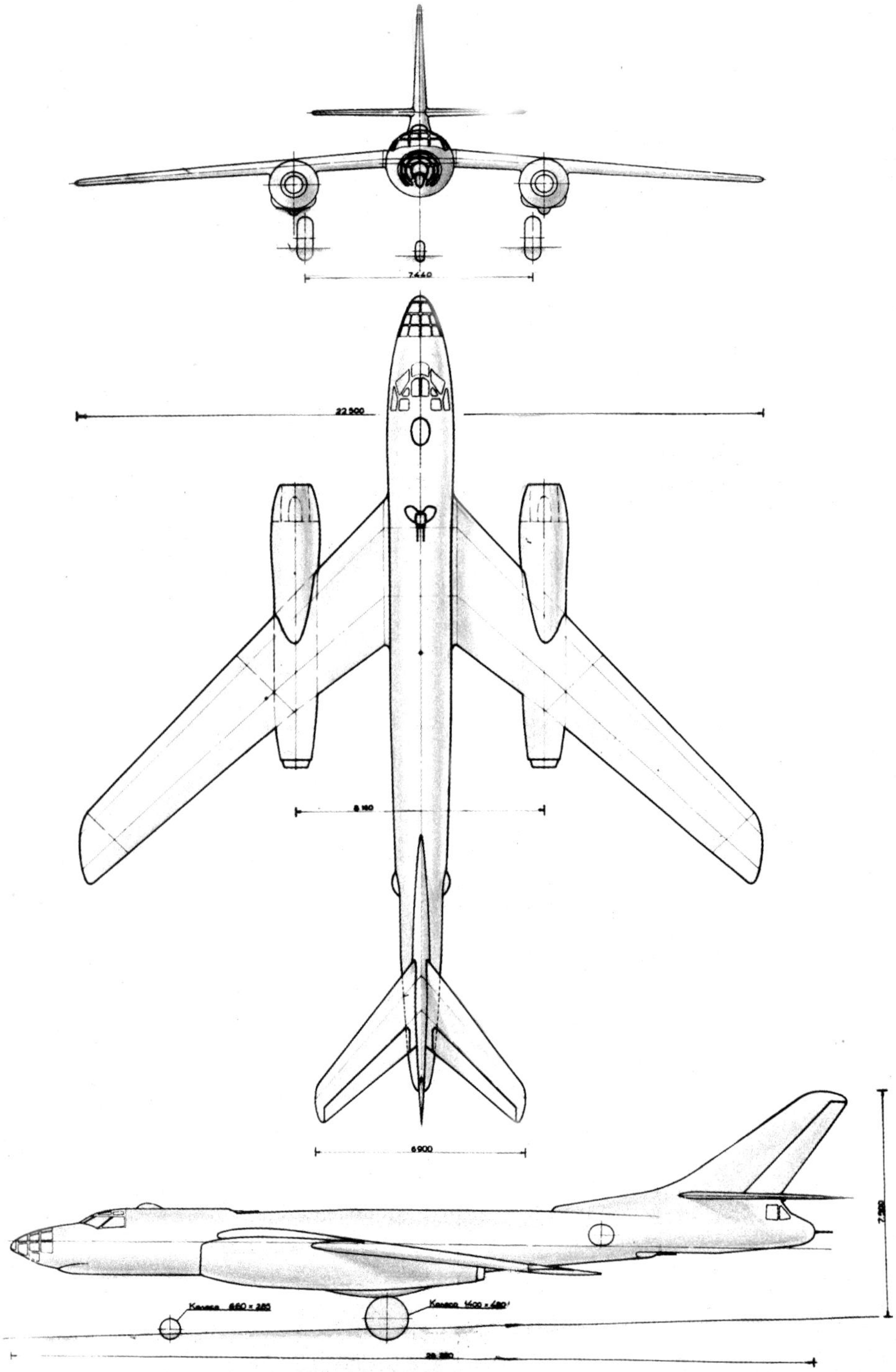

General arrangement of the Tupolev 'Project 491'.

liminary layout and calculations were completed and these suggested that the aeroplane's maximum speed at 21,325ft (6,500m) would be 674mph (1,085km/h), service ceiling 44,291ft (13,500m) and range 3,108 miles (5,000km). That was as far as the work on the '491' progressed but the data obtained proved most valuable and was incorporated into Tupolev's first transonic aircraft projects, described in the next chapter.

Tupolev 'Aircraft 87'

Almost alongside its work on the AM-02-powered 'Aircraft 86', the Tupolev OKB also completed its evaluation of the version fitted with TR-3s and this was given the designation 'Aircraft 87'. The engines were the only real difference, the structure was near identical, and the OKB's preliminary project was completed on 18th July 1949. Sometime later, when the '86' had reached a stage which had no clear objectives, the OKB switched entirely to the TR-3-powered aeroplane, and then after the '86' was cancelled it applied the 'Aircraft 87' results to the initial work that led to 'Aircraft 88'. 'Aircraft 87' had an estimated service ceiling of 43,963ft (13,400m) and a range of 2,952 miles (4,750km).

Tupolev Tu-16 Family

Tupolev 'Aircraft 88'/Tu-16

By 1948 one of the most critical requirements for the Soviet Air Force, and a subject which was to receive top priority, was to find and develop a long-range jet bomber that could replace the piston-engined Tu-4 in front-line service. The Air Force wanted a bomber offering a speed of approximately one and a half to two times that of the Tu-4 combined with the piston aircraft's load-carrying capacity. It was felt that such an aircraft could contain American ambitions since it could hit both that country's military bases in Europe and Asia plus her allies' political, economic and military centres. This bomber would also be able to deal with American and British shipping, in particular their aircraft carriers (which presented a strategic threat to the USSR) and merchant fleets sailing near the USA, Europe and Asia. For the Western allies any long-term conflict with the Soviet Union would be impossible to sustain without the extensive use of merchant shipping.

Wind tunnel models of five of the designs embraced by Tupolev's 'Project 494' designation.

The accounts of those aircraft projects described so far show that, soon after the close of the war, the effort put into jet bomber design in the Soviet Union started to grow very rapidly. Research into high-speed jet bombers began at the Tupolev design bureau as early as 1947-48 and it was the OKB's Projects Department, led by B M Kondorski, that completed an examination into the flight characteristics of such aeroplanes, including the use of swept wings. This work was carried out as a joint study with TsAGI and resulted in a series of projects for heavy swept wing jet bombers which were to act as a base for an examination of both the theoretical and the practical problems of developing such a type. The completion of this work, which solved many of the aerodynamic problems inherent in producing a heavy swept wing aircraft, brought a big step forwards in that it gave Tupolev the capability to produce the type of bomber that the Air Force wanted.

It was during this period that the Air Force ordered Ilyushin's Il-46 jet bomber but the existence of this new rival gave Tupolev the

Artist's impression of the Tupolev 'Aircraft 103'.

aim of producing a bomber that offered a better performance than both the Il-46 and its VVS specification. The first preliminary studies showed an aircraft with a 35° swept wing and an aspect ratio of 7 to 9 and in 1949 the work became a full project under the designation 'Aircraft 88'. Kondorski and his team, including some young engineers who had just graduated from the Moscow Aviation Institute such as A A Tupolev (Andrei N Tupolev's son), G A Cheryomukhin, A A Judin, I B Babin and V A Sterlin, completed a set of preliminary projects and then assessed whether these designs could reach the following performance parameters – maximum speed 590mph to 622mph (950km/h to 1,000km/h) at 32,808ft (10,000m), maximum bomb load 13,228 lb to 26,455 lb (6,000kg to 12,000kg), range with a normal bomb load 4,661 miles (7,500km) and service ceiling 39,370ft to 42,651ft (12,000m to 13,000m).

This series of designs was called collectively 'Project 494' and embraced a mix of alternative powerplant arrangements with either two or four AL-5 or AM-3 engines. A A Tupolev also suggested a layout which had the engine nacelles set at the wing roots adjoining the fuselage and this was selected for further study as 'Project 495' or '495-88'. When it was completed, Andrei N Tupolev officially proposed this design as 'Aircraft 88', together with a guarantee that it would have a better all-round performance than the Il-46. Finally, on 10th June 1950 the USSR Council of Ministers issued a Decree that secured for the bureau the development and manufacture of the '88' with two AL-5 engines. In due course however, the powerplant was changed to two 17,637 lb (78.3kN) thrust AM-3s.

In June 1949 the Mikulin design bureau in Moscow had begun designing the AMRD-03 engine which, for its time, had a very high static thrust of 18,078 lb (80.3kN). The first example was completed at the beginning of 1951 and bench testing began in the following spring. Once various improvements had been incorporated the engine was put into series production as the AM-3 with a maximum thrust of 19,290 lb (85.7kN). Besides Tupolev's bomber, it was to be used by several other famous aircraft.

On 15th June 1950 the VVS completed its Technical Requirement while, as the head of the OKB, Andrei Tupolev paid very close attention to the progress of 'Aircraft 88' as it passed through its detail design stage in the hands of a team led by S M Yeger; this was completed on 20th April 1951. In July the version with AM-3s was approved by the VVS together with the '88' mock-up. In its design 'Aircraft 88' was quite different to anything that had been handled previously by the OKB and its aerodynamic shape almost fell within the 'area rule' (before this law had been officially discovered). The empennage had been given more sweep than the wing which meant that the problems of 'shock stall' began to form on the empennage at higher speeds than on the wing, and this gave the aircraft good stability and controllability up to relatively high Mach numbers.

The first 'Aircraft 88' flew on 27th April 1952 and the type entered service as the Tu-16. It received the Western codename *Badger* and around 1,800 examples were built. It proved to be one of the finest jet bombers of its era and several versions were to be developed during its service career including a reconnaissance type and a dedicated air-refuelling tanker.

Tupolev 'Aircraft 90'

As noted 'Aircraft 88's original powerplant was to have been two AL-5 engines and, after the AM-3 became the preferred choice, it was decided to proceed with an alternative version with AL-5s as a fail-safe back-up, just in case the AM-3 proved to be a failure. The OKB's technical proposals were labelled 'Aircraft 90' and the aircraft was practically identical to the '88' except for the powerplant itself. The AL-5 supplied little more than half of the thrust of the AM-3 and so the '90' had four engines in two alternative arrangements. The first had two AL-5s in nacelles adjoining the fuselage (as per 'Aircraft 88') plus two more beneath the wing at two-thirds distance across the span; the second had the engines mounted in pairs away from the fuselage under the wing at one-third span. On the first version the main landing gears retracted into special wing pods similar to the '88's but in the second they folded into the engine nacelles; the underwing outrigger legs were retained by both types and the fuselage was near identical to the '88'. The success of the AM-3 powerplant meant that 'Aircraft 90' was never required.

Some time later, at the initiative of S M Yeger, the OKB evaluated an 'Aircraft 88'/Tu-16 development fitted with two TV-12 turboprops. This too received the in-house designation 'Aircraft 90' but the redesign needed to accommodate the new powerplants proved to be quite difficult. The large diameter propellers would necessitate the complete re-engineering of the landing gear joints, a redesign of most of the wing and some further changes to the fuselage. As a consequence the project did not prove to be viable, despite the initial estimates suggesting an increase in range at the expense of a slight decrease in the cruise and maximum speeds.

Tupolev 'Aircraft 97'

At the beginning of the 1950s the Tupolev OKB began work on studies for large jet aircraft which were to be capable of supersonic or high transonic speeds and the first projects to come from this, 'Aircraft 97' and '103', were both developments of 'Aircraft 88'. For the former the OKB evaluated the installation of 28,660 lb (127.4kN) VD-5 engines which, besides offering a considerable increase in power over the AM-3, were also more fuel-efficient; this meant that the '88's range could be maintained or even increased. To ensure that the additional thrust was used to the maximum effect, a new wing swept 45° at quarter chord was designed which was expected to give 'Aircraft 97' transonic performance. This project did not progress beyond the technical proposal stage but it eventually evolved into 'Aircraft 103' and then the supersonic 'Aircraft 105' (Chapter 4).

Tupolev 'Aircraft 103'

One version of 'Aircraft 97' was fitted with either four 24,250 lb (107.8kN) VD-7 engines or four AM-13s, the latter another new engine which was created by coupling together two AM-11 units. This project was eventually redesignated 'Aircraft 103' and retained its '88'-type fuselage; it also used the 45° sweep wing and had the engines mounted in pairs in wing root nacelles. 'Aircraft 103' was another design not to leave the drawing board.

Tupolev Tu-95 Family

Tupolev 'Aircraft 95'

Following the rather swift obsolescence of piston bomber aircraft the world's more powerful nations began to build big jet bombers. Britain started its V-bomber programme and the Americans produced first the Boeing B-47 Stratojet and then the massive B-52 Stratofortress. The Soviet Union tried to match these aeroplanes but also had to decide, early on, whether to use turbojet engines or turboprops to power them. In the end the Soviet's intercontinental bomber force employed one of each format. The need to replace the piston Tu-85 (Chapter 1) became so critical that the new bomber had to be in VVS service by 1954, the year that the Kremlin had set as a possible starting date for a nuclear conflict with the United States. The competitors were Myasishchev, whose M-4 is described shortly, and Tupolev.

In 1951 Tupolev sent a memo to Stalin explaining his theories of big bombers. The first stage was to find what the highest maximum speed was likely to be for an aircraft with a swept wing, regardless of the powerplant. With TsAGI research it became clear that speeds around 594mph (955km/h) at 26,247ft (8,000m) and 553mph (890km/h) at 39,370ft (12,000m) were possible with current knowledge and technology. The relationships between type of powerplant, the power that it supplied and the aircraft's size and weight were balanced against the aircraft's estimated speed, ceiling and range and these showed that a turbojet could give the bomber these speeds, but ensuring that a range in excess of 6,837 miles (11,000km) could go with them would be difficult and require an exceptionally large and heavy aeroplane. However, the new turboprop engines designed by Kuznetsov made the objective rather easier to achieve and suggested that an aircraft of 286,596 lb to 352,734 lb (130,000kg to 160,000kg) maximum weight could have a range of 9,323 miles (15,000km) and, quite possibly, rather more.

The Tu-95 heavy bomber programme was started in the spring of 1950. Preliminary studies embraced designs having swept flying surfaces and powerplants of four AM-3 jets, four TV-10 turboprops plus two TR-3A jets, four TV-4 turboprops plus two AM-3s, or four TV-10 turboprops only. The final results suggested that there were two possible alternative solutions, one with four turboprop engines and another with four turbojets. For an aeroplane weighing 440,917 lb (200,000kg), four AM-3s were expected to give a range of 6,215 miles (10,000km) at a speed of 559mph to 590mph (900km/h to 950km/h), when four turboprops offered 8,080 miles (13,000km) at around 497mph (800km/h).

Tupolev himself understood that range was the primary consideration for this aircraft (that is, the ability to reach America) and consequently accepted the turboprop despite the fact that VVS Command, and some of the heads of the Aviation industry, had insisted that his 'Aircraft 95' should have four jets. However, Tupolev's case was proved after flight testing had begun – 'Aircraft 95' showed a practical range of over 8,701 miles (14,000km) when the Myasishchev M-4 jet-powered prototype could only cover 5,594 miles (9,000km).

The Bureau completed its preliminary design in October 1951 with four Kuznetsov 2TV-2F turboprops. During this phase the only existing turboprop offering anything like enough power had been the 6,250hp (4,661kW) Kuznetsov TV-2F, developed out of the Jumo-022, but to function properly the '95' actually needed something like 12,000hp

Tupolev Tu-95.

(8,948kW) and so it was decided to use coupled 2TV-2Fs in each nacelle, with the prospect of replacing them with single, more powerful, TV-12 units later. Developed to go with them was a reduction gear to drive two huge counter-rotating four-blade propellers on each nacelle. This choice of powerplant, however, meant that the Tu-95 could only be re-engined with units mounted on or beneath the wings and the result was that this aircraft's configuration, swept wing with four tractor turboprops, has never been repeated by any other type except the follow-on 'Aircraft 96'.

'Aircraft 95' or Tu-95 had been formally launched by a SovMin directive on 11th July 1951 which covered both the 2TV-2F and TV-12 versions, with the prototypes to be ready for flight testing in September 1952 and September 1953 respectively; production was to start at Factory No 18 at the beginning of 1953. The OKB's Strength Department, led by A M Cheryomukhin, had a key task in that the aircraft had to possess a flexible swept wing with a high aspect ratio and this required the design of a light, strong and durable structure. A joint research programme was carried out with TsAGI and the Myasishchev OKB (for the M-4) and resulted in a substantial reduction in wing weight. A string of other difficult development problems also had to be solved and to co-ordinate the task Tupolev appointed N I Bazenkov, his closest assistant, to lead the full programme.

The first prototype was completed in autumn 1952 and made its maiden flight on 12th November; however, it was tragically lost in a fatal crash on 11th May 1953 after a reduction gear pinion in engine number three failed (it is believed that after the crash Tupolev ordered all photographs of the first aeroplane to be destroyed). The situation was close to disaster with the Ministry of Aviation Industry considering the abandonment of the whole programme and, for Factory 18, a switch in production to the Myasishchev M-4. However, Andrei Tupolev and his colleagues worked hard to save the project and, after introducing a series of technical and organisation changes, they succeeded. Work continued on the second prototype (the 'doubler') with four TV-12 (NK-12) turboprops but the delays caused by these problems and other refinements meant that the aircraft did not get airborne until 16th February 1955. In mid-1956 this machine and the first two production aeroplanes carried out the type's state testing, when the 'doubler' achieved a range of 9,347 miles (15,040km).

The State assessment resulted in a new engine being fitted, the NK-14M with more take-off power, and together with more internal fuel, the resulting Tu-95M became the standard production model. At last the Soviet Union had a modern nuclear weapon delivery system and the type was eventually built in many versions, including the anti-submarine Tu-142 described in Chapter 9, with different engines and weapons including cruise missiles. The Tu-95 received the Western codename *Bear*.

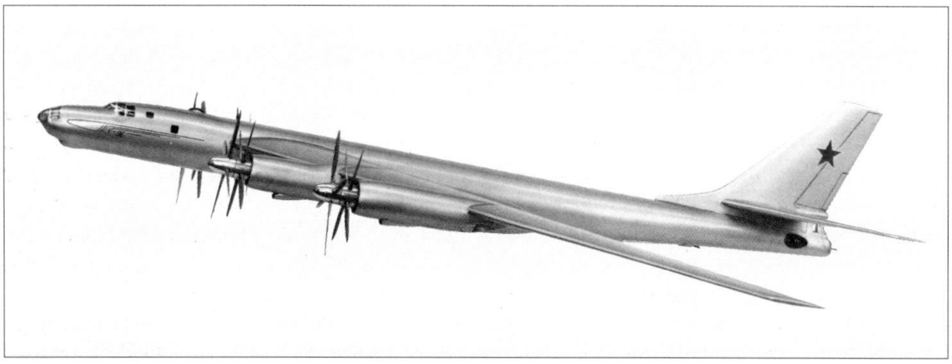

Impression and model of the Tupolev 'Aircraft 96'.

Tupolev 'Aircraft 96'

By the time 'Aircraft 95 had reached the preliminary design and evaluation phase Tupolev was considering a high-altitude version called 'Aircraft 96' which was designed to operate at heights of 52,493ft to 55,774ft (16,000m to 17,000m). It was to be similar to the basic '95' but would introduce high-altitude TV-16 turboprop engines (a development of the TV-12) and a much greater wing area with a new wing centre-section; in addition the fuselage and the nose crew cabin were also new. The SovMin resolution covering the programme, dated 29th May 1952, stated that two prototypes were to be built and the preliminary design had to be completed in March 1953, a target that was achieved.

However, the loss of the first Tu-95 plus the difficulties experienced with the development of the TV-12 brought the '96' close to cancellation. With most of the Tupolev and Kuznetsov OKB's energy going into dealing with the Tu-95 and TV-12, progress on 'Aircraft 96' automatically slowed down and it was not until the second Tu-95 entered flight test that the later aircraft got moving again. The prototype was built in 1955 but problems were then encountered with the TV-16 which was even heavier than the TV-12 and suffered several failures on the bench. Kuznetsov was unable to bring the engine to an airworthy state and no examples ever reached an airframe.

Instead Tupolev had to fit TV-12s (later four NK-12Ms were installed) and factory testing began in the summer of 1956, but the rapid advances made in the development and introduction of high-altitude supersonic jet fighters and surface-to-air missiles prompted a decision made in March 1956 to halt the programme. Such an aeroplane would have a minimal chance of surviving in an environment defended by such systems and so the VVS opted for aircraft carrying strategic cruise missiles, where the airframes did not have to enter defended airspace.

With TV-12s 'Aircraft 96's service ceiling was 40,682ft (12,400m) and range 9,323 miles (15,000km); the equivalent estimated figures for TV-16s were 55,118ft (16,800m) and 10,441 miles (16,800km) and the later units also offered a maximum speed of 561mph (902km/h). The only aircraft to be completed was used for research and experimental purposes until the end of the decade but the bureau did assess a variant intended to carry and launch the 'Aircraft 100' bomber (Chapter 5). After release from its carrier aircraft, the '100' would proceed to a point where it could launch a nuclear weapon at its target and then return to base under its own power.

Another engine assessed for 'Aircraft 96' was the RVD turboprop designed by the Shvetsov OKB. It was calculated that the '96's ceiling with this engine would be 52,493ft (16,000m) and speed at that altitude 622mph (1,000km/h). The total thrust from one RVD unit was estimated to be 10,000lb (44.4kN) which was split almost equally between the propeller and exhaust; the propeller thrust was 5,200lb (23.1kN) with an efficiency of 0.8, the jet thrust 4,800 lb (21.3kN). During ascent to cruise the power would remain relatively constant, although the efficiency of the propellers would decay with height. At cruise speed the engine was expected to provide a continuous 12,000shp (8,948kW) but, above 52,493ft (16,000m), horsepower would be lost proportional to the increase in altitude and loss of air pressure. On the other hand the pure jet thrust was expected to increase linearly from sea level to 52,493ft (16,000m) and then stabilize at 4,800 lb (21.3kN). Above this height the jet thrust would also decay but, until then, the RVD's total thrust actually increased with altitude.

Tupolev 'Aircraft 99'

Another outcome of the loss of the Tu-95 prototype was that Tupolev had to assess some alternative engines for the basic airframe. One solution was to introduce the D-19 currently being developed by the Soloviev design bureau which offered the same power as the Kuznetsov 2TV-2F and TV-12 turboprops. At the same time evaluations were also concluded for four Dobrynin VD-5 turbojets but difficulties associated with the design, construction and static testing of this new jet engine eventually led to a decision to install four 24,250 lb (107.8kN) VD-7s (a developed version of the VD-5) or six 14,330 lb (63.7kN) Lyulka AL-7 turbojets; it was intended to put these inside nacelles hung under the wings on pylons in an arrangement similar to that used on the American B-47 bomber.

All of this work was brought together under the designation 'Aircraft 99' and the configurations examined also included buried engines, but the version with four pylon-mounted units proved to be the most efficient. Using the VD-7, the '99's estimated range was between 7,458 and 8,080 miles (12,000km to 13,000km), which was superior to that claimed for the Myasishchev M-4 jet bomber (below), and estimated maximum speed 590mph to 622mph (950km/h to 1,000km/h).

Also during 1953 Tupolev assessed the value of several alternative aerodynamic configurations and propulsion systems including changes to the wing area and take-off weight and changes to the wing sweep. The purpose of this research was to determine further the optimum characteristics for pure turbojets versus turboprops. The first successful indigenous Soviet turboprop engines were produced during the early 1950s by the two OKBs supervised by A D Shvetsov and N D Kuznetsov. The core of the Kuznetsov turboprop, which was given the designation P-8, had originally been created in the early 1950s for Tupolev's stillborn supersonic strategic carrier designated 'Aircraft 108' (Chapter 5); as noted, Shvetsov's RVD turboprop was designed for 'Aircraft 96'. The 'Aircraft 99' studies included wing configurations with areas varying from 3,226ft² to 4,839ft² (300m² to 450m²) and variations in sweep angle between 45° and 55°. The take-off weight was also altered with the smaller designs offering gross weights of 300,000lb (136,080kg) and the largest 500,000lb (226,800kg). Some of the configurations were tunnel tested by the TsAGI.

Concurrent with this Tupolev upgrade, the Myasishchev OKB moved ahead with an improved M-4 equipped with four VD-7s (the M-6 described below) and with these new powerplants a significant improvement in range over the M-4 was claimed. Myasishchev's new bomber successfully completed its first flight during 1956 and entered full-scale production, and so 'Aircraft 99' was abandoned. In addition, after the Kuznetsov design bureau introduced modifications to its TV-12 turboprop which improved the engine's reliability, the requirement for an alternative powerplant for the Tu-95 was dropped.

Myasishchev M-4 and 3M Family

Myasishchev VM-25/M-4

When Vladimir Myasishchev officially returned to designing his own aircraft in 1951 one of his first assignments, in collaboration with G N Nazarov from OKB 22 and TsAGI, was to assess the feasibility of building a strategic long-range bomber (*strategichesky dahl'ny bombardirovschchik* or SDB). In fact Myasishchev himself had actually been working on such a type since 1948 but the new task embraced many parametric studies which, as ever, were a painstaking task requiring a vast quantity of calculations and graphs with the slide rule still the principal tool. However, the results were favourable and a meeting with the TsAGI's technical committee concluded that such a bomber could be produced.

Myasishchev's first full SDB proposal was made in February 1951 while he was still based at the Moscow Aviation Institute (the MAI) as head of fixed-wing aircraft design, and he observed that the main problem for such an aeroplane was to blend the required high speed with the maximum range. One of the preliminary studies showed a design powered by six VK-5 engines which, at a take-off weight of 198,413 lb (90,000kg) and flying at a speed between 466mph and 497mph (750km/h and 800km/h), was expected to be able to deliver a 6,614 lb (3,000kg) bomb load across a range of 7,458 miles (12,000km). A full SDB, however, would be even bigger with a span of 164ft (50m), length 144ft 4in (44m), wing area 3,226ft² (300m²), maximum take-off weight 242,504 lb (110,000kg), maximum bomb load 44,092 lb (20,000kg) and range with 11,023 lb (5,000kg) bombs 7,458 miles

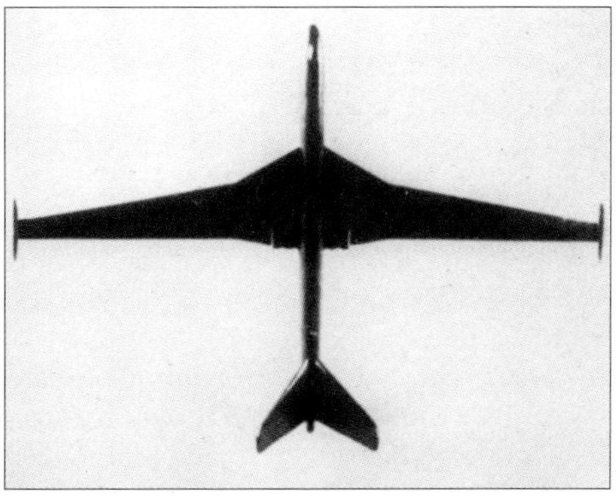

The first version of Myasishchev's Aircraft M had straight wings (1951).

Model of the Myasishchev M study which had swept wings and a butterfly V-tail (1951).

(12,000km). This design had swept flying surfaces with shoulder-mounted wings, a bicycle undercarriage and four Mikulin AMRD-03 engines; the wing itself had a cranked leading edge and two kinks in the trailing edge.

This project was the largest and heaviest aircraft yet designed in the Soviet Union and was thought to be quite attractive. The SDB programme was officially authorised by a SovMin directive dated 24th March 1951 and it was this move which allowed Myasishchev to re-open his OKB (at Factory No 23 at Fili near Moscow), together with permission to recruit as many specialist engineers as he required; as a result the majority of the 1951 graduates from the Moscow Aviation Institute went to OKB-23. Myasishchev called his bomber the VM-25 and it was later officially designated Aircraft M (and then M-4 after service acceptance). The Soviet Government took the project very seriously and the rapid increase in the number of technical staff allocated to the new facility ensured that its progress was swift.

Leonid Selyakov was entrusted with the preliminary project, the design of which had to be completed in just four months so that the large lead which America held over the Soviet Union in the development of this type of aeroplane could be drastically reduced. It was Selyakov who actually drew the first alternative general arrangement for Aircraft M which had straight wings, wing root engines and a swept V-tail. The next layout had the wings swept, the third had a swept wing but introduced a conventional swept fin and tail with wing root engines and the fourth was the same as the third except that the engines were moved to underwing pods, one per side with two units in each nacelle.

Design number five had eight 6,250ehp (8,381kW) Kuznetsov TV-2F turboprops driving eight-blade contra-rotating propellers which were installed back to back in nacelles going clean through the wing torsion box; the forward engine in each pair was in tractor configuration while the aft engine acted as a pusher. The aircraft was 165ft 8in (50.5m) long, had a span of 169ft 1in (51.54m), wing area 3,269ft^2 (304m^2) and a maximum take-off weight of 385,802 lb (175,000kg). Version six was a swept-wing aircraft with six Lyul'ka AL-5 turbojets, four of which were buried in the wing roots and received their air from elliptical intakes (as per the future M-4) while the other two were mounted further outboard in underwing nacelles. The external dimensions and wing area were identical to the turboprop M but the maximum take-off weight had risen to 392,416 lb (178,000kg).

A seventh design was almost identical but was powered by four AM-3 turbojets all mounted in the wing root position (maximum take-off weight was 390,212 lb [177,000kg]) while an eighth design was very similar but had a slightly lower span of 151ft 11in (46.3m), wing area 3,010ft^2 (280m^2) and maximum take-off weight 410,053 lb (186,000kg). Some even bigger designs were also considered which included engines carried on underwing pylons. Towards the end of the studies the powerplant choice was narrowed to either four 11,032 lb (49.0kN) Lyul'ka AL-5s or 17,634 lb (78.4kN) Mikulin AM-3s.

Finally, after careful consideration, the engineers chose what was considered to be the optimum layout – a swept-wing aircraft with a conventional swept tail unit powered by four AM-3 engines, the most powerful Soviet jet engine at that time; a 'zero-track' bicycle undercarriage was also adopted. Myasishchev opted for AM-3s because, despite giving a shortfall in take-off performance and ceiling, there was no better engine available while an M fitted with AL-5s did not meet any part of the requirement, particularly in regard to field performance where the take-off run was estimated to be a totally unacceptable 11,483ft (3,500m).

Alongside the design of the aircraft itself, hundreds of changes in manufacturing practice had to be introduced by the sub-contractors as part of the VM-25's development programme. For example an entire section (or brigade) at the VIAM (Aviation Materials Institute) concentrated on developing advanced light alloys while new facilities such as heat treatment furnaces had to be constructed to deal with much larger components. To save weight, for the first time 6ft 7in (2m) wide sheets of dural were used when, formerly, the standard width had been 5ft 0in (1.5m).

By 1st April 1952 prototype drawings were being prepared and on 15th May the prototype itself began final assembly. In November this aircraft was cleared for its manufacturer's flight testing and then in December it was rolled out and dismantled for shipping down the Moscow river to Zhukovsky. On 20th January 1953 the M successfully completed its maiden flight and early testing soon revealed that the aircraft was capable of delivering 11,023 lb (5,000kg) of ordnance to a target 6,650 miles (10,700km) from base. Various modifications had to be made, some of which were delayed by the factory's own production of the Il-28, and this put back the state acceptance testing until May 1954. Participation in the 1954 Mayday flypast also held up proceedings and, after having been seen there by Western observers, the aircraft was codenamed *Bison*; in the West it was also called, for a period at least, *Molot* (Sledgehammer).

The trials found that generally the aircraft was good and should be put into VVS service, but its maximum speed at 29,528ft (9,000m) and cruise speed for maximum range did fall short of requirements. In comparison to the Tu-95, when carrying an 11,023 lb (5,000kg) war load the M-4 could fly 35mph (57km/h) faster but its range was 3,108 miles (5,000km) less. Andrei Tupolev, who seems to have been something of a grumpy character, was critical of Myasishchev's bomber when comparing it to his own turboprop-powered Tu-95 but officials in the SovMin countered this by giving their support to Myasishchev. However, for comparison, Myasishchev assessed a separate version of the M powered by four NK-12 turboprops driving eight-blade contra-rotating tractor propellers but this was never built.

In due course the M-4 was adopted for production, the decision in part being linked to the loss of the first Tu-95, and this began in 1954. However, the shortfall in performance would have to be addressed and in 1953 the decision was taken to carry out further developments. There were two possible solutions – improve the aerodynamics and fit newer engines with better fuel consumption or introduce in-flight refuelling (IFR). The first major Soviet developments in the IFR field followed a SovMin resolution made on 18th December 1953 and during 1954 this was followed by further resolutions requesting the development of a Tu-16 tanker to refuel MiG-19 fighters plus other Tu-16s and an M-4 tanker to refuel M-4 bombers. This represented another big step for the Soviet aircraft industry to take and there were problems in seeing it through; as a result the M-4 tanker's state acceptance trials were not concluded until July 1958. A total of 35 M-4s was eventually built (including two prototypes and a static test airframe) and from 1958 production machines were converted into tankers, but the appearance of this new strategic bomber represented a substantial threat to the United States.

Myasishchev M-26

A projected derivative of the basic VM-25 was called the M-26 (the 'V' prefix was dropped at about this time) which featured a redesigned inner wing to accommodate a new powerplant – two Dobrynin VD-5 axial-flow turbojets with a take-off rating of 28,660 lb (127.4kN) and a nominal rating of 24,250 lb (107.8kN). Like the production M-4, the M-26 had a crew of eight, an identical defensive armament of six 23mm cannons in three powered barbettes and provision was made for rocket-assisted take-off (RATO) boosters with two 19,841 lb (88.2kN) rockets mounted under the wing roots. Span was 173ft 11in

Two views of the Myasishchev M-4 prototype.

Model of the Myasishchev M-26 development of the M-4.

The first Myasishchev's M-28 DVB bomber had one engine in each wing root and another in an underwing pod half way along the inner wing. Russian Aviation Research Trust

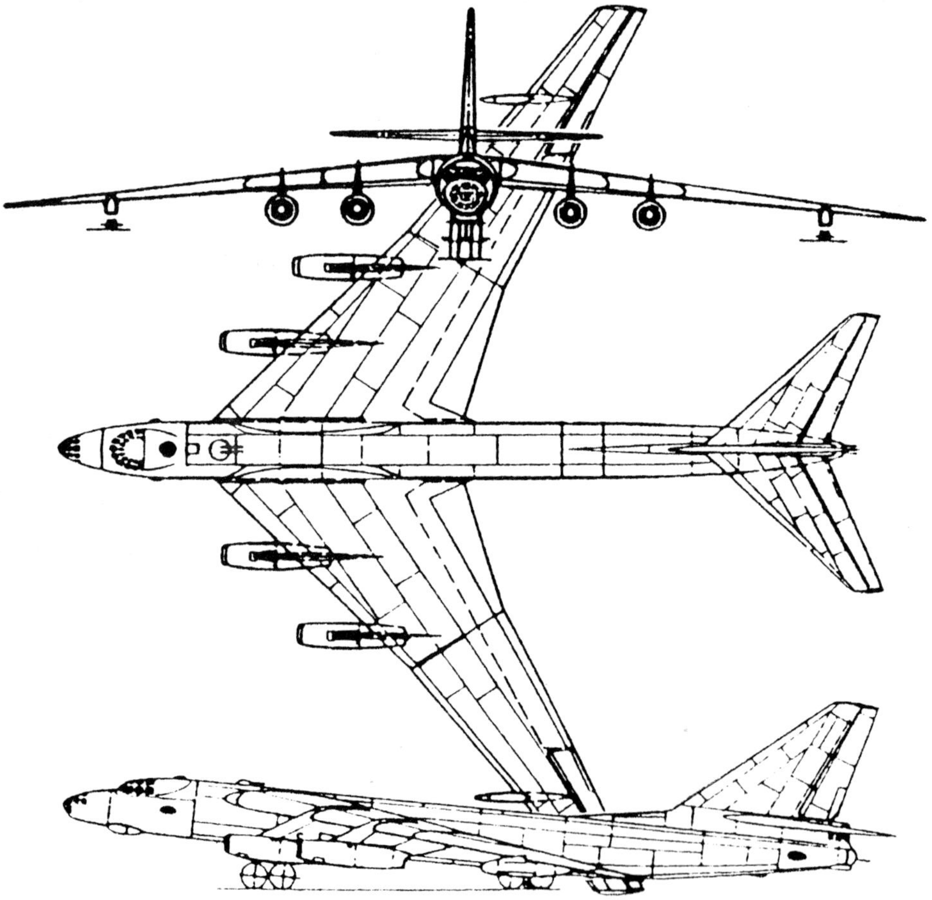

The third Myasishchev M-28 configuration (1952/53). Russian Aviation Research Trust

(53.0m), length 160ft 0in (48.75m), wing area 3,817ft² (355m²), maximum take-off weight either 231,481 lb or 265,550 lb (105,000kg or 120,000kg) and top speed 566mph (910km/h) at 29,528ft (9,000m).

Myasishchev M-28

The big weakness of the SDB/VM-25/'M' concept, which left it well behind the contemporary American B-52, was that its maximum range could only be achieved when flying above 29,528ft (9,000m). In addition its ceiling was only 45,932ft (14,000m) and the wings were incapable of supplying sufficient lift to take the aircraft any higher and thus put it out of the reach of anti-aircraft fire. Myasishchev's first answer was to produce the M-28 long-range high-altitude bomber, which was initially called the DVB, and this new aircraft's development, under the official label 2M, was subsequently covered by a SovMin resolution of 29th March 1952. It had to be able to carry 11,023 lb (5,000kg) of bombs over a range of 7,769 miles (12,500km) and, when flying over its target, reach a height of 56,759ft (17,300m). Maximum fuel capacity, including external carriage, would be 235,891 lb (107,000kg) and there were six crew. The offensive load could also include sea mines or 45-36AV torpedoes specifically designed to be launched at high altitude.

In general the layout would be similar to the M-4 with the same 23mm defensive turrets and the initial studies stuck to the M-4's engine arrangement; however, by the time the full-scale mock-up was officially inspected by a VVS team in January 1953, the four jets had been mounted on underwing pylons. The first idea had been to have the engines in a paired nacelle under a single pylon on each wing, then a switch was made to having two individual nacelles with their own pylons under the inner half of each wing, the latter giving a planview reminiscent of the first generation of American four-jet airliners and which appears to have been the configuration first produced in mock-up form. Later, and quite unusually, the outer engines were pushed much further out towards the wingtips. The bicycle undercarriage was retained and one benefit from having the engines further out was that the outrigger wheels could now be housed in the outer engine nacelles rather than in their own individual pods.

The flight test programme, as first laid down by the SovMin resolution, declared that the aircraft should be ready for its state testing in May 1954. This was then postponed to June 1955 and then the project was cancelled by another SovMin dated 16th October 1954. This document reported that 'Due to the insufficient 4,786 miles (7,700km) range of the 2M long-range high-altitude bomber … and new MAP instructions to develop new versions of the M-4 with a range of 8,701 miles (14,000km), the Ministries of Aircraft Production, of Defence and of Agricultural Engineering (the latter was also involved in some armament production) had agreed to terminate the 2M's development'. In fact, as the

Full-scale mock-up of the Myasishchev M-28/2M.

design work on the M-28 proceeded, more detailed estimates had revealed that the bomber's take-off weight would increase and there would be a consequent decrease in speed and range. The M.28 was nicknamed *Visotniy* (high-altitude climber).

Myasishchev M-36

After the M-4 and the Tupolev Tu-95 had entered service, the former offering the higher speed and bomb load while the Tu-95 had the range, it was clear that the Soviet Union had not managed to develop a bomber that could equal the B-52; nevertheless, the research and development effort continued. MAP ordered TsAGI, TsIAM, Myasishchev and Tupolev to find a way of increasing the range of heavy jet bombers and in April 1954 Minister P V Dement'yev was able to tell the SovMin that the fuel consumption of current jet engines could be reduced by 20 to 25% and this, coupled with improved airframe aerodynamics, should increase range by a substantial figure.

Dobrynin, whose OKB had produced the VD-5, had proposed the VD-7 which, besides having a better fuel economy, was expected to offer 22,046 lb (98.0kN) of thrust. As a result Tupolev calculated that a bomber with four VD-7s could reach a top speed of 572mph to 590mph (920km/h to 950km/h) and achieve a range of up to 8,701 miles (14,000km) while Myasishchev's data gave the same range but a speed of 590mph to 622mph (950km/h to 1,000km/h); both also proposed to include IFR in their aeroplanes which would increase

Model of the Myasishchev M-28/2M. George Cox

Model of the M-28 in the form that was to be built. George Cox

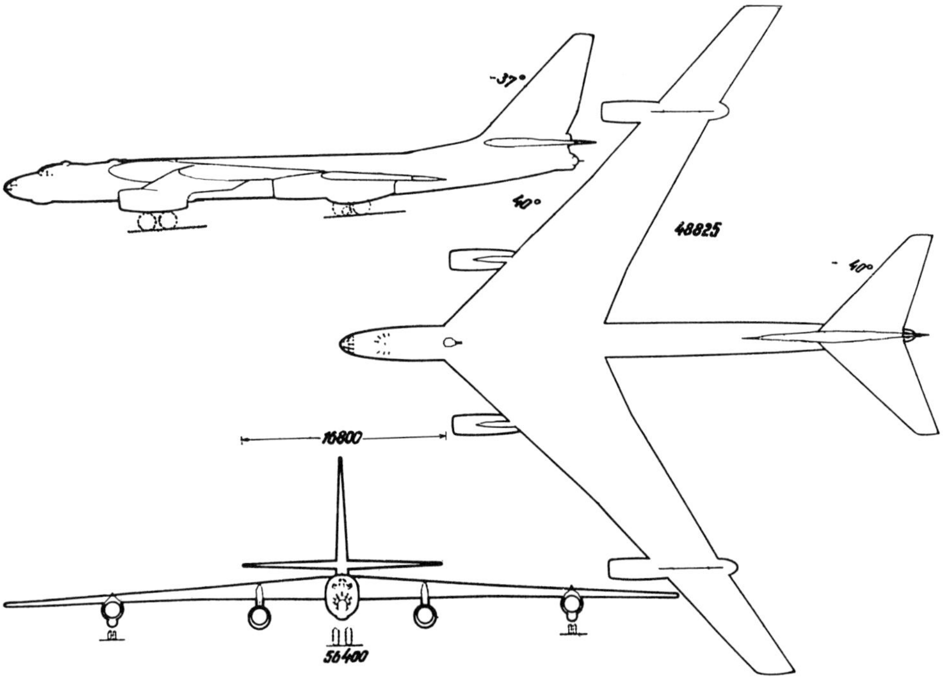

The last Myasishchev M-28 configuration (1952/53).

Artist's impression of the M-28 in its ultimate form.

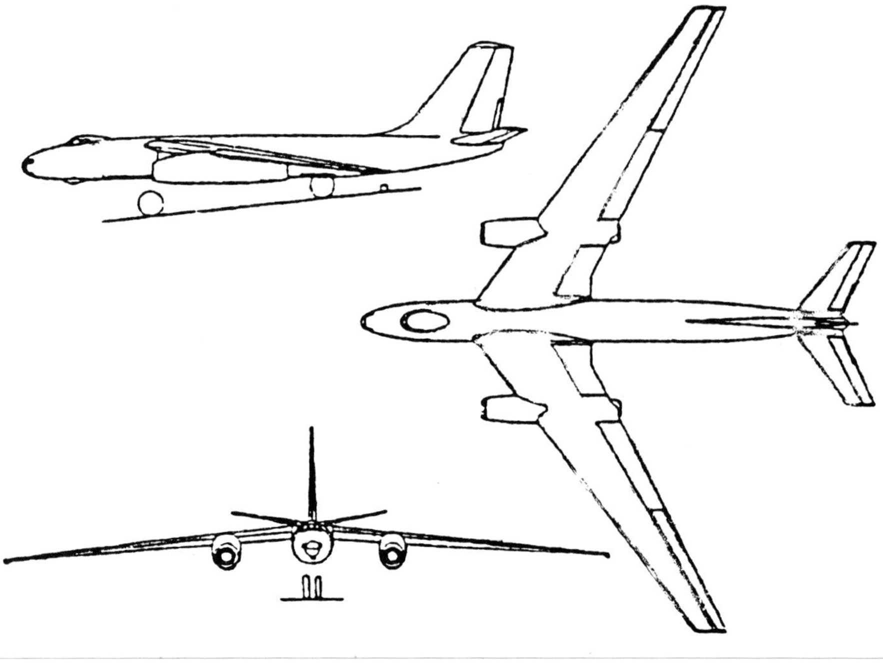

Myasishchev M-30 (1953). Russian Aviation Research Trust

the range to between 9,944 miles and 10,566 miles (16,000km and 17,000km).

Three months later the SovMin authorised the development of a modified M-4 with VD-7s called the M-6, but with the Klimov VK-9 or Mikulin AM-13 held as a back up because the M-6 was essentially a new aircraft; it was also called the 3M. There was the possibility that TV-12 turboprops might be fitted which would give an unrefuelled range of 8,701 miles (14,000km). The M-4's basic design was revised with a new forward fuselage, the new lightweight engines and a lighter structure which overall gave 14,330 lb (6,500kg) less weight. There was no change to the maximum bomb load although service aircraft would rarely carry more than a single 19,841 lb (9,000kg) or two 13,228 lb (6,000kg) stores.

The prototype 3M was a converted M-4 (whose original serial number is unknown) and flew in its new form for the first time on 27th March 1956, but with AM-3A engines fitted because the VD-7s were not yet available. Compressor surge problems delayed the state testing but, after some doubts as to whether any production machines would be ordered (in March 1956 the SovMin ordered GAZ No 23 to follow its M-4 production with the Antonov An-8 transport), in August a resolution was finally passed covering the 'Production of 3M strategic bomber by Myasishchev'.

So acute was the need for this bomber it was actually included in the Soviet Air Force inventory before its state acceptance trials were even under way. Back in 1954 a SovMin resolution had also outlined a requirement for the aircraft to carry the new Kh-20 cruise missile, thus forming the K-20 (MK) weapon system. This step was confirmed in 1956 and, to achieve this, the 3M's wing had to be extensively modified and new equipment fitted. There was also a 3MT IFR tanker and eventually a total of 85 production 3Ms were built, the last appearing in 1960. The 3M was a long-lived aeroplane – the final operational flight was made in March 1994 and during August 1997 the surviving 3Ms were removed from store and ferried to a designated airbase for scrapping.

During the design and development phase a 3M-M flying boat variant was also suggested which, in appearance, showed the 3M's fuselage and wing blended onto an additional flying boat hull underneath. This was expected to offer a cruise speed of 497mph (800km/h) and a very long range, but it does not appear to have been examined in depth.

Myasishchev M-30

An unrelated design, but one which appears to have used a good deal of M-4 experience, was this high-altitude medium reconnaissance bomber project of 1953. This aircraft had the appearance of a scaled-down M-4 and employed a bicycle undercarriage, a pressurised forward fuselage for four crew, a lower fuselage internal bomb bay with downward-facing cameras directly behind, and no defensive armament. Power was supplied by two 18,739 lb (83.3kN) Mikulin AM-9 engines in underwing nacelles, span was 121ft 5in (37.0m) and length 98ft 5in (30.0m), top speed between 584mph and 603mph (940km/h and 970km/h), ceiling 56,594ft (17,250m) and range 3,325 miles (5,350m). It was not built

Heavy / Medium Bombers – Data / Estimated Data

Project	*Span ft in (m)*	*Length ft in (m)*	*Gross Wing Area ft² (m²)*	*Max Weight lb (kg)*	*Powerplant Thrust lb (kN)*	*Max Speed / Height mph (km/h) / ft (m)*	*Armament*
'Baade' EF 132 (early design)	112 10 (34.4)	129 3 (39.4)	?	192,901 (87,500)	6 x Jumo 012 6,615 (29.4)	590 (950)	4 x 15mm guns Normal 8,818lb (4,000kg) bombs
'Baade' EF 132 (later design)	119 5 (36.4)	126 3 (38.475)	2,581 (240.0)	198,743 (90,150)	6 x AM-TKRD-01 7,275 (32.3)	609 (980)	6 x 20mm cannon, max 39,683lb (18,000kg) bombs
Baade Type 140 (flown)	63 8 (19.4)	64 7.5 (19.7)	628 (58.4)	55,600 (25,220)	2 x AM-TKRD-01 7,275 (32.3)	562 (904) recorded at 9,843 (3,000)	2 x 2 x 23mm cannon; on 140B/R up to 9,921lb (4,500kg) bombs
Baade RB-2	?	?	?	90,388 (41,600)	2 x TR-3 11,025 (49.0)	622 (1,000) at 15,420 (4,700)	1 x 1 + 2 x 2 23mm cannon, up to 13,228lb (6,000kg) bombs
Baade Type 150 (flown)	79 1 (24.1)	87 9 (26.75)	1,346 (125.2)	119,048 (54,000)	2 x AL-5 10,140 (45.1)	528 (850) at S/L, 578 (930) at 32,808 (10,000)	2 x 2 x 23mm cannon, up to 13,228lb (6,000kg) bombs
Ilyushin IL-46 (flown)	95 2 (29.0)	83 1 (25.325)	1,108 (103.0)	118,580 (53,788)	2 x TR-3A 11,025 (49.0)	577 (928) max achieved	1 x 2 + 2 (fixed) 23mm cannon, 13,228lb (6,000kg) bombs
Ilyushin Il-46S	95 2 (29.0)	83 1 (25.325)	In excess of 1,108 (103.0)	116,376 (52,788)	2 x AL-5 11,025 (49.0)	622 (1,000)	1 x 2 + 2 (fixed) 23mm cannon, 13,228lb (6,000kg) bombs
Tupolev 'Aircraft 86' (first swept wing layout)	85 4 (26.0)	74 7 (22.75)	892 (83.0)	57,319 (26,000) (normal t/off weight)	2 x AM-02 10,540 (46.8)	634 (1,020) at 19,685 (6,000m)	1 x 23mm + 3 x 2 x 20mm cannon, max of 6,614lb bombs
Tupolev 'Aircraft 86' (second swept wing layout)	84 2 (25.66)	79 3 (24.15)	895 (83.2)	66,138 (30,000) norm 92,593 (42,000) o'load	2 x AM-02 10,540 (46.8)	609 (980) at 19,685 (6,000m)	4 x 2 x 23mm cannon, max 13,228lb (6,000kg) bombs
Tupolev 'Aircraft 86' (third swept wing layout)	90 2 (27.49)	90 2 (27.48)	1,075 (100.0)	66,138 (30,000) norm 92,593 (42,000) o'load	2 x AM-02 10,540 (46.8)	590-622 (950-1,000) at 19,685 (6,000m)	3 x 2 + 1 x 1 x 23mm cannon, max 13,228lb (6,000kg) bombs
Tupolev 'Aircraft 87'	90 2 (27.49)	90 2 (27.48)	1,075 (100.0)	66,138 (30,000) norm	2 x TR-3	603 (970) at 13,123 (4,000)	3 x 2 + 1 x 1 x 23mm cannon, max 13,228lb (6,000kg) bombs
Tupolev Aircraft 88' (Tu-16 prototype – flown)	108 2.5 (32.977)	113 6 (34.6)	1,770 (164.6)	170,525 (77,350)	2 x AM-3 17,637 (78.3)	634 (1,020) at 16,404 (5,000)	3 x 2 + 1 x 1 x 23mm cannon, max 19,841lb (9,000kg) bombs
Tupolev Tu-95M (flown)	164 2 (50.04)	151 6 (46.17)	3,051 (283.7)	401,235lb (182,000)	4 x NK-12M 15,000hp (11,186kW)	548km/h (882)	6 x 2 x 23mm cannon, up to 11,023lb bombs at max range
Tupolev 'Aircraft 96'	167 5 (51.04)	151 7 (46.2)	3,715 (345.5)	341,711 (155,000) normal	4 x TV-12	547 (880) at 20,998 (6,400)	3 x 2 x 23mm cannon, 11,023lb (5,000kg) bombs
Myasishchev VM-25 (M-4) (early production)	165 9 (50.53)	156 4.5 (47.66) (less probe)	3,509 (326.35)	400,132 (181,500)	4 x AM-3A 19,290 (85.7)	589 (947)	3 x 2 x 23mm cannon, max bomb load 52,910lb (24,000kg) incl 3 x 13,228lb (6,000kg) or 2 x 19,841lb (9,000kg)
Myasishchev M-28 (final form)	160 1 (48.8)	150 3 (45.8)	3,978 (370)	407,848 (185,000)	4 x VD-5 28,660 (127.4)	584-603 (940-970)	3 x 2 x 23mm cannon, max bomb load 52,910lb (24,000kg) incl 11,023lb (5,000kg), 13,228lb (6,000kg) & 19,841lb (9,000kg) bombs, + mines and torpedoes
Myasishchev M-36 (3MN-1) (early production)	174 4 (53.14)	159 11.5 (48.76) (less probe)	3,783 (351.8)	425,485 (193,000)	4 x VD-7B 20,945 (93.1)	575 (925)	3 x 2 x 23mm cannon, max bomb load 52,910lb (24,000kg) incl 11,023lb (5,000kg), 13,228lb (6,000kg) & 19,841lb (9,000kg) bombs

Chapter Four

The First Supersonic Bombers

This aircraft was the only example of the M-50 to fly.

By the start of the 1950s those design bureaux in the Soviet Union that specialised in the design of bombers, Ilyushin, Myasishchev and Tupolev, had all moved on to the study of types possessing transonic or supersonic capability.

Ilyushin OKB

Ilyushin Il-54

In December 1952 Ilyushin started working on a transonic tactical bomber which it called the Il-54. On 29th December an official go-ahead for the project was given by a Council of Ministers directive and this requested a maximum speed of Mach 1.15 at 15,584ft (4,750m) and, with a 6,614 lb (3,000kg) bomb load, a range of 1,367 to 1,554 miles (2,200km to 2,500km). The aircraft had to be submitted for its state acceptance trials in July 1954, but this estimate was to prove optimistic. The governing factor for the choice of layout was the stipulated transonic maximum speed and, in the opinion of the OKB's aerodynamics experts, thin wings of considerable sweepback, up to 55°, would be needed to achieve this. While possessing the necessary properties for transonic cruise, when compared to straight or moderately-swept wings such sweepback would bring a deterioration in the aircraft's range and field performance – a longer runway would be required to accommodate the higher take-off and landing speeds.

As usual, several configurations were studied and one (according to a published drawing) showed a low-wing monoplane with the engines mounted in small nacelles on the fuselage above the wing root trailing edges and a tailplane set in the mid position on the vertical fin. However, the first version of the Il-54 to be officially endorsed by Ilyushin himself, on 23rd March 1953, was quite different. This was an aircraft with mid-fuselage wings, engines in the wing roots and a sharply swept T-tail. The accommodation for its three crew members and the layout for the defensive guns followed the well-established pattern used on the Il-28 and the main units of the tricycle undercarriage retracted forwards into the wing roots, swivelling through 90° to permit the wheels to lie flat inside the wings.

The powerplant comprised two Lyulka TRD-I axial-flow non-afterburning turbojets (later renamed AL-7) which in their original form possessed a maximum thrust of 16,975 lb (75.4kN). However, the Il-54 project documents mention an 'operating mode' for the engine 'with thermal control', which may refer to the exhaust temperature and was probably a proposed engine development, and this was expected to increase the thrust to 18,959 lb (84.3kN). The estimated performance included a maximum speed of 746mph (1,200km/h) at 16,404ft (5,000m) with the engines operating in normal condition, but the 'thermal control' was expected to push this up to 780mph (1,255km/h) at the same altitude. An effective range of 1,492 miles (2,400km) on internal fuel was envisaged but additional drop tanks would increase this to 1,709 miles (2,750km).

Besides the basic bomber version, there were proposals for a torpedo-bomber, a trainer and a photographic intelligence

The first Il-54 proposal to be endorsed by Ilyushin (23.3.53).

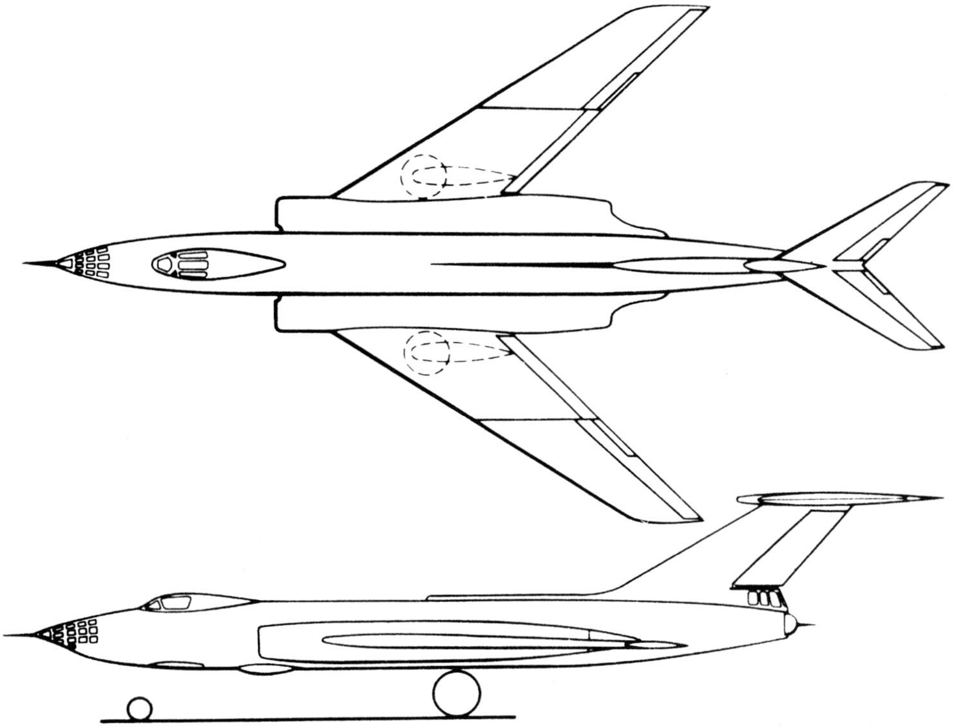

The prototype Il-54. The enormous main wheels are very evident.

(PHOTINT) aircraft. The AL-7 was the first Lyulka engine to go into mass-production, the initial examples being built in March 1953, and different versions were eventually produced for many types including fighters, the Beriev Be-10 (Chapter 9), Il-54 and Tupolev's Tu-98 (below).

After military representatives had studied the advanced development project and inspected a full-scale mock-up of this aircraft they voiced some criticism concerning the undercarriage. In the opinion of the Air Force specialists, wheels of the size quoted in the brochure would not allow the Il-54 to operate routinely from dirt strips. They insisted that the runway loading should be reduced to the same level as the Il-28 and this resulted in an increase in the size of the wheels which, in turn, made it impossible to stow them in the space between the spars of the thin swept-back wing. Consequently, this Il-54 layout, chosen by Ilyushin, had to be discarded.

A new design, endorsed by Ilyushin on 16th November 1953, showed an aircraft transformed into a shoulder-wing type with a low-set tailplane mounted at the root of the vertical tail. The wings had 55° sweepback at the leading edges, two wing fences on each wing and retained the AL-7s, but these were now placed in pylon-mounted underwing nacelles, an arrangement adopted after wind tunnel testing had demonstrated that pylon-mounted engines offered less drag during transonic cruise. To deal with the demand for rough-field operation much bigger wheels were fitted but these had to be arranged as a widely spaced bicycle undercarriage and so could not be stowed in the thin wings or slim nacelles. The defensive armament comprised three 23-mm Afanasyev/Makarov AM-23 cannon which, in terms of rate of fire and weight of salvo, were considerably superior to the NR-23 used on the Il-28 and Il-46. One cannon was mounted in a fixed installation on the port side of the forward fuselage; the other pair formed part of the DK-35A remote-controlled tail turret.

The first Il-54 made its maiden flight on 3rd April 1955 and displayed good handling qualities, but on the ground the bicycle undercarriage proved tricky, especially during landing. Some improvements had to be made to the undercarriage after the test pilot lost control on the runway in an incident which resulted in minor damage to the airframe. By the spring of 1956 the second prototype had joined the programme having been fitted with two AL-7Fs producing 22,046 lb (98.0kN) of thrust with afterburner. However, for a number of reasons the Il-54 failed to reach production. One problem was the competition coming from the Yakovlev design bureau which was currently working on the Yak-26 (Chapter 8), a derivative of the Yak-25 fighter intended for the tactical bomber role. Aleksandr Yakovlev argued that this smaller and lighter machine could accomplish most of the missions envisaged for the Il-54. Another important point was the tendency of Soviet political and military leaders during this period to overestimate the value and capability of unmanned rocket-powered weaponry, to the detriment of aviation, which resulted in less official interest in conventional bombers; this point is discussed later.

It was intended to put the Il-54 at the head of a column of new prototype aircraft to be displayed during the 1956 annual show at Moscow-Tushino; however, shortly before this event took place the Il-54's participation was cancelled. Instead, on 30th June 1956, the aircraft was shown to a US military dele-

Unbuilt Ilyushin Il-54T project (1954).

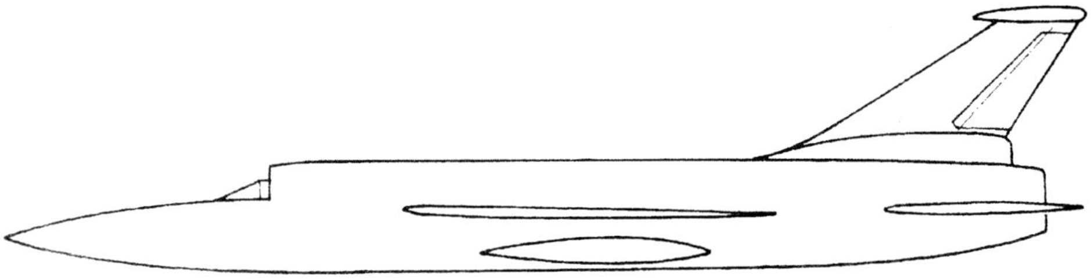

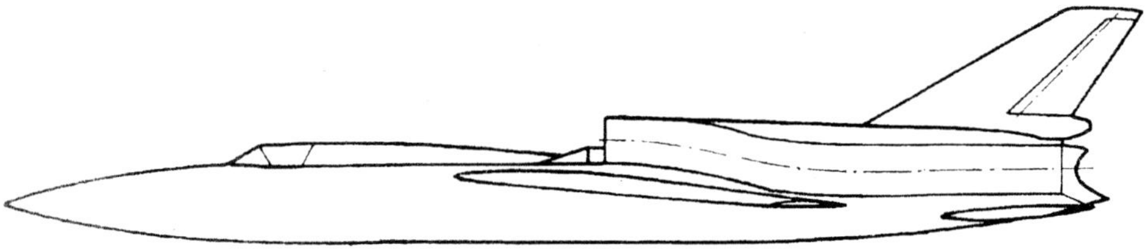

Two versions of the Ilyushin Il-56 (1955).

gation (on the ground) at Kubinka airbase west of Moscow, together with a selection of other prototype aeroplanes. This delegation was led by the USAF Chief of Staff, General N F Twining, and the demonstration aroused much comment in the Western aeronautical press. Ilyushin's Il-54 was assessed by Western observers as a transonic tactical bomber on a par with the current state of the art although, at the time, the aircraft's correct designation was unknown. Consequently it was given the NATO codename *Blowlamp* and for some time was inaccurately designated Il-140. A source close to the Ilyushin OKB claims that the Il-54 was demonstrated in Kubinka under the designation Il-149.

The Il-54 was to be the last manned bomber to be designed and built by Ilyushin. There were, however, several proposed developments of the basic aircraft produced against a special Government resolution of 2nd March 1954, and these were to be submitted for state acceptance trials in May 1955 (this deadline was later revised). The new variants were the Il-54T torpedo bomber (which was to be capable of carrying all types of air-launched torpedo in a longer weapon bay), the Il-54R photo reconnaissance aircraft and Il-54U trainer (with an additional instructor's cockpit). None of these types were built.

Ilyushin Il-56

Alongside its work on the Il-54, the Ilyushin OKB also undertook preliminary studies into a supersonic tactical bomber designated Il-56. This was in response to a joint directive of the Communist Party Central Committee and the Government, dated 28th March 1955, in which Ilyushin and MAP were to present their proposals in June against a deadline which covered prototype construction and submission for state acceptance trials. The basic specification requested a maximum Mach 1.75 to 1.9, a 4,409 lb (2,000kg) bomb load and a range of 1,367 miles (2,200km). An additional MAP order contained such figures as 1,057mph to 1,119mph (1,700km/h to 1,800km/h) speed, service ceiling 45,932ft to 49,213ft (14,000m to 15,000m) and all-up weight 50,705 lb to 52,910 lb (23,000kg to 24,000kg). Various layouts were considered, some of them based on a single AL-7F or two Mikulin AM-11 engines, the latter a new power unit currently under development. In due course the designers opted for the twin-engined configuration.

The Il-56 was a two-seat high-wing monoplane with a low-set tail and a bicycle undercarriage. Semi-circular air intakes, featuring movable centrebodies in the form of a half-cone, were placed on top of the fuselage in the area of the wing centre section and different versions of the basic aircraft had their wings swept at angles of either 45° to 55°. During 1955 and into 1956 the TsAGI wind tunnels conducted experiments with scale models of the Il-56, but in early 1956 the OKB was told to terminate all work on the bomber because the OKB's resources were to be concentrated on a new task. This was the development of a cruise missile or, as it was termed at that time, an 'unmanned winged missile'.

Myasishchev '30' Series

During 1952 Myasishchev produced a set of three transonic or supersonic bomber projects which helped pave the way for the M-50 described later. Work on these proceeded concurrently with the development of some versions of the M-4 described in Chapter 3.

Myasishchev M-31

The development of supersonic fighters meant that any new strategic bomber would itself need transonic or even supersonic performance to try to avoid being intercepted. However, the aerodynamics and structures needed in a bomber airframe to make it capable of such speeds would require a considerable amount of new research and experimentation and the Myasishchev OKB got moving on this in 1952. It was confirmed that increasing speed from Mach 0.7 to about Mach 1.1 brought a big drop in range because of both the rise in fuel consumption and the reduction in lift delivered by a sweptback wing. Increases in wing sweep reduced the take-off performance while a thinner wing pushed up the structure weight. Nevertheless, in co-operation with TsAGI, the design bureau began work on a heavy transonic bomber called 'Article 31' or M-31.

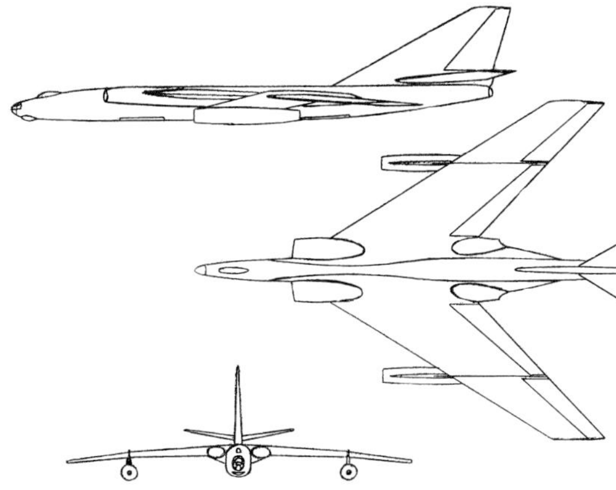

Myasishchev M-31 (1952). Russian Aviation Research Trust

This M-31 model gives an indication of the design's considerable bulk. John Hall

A 55° sweep wing was selected after wind tunnel results for alternative high sweepback or rhomboid shaped wings had been analysed. The M-31 was to have four Dobrynin VD-5 engines and the possible alternative arrangements would have all four of these in the fuselage or to put two under the wings on pylons with the other pair in the wing roots. However, the M-31's wing thickness/chord ratio was 6% maximum, which meant that the wing skins would be relatively thick, so the second format offered a saving in weight while keeping down the size of the fuselage; in addition using pylon-mounted engines would improve the wing's flutter characteristics by pushing up the critical speed at which this phenomenon appeared. Such a thin wing also prevented the use of a tricycle undercarriage, a zero-track bicycle type being fitted instead, while RATOG equipment supplying up to 44,000 lb (195.6kN) of additional thrust would be used on take-off.

The M-31 could carry one 19,841 lb (9,000kg), two 13,228 lb (6,000kg), three 6,614 lb (3,000kg), six 3,307 lb (1,500kg), twenty 1,102 lb (500kg) or thirty-six 551 lb (250kg) bombs, or two 3,527 lb (1,600kg) guided bombs or six torpedoes or sea mines; for rear defence a battery of two or four 23mm cannon was provided at the end of the fuselage and four crew were carried. The airframe structure was all-metal, mainly V-95 light alloy, with a semi-monocoque oval-shaped fuselage and a central bomb bay. The wing used box-spar construction and the fuel (216,049 lb [98,000kg] in total) was housed in both wings and fuselage. A Rubin bombing radar was carried and the aircraft was expected to offer a range of 4,972 miles (8,000km) and a service ceiling of 49,213ft (15,000m). A preliminary project was presented to MAP for review in August 1952 (together with material on the M-32 below) but Myasishchev's designers recognised that much more research would probably be needed before they could begin to design and build a bomber of this size for hitting major military and industrial targets, particularly one that offered even better performance.

Myasishchev M-32

Alongside the M-31, the bureau was also working on the 'Article 32' or M-32 supersonic bomber, with A K Batukhov in the role of chief designer. TsAGI was again involved and that extra step of making the aircraft a pure supersonic design meant that another new aerodynamic shape needed to be explored – a 60° delta wing, which offered better aerodynamic efficiency above Mach 1 than a swept wing, plus a T-tail. In addition, the delta had inherent structural strength which reduced the overall weight and improved the offensive load/weight ratio. A semi-monocoque structure was employed in an oval fuselage and a tricycle undercarriage was fitted with twin-wheel nose gear and paired main wheels. Together with the M-31, the M-32 preliminary project was sent to MAP on 15th August 1952 and it was the delta wing project that generated the most interest. A Deputy MAP Minister issued an order on 4th February 1953 authorising a more detailed preliminary project to be

Model of the Myasishchev M-32 (1952). John Hall

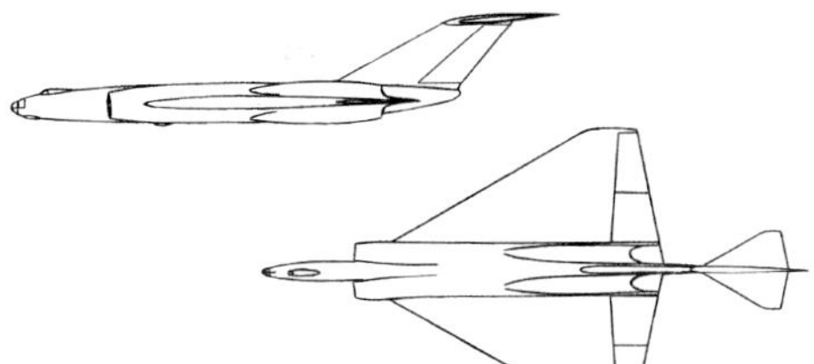

Myasishchev M-32 (1952). Russian Aviation Research Trust

The redesigned M-32 with side-by-side engines (c2.53).

made and, during this process, Myasishchev changed the design quite considerably.

Four (unspecified) turbojets had originally been mounted in pairs at the wing roots, one above the other with split intakes and jet pipes to serve each unit, but, to reduce drag, the new format had side-by-side wing root engines served by individual intakes, the inner pair more forward than the outer intakes; in addition a radar replaced the navigator's nose cockpit. Wing area was increased (to 5,484ft² [510m²]), which brought more space for fuel, and efforts were made to reduce the weight overall. The detailed preliminary project was submitted to MAP on 10th March 1953 but neither the M-31 or M-32 progressed beyond the design and model stage because their estimated maximum speeds were thought to be insufficient to meet future needs. The M-32 was intended to attack strategic targets and had an estimated cruising speed of around 839mph (1,350km/h), a range of at least 5,220 miles (8,400km) and ceiling 59,055ft (18,000m). There was no defensive armament but the offensive load could include a single 19,841 lb (9,000kg) bomb. (Note: in 1951 the Myasishchev OKB proposed a fighter-size research aircraft called the M-33 to examine the delta wing in flight, but it was not built).

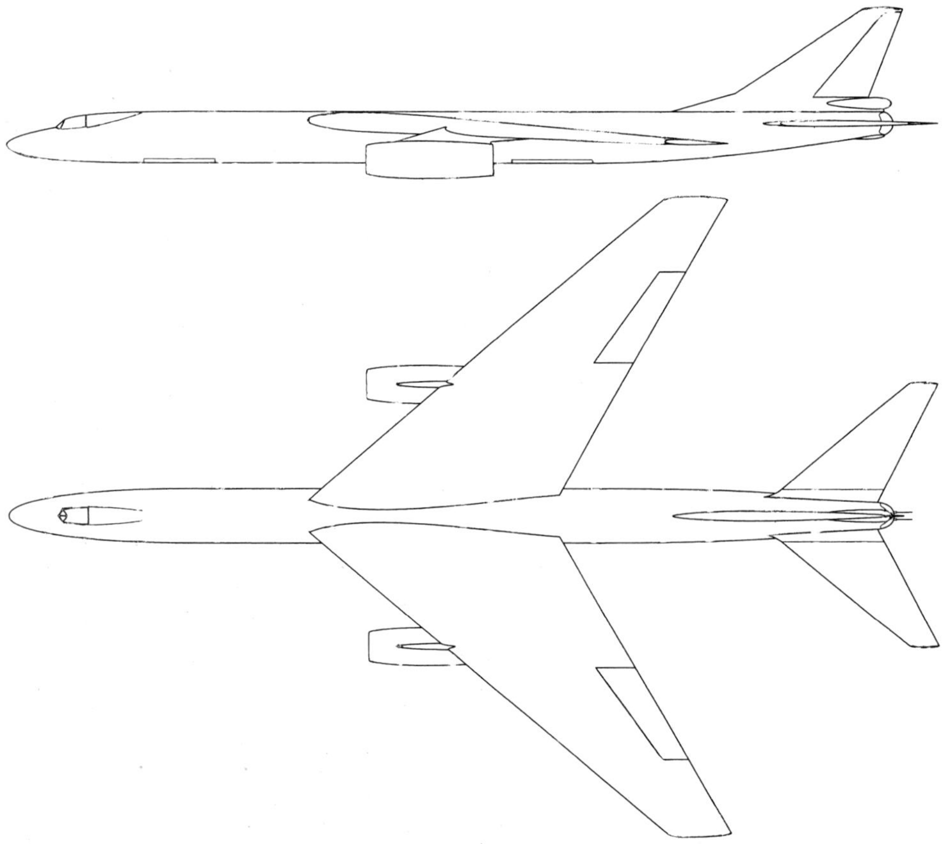

Myasishchev M-34 (1953). A large cruise missile was carried beneath the forward part of the centre fuselage. Russian Aviation Research Trust

Myasishchev M-34

During the early 1950s bomber design had to match even more the rapidly increasing top speed of new fighters. Yet again any improvement in speed was matched by a loss in range and yet the official requirements asked for a range in the region of 8,080 miles (13,000km). Myasishchev studied the problem carefully and concluded that this range was possible for an average cruise speed of 656mph (1,050km/h), with a dash over the target at about 777mph (1,250km/h), by making several important changes to the aircraft. As one might expect these included better aerodynamics and fuel consumption, but also included a smaller crew, the removal of the fuel tank self-sealing covers, the introduction of special take-off trolleys and, from a strength point of view, a much more accurate design of the structure; the latter was backed up by using titanium and other alloys that had high strength/weight ratios.

In appearance a relatively simple design, with beautiful swept wings and tail surfaces, the resulting M-34 was a logical continuation of the M-31/M-32 and had one Mikulin AM-13 or Dobrynin VD-7 engine hung in a nacelle on a pylon under each wing. These were to confer a maximum speed of around 746mph (1,200km/h) at sea level and a ceiling approaching 49,213ft (15,000m) and the estimated range was expected to be in excess of 8,080 miles (13,000km). Myasishchev's calculations were aimed at keeping the take-off weight to 176,367 lb (80,000kg), roughly the same as the Tupolev Tu-16 (Chapter 3), and the estimates suggested that the M-34 would have twice the range of that aircraft with 20% more speed. The structure was all-metal with a circular fuselage of semi-monocoque construction plus box spar wings, and a bicycle undercarriage was fitted. One 15,432 lb (7,000kg) Kh-20 cruise missile was carried

The sole example of the Tupolev Tu-98.

semi-recessed beneath the middle fuselage, conventional bombs of up to 3,307 lb (1,500kg) weight could be carried and two 23mm cannon were mounted in the end of the fuselage; there were three crew and the chief designer was Ye D Mochalin. A preliminary project was submitted to MAP in 1953 but again this aircraft never progressed to the hardware stage.

Tupolev OKB

Tupolev Tu-98

Andrei Tupolev's OKB first entered the arena of supersonic bomber design with 'Aircraft 97', a variant of the Tu-88 described in Chapter 3, but the next number in the series covered an all-new design which was eventually taken to prototype status. From the end of 1949 the design bureau, working together with TsAGI, undertook a series of theoretical studies into the main parameters for a future heavy aircraft designed to fly at high transonic and supersonic airspeeds. From the beginning of the 1950s the results of this research found their way into the OKB's first supersonic designs, which included 'Aircraft 98'.

Preliminary work on 'Aircraft 98' began in January 1953 within the Technical Projects Department headed by S M Yeger. On 12th April 1954 a USSR Council of Ministers Decree was issued which ordered the bureau to produce a high-speed front-line bomber with two AL-7F turbojets, giving 14,330 lb (63.7kN) of dry thrust and 20,940 lb (93.1kN) in reheat, and capable of a top speed of 808mph to 870mph (1,300km/h to 1,400km/h). It was intended that these engines should eventually be replaced by Mikulin AM-15s (paired AM-11s), which offered a thrust of 25,130 lb (111.7kN) each, or with 26,450 lb (117.6kN) VK-9s. Engineering development and prototype manufacture got moving in November 1954 and by July 1955 development was complete. Among the pieces of equipment produced specially for the aircraft was a new twin AM-23 cannon mounting to be guided by the navigator/operator using the '98's PRS-1 Argon radar. The Chief Designer on 'Aircraft 98' was D S Markov while A I Zalesski was responsible for the team building the prototype.

Due to delays with the AL-7Fs, the '98 prototype was not completed until February 1956. The first flight was made on 7th September and in 1957, during its test programme, the bomber reached a speed of 769mph (1,238km/h) at 39,370ft (12,000m). It was eventually used extensively on research into the development of heavy supersonic aeroplanes, with special attention being given to the flight control system and the functioning of the powerplant. In the event, 'Aircraft 98' was never transferred for state testing because, during the second half of the 1950s, the decision was made to re-equip the Soviet's front-line aviation with supersonic Sukhoi Su-7B fighter-bombers (Chapter 8). From this point, until the 1960s, no new types of front-line bomber entered development for the USSR Air Force, that is until Sukhoi began work on its Su-24 (Chapter 8).

Tupolev produced several projects based on 'Aircraft 98'. The '98A', to be called Tu-24 in service, was to have been a front-line bomber and missile carrier powered by AL-7F-1 engines. It was expected to be lighter than the '98' and to introduce better aerodynamics (with a more extensive application of the 'area rule'), have an estimated top speed of 1,119mph to 1,243mph (1,800km/h to 2,000km/h) and was to be armed with one or three P-15 missiles. After this came the '98B' with more powerful engines based on the AL-7F2, VD-15 or AL-9 turbojets, a bigger wing and a new undercarriage. A much more extensive development was 'Aircraft 122', but in February 1958 the Council of Ministers cancelled all further work on the '98' and every one of these projects were terminated along with it. However, the line of development continued into the Tu-128 long-range supersonic patrol fighter-interceptor ordered in 1958. 'Aircraft 98' itself continued its research duties until 1959 and this work later included flight testing the Tu-128's RP-7 Smerch on-board targeting radar.

Tupolev 'Aircraft 122'

There was so much redesign introduced by this project over the original 'Aircraft 98' that a new number was thought to be necessary. It was prepared as a preliminary proposal during 1957 and power was to be supplied by two AL-11 engines. The estimated maximum speed was 1,243mph to 1,367mph (2,000km/h to 2,200km/h), service ceiling 65,617ft (20,000m) and range 1,865 miles (3,000km). No further progress was made with 'Aircraft 122' because the Air Force eventually selected the Yakovlev Yak-28 as its frontline supersonic bomber; however, the '122's development work, layout and design were later used in the Tu-128 interceptor.

Tupolev 'Tu-105'

By now the Tupolev OKB was aiming at a more capable aircraft. Following its proposals for 'Aircraft 97' and '103' (Chapter 3), in early 1954 Andrei Tupolev suggested to the Ministry of Aircraft Industry the development of a supersonic long-range bomber based on the Tu-16 (Chapter 3) and which, in due course, would replace the Tu-16 in service. His proposal was accepted and given support by the Government and on 30th July 1954 the Council of Ministers issued a decree covering its development. Designated 'Aircraft 105' its two Dobrynin VD-5F turbojets, giving 40,810 lb (181.4kN) of thrust in reheat, were expected to bestow a speed of 870mph to 932mph (1,400km/h to 1,500km/h) and the first prototypes were expected to be ready in 1956.

Tupolev '105' as first proposed in November 1954 with wing root engines.

Tupolev '105' first prototype.

Production Tupolev Tu-22KD missile strike aircraft.

By November 1954, as so often with a new advanced aeroplane like this, a selection of different aerodynamic schemes had been completed, the basic approach being to begin with the proven Tu-16 layout and then introduce suitable alterations to make the aeroplane capable of supersonic flight. At this stage the bomber still looked much like its subsonic predecessor but it did have new 55° sweep wings and no main landing gear fairings extending beyond the trailing edge (which were so typical of Tupolev designs of this period). The VD-5s were housed in nacelles placed along the fuselage sides which stretched right back to the tailplane. After this some more TsAGI recommendations had to be worked in which brought further changes, principally in moving the engines to the dorsal position above the rear fuselage. Several ways of 'squeezing' the fuselage were tried during the OKB's attempts to introduce area rule, and then in the summer of 1955 the VD-5F powerplant was replaced by a pair of VD-7Ms which offered 35,270 lb (156.8kN) of reheated thrust. A definitive configuration was established in August.

The preliminary studies and choice of the best design was the task of the OKB's Technical Department under S M Yeger, while D S Markov headed the group that took on the detail design and the development of further versions. The '105's engineering development got moving on 15th August 1955 and the Experimental Factory began to build the prototype in November; it was completed in December 1957 but frequent changes in the design had to be incorporated during this stage to cure the many problems which were encountered. In appearance the bomber was very different from anything the Tupolev Bureau had created in the past with a long slim fuselage, sweptback rear-mounted wing and an unusual engine installation, all of which suggested outstanding flying characteristics.

The first flight was made on 21st June 1958 and the prototype completed about twelve test flights before receiving some damage during a landing accident. It was not repaired because the second prototype (now called the '105A') took its place and this became the model for the production aeroplanes that followed. The first machine was never taken above subsonic speeds and little of its performance data was recorded. After lying derelict for a number of years, the prototype '105' was scrapped.

Tupolev 'Tu-105A'/Tu-22

'Aircraft 105A' was a further improvement of the basic '105' which the OKB introduced during the original '105's design. Eventually there were to be two lines of development, powered by VD-7M and NK-6 turbojets respectively, and the second, 'Aircraft 106' described shortly, was expected to achieve Mach 2 speeds. The most important change in the VD-7M version was the application of the 'area rule', represented by a 'thinning' of the fuselage, while the wing saw the return of typical Tupolev-style nacelles to house the main landing gear (which fitted the requirements of 'area rule' like a glove). Other changes included a small extension to the wing leading edge near the wing root to help improve the bomber's stability in supersonic flight while the number of cannon in the tail mounting was cut to just one. In addition, the engine nacelles were extended aft and given a small 'toe-in' angle; from the start the aircraft was to have a nuclear capability.

The USSR Council of Ministers Decree covering the '105A', which in service was to be designated Tu-22, was issued on 17th April 1958 and described not just the bomber but also the Tu-22K cruise missile carrier armed with the Kh-22 air-to-ground missile. With construction beginning in January 1958, the second prototype, (that is, the first '105A' – a static test '105A' airframe was also built) was completed in mid-1959 and became airborne on 7th September; however this machine's factory testing was never completed because on 21st December it was lost in a fatal crash. The test programme was finished using the first production aeroplanes which, in the aftermath of the crash, were quickly modified. These had their airframe strengthened, the elevator on the all-moving tail was deleted, the edges of the tailplane itself were clipped to improve the flutter characteristics and anti-flutter weights were added to the wingtips. Finally the engine nacelles were moved upwards a little, above the fuselage, to prevent any possibility of engine surge.

During 1960 a total of twenty aircraft rolled off the production line (nine of which flew over Tushino during that year's air show) and for several years all of these were employed on test programmes, improvements being introduced right through the period of production and delivery. The Tu-22 was eventually procured in many versions with deliveries continuing until 1969, but the type's teething troubles actually lasted into the 1970s. The West codenamed the aircraft *Blinder* and it entered service in 1962; by the mid-1990s only a few remained in service in the Ukraine.

Tupolev 'Aircraft 106' Series

It was realised that to improve the '105'/Tu-22 formula as a long-range bomber and missile strike aircraft it would be necessary to make the basic airframe cleaner from an aerodynamic point of view and give it a better engine, more powerful weapons and more capable avionics. The question of developing the '105' into a faster aircraft was already being considered by the Tupolev OKB when work on the original '105' was still in its early stages. During 1954, when the bureau received its orders to develop the '105', plans were formulated for 'Aircraft 106' based on the '105' but with more powerful engines to give a speed of Mach 2. Accordingly another Council of Ministers Decree was received, dated 30th July 1954, which ordered the development of 'Aircraft 106' along these lines using either two AM-17 or VD-9 engines, which gave a thrust for take off of between 33,070 lb to 37,480 lb (147.0kN to 166.6kN).

It was expected that such an aircraft would exhibit a maximum airspeed of 1,057mph to 1,119mph (1,700km/h to 1,800km/h), have a range of 3,605 miles (5,800km) when flying at 590mph to 622mph (950km/h to 1,000km/h) and carrying 6,613 lb (3,000kg) bombs and, when flying supersonically, have an altitude over the target of 49,213ft to 52,493ft (15,000m to 16,000m). The offensive and defensive armament, and crew, would be the same as the '105' and it was thought that a preliminary project should be prepared for this aircraft by

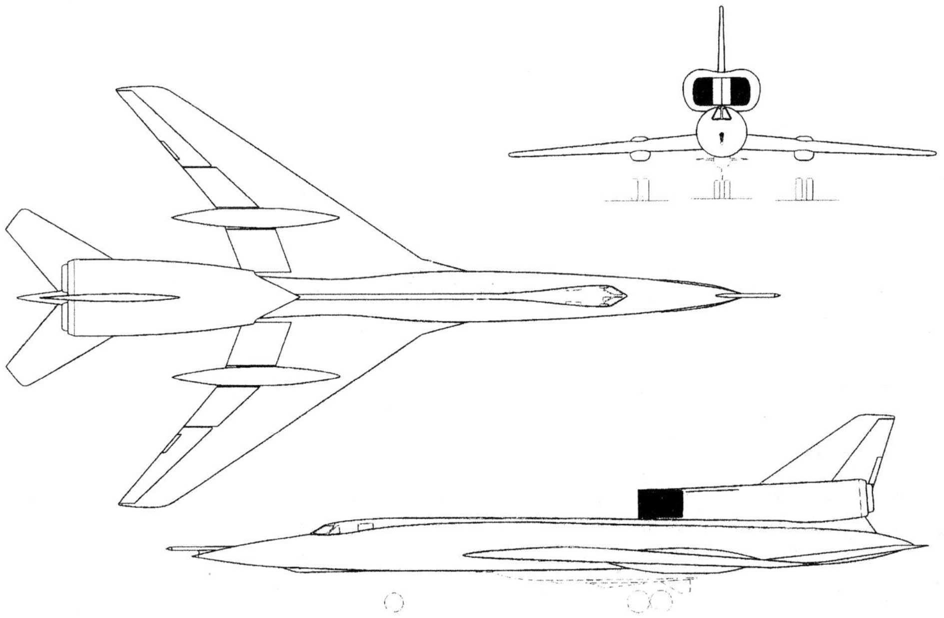

Tupolev '106K' as revamped in 1955 with a box nacelle for two NK-6 engines and a cruise missile beneath the fuselage. Russian Aviation Research Trust

Model of the '1955' Tupolev '106K'. George Cox

Artist's impression of the '106K' cruise missile carrier.

Cutaway of the '106K' showing the stowage position for the Kh-22 air-to-surface missile.

February 1955. If a go-ahead was given in March, the '106' should be flying by 1958. However, when the initial '105' and '105A' were still on the drawing board, someone proposed fitting Kuznetsov NK-6 afterburning turbofans to the airframe and in fact this idea was floated on several occasions, but each time the designers rejected it in the knowledge that you cannot squeeze Mach 2 out of a Mach 1.5 aircraft merely by putting in more powerful engines; any substantial improvement in performance required more radical changes.

The NK-6 was, by 1961, the most powerful Soviet-designed turbofan engine created so far and the very high level of thrust that it was initially expected to provide (possibly as much as 48,500 lb [215.6kN] with afterburning) was based on raising the temperature in the turbine to 1,130°C. Bench tests undertaken in November 1960 demonstrated a maximum reheated thrust of 49,380 lb (219.5kN) but also revealed that to establish a period of fifty hours between overhauls, the maximum thrust would have to be cut to 41,890 lb (186.2kN), with the maximum dry thrust 28,660 lb (127.4kN).

In 1955, after the VD-7M-powered '105A' had become reality, the '106's layout was changed and in due course Tupolev's designers produced a thoroughly revamped version. New features were mainly the powerplant plus the wings now swept back 60° at quarter-chord (rather than 52°14'30", although initially the wings were identical to those of the Tu-105A); these were also thinner, with an average thickness-chord ratio of 3.5% instead of 6%. Two NK-6s, giving 47,400 lb (210.7kN) thrust, were housed in a common box nacelle placed at the fin root which had a V-shaped front that looked very similar to the underwing nacelles later used by the Tu-160 *Blackjack* (Chapter 11). The vertical splitter dividing the supersonic air intake into port and starboard engine ducts incorporated variable ramps.

From the outset the new bomber lacked defensive armament, a Siren ECM pack replacing the Tu-22's tail barbette (the ECM pack was put through its paces on a production Tu-22P specially adapted as an avionics test-bed). Another redesign followed in 1956/57 when 'area rule' considerations were introduced along the lines of the changes that had turned 'Aircraft 105' into the '105A'.

As the bureau's involvement with 'Aircraft 106' increased, with input from TsAGI and other related organisations, a Decree was issued on 17th April 1958 which brought both the '105' and '106' programmes together as a combined Tu-22 and K-22 missile system. The NK-6-powered '106' was considered as the main aircraft with the VD-7M '105A' serving as a reserve back-up in case the NK-6 failed to reach fruition. With the NK-6's thrust now increased to 48,500 lb (215.6kN), the '106' was expected to have a top speed of 1,119mph to 1,243mph (1,800km/h to 2,000km/h), a range when flying at 590mph to 622mph (950km/h to 1,000km/h) and carrying 6,613 lb (3,000kg) bombs of 3,729 miles (6,000km) and, when flying supersonically, an altitude over the target of 52,493ft to 55,774ft (16,000m to 17,000m). A variant carrying the Kh-22 missile would release its weapon at heights between 32,808ft to 45,932ft (10,000m and 14,000m) and further increases of thrust from the NK-6 were expected to give a top speed approaching 1,553mph (2,500km/h). The development programme stated that the aircraft should be ready for joint state testing (with the VVS) during the third quarter of 1960.

However, the lengthy teething problems experienced by the '105A'/Tu-22 above (particularly with some elements of the K-22 system), plus delays to the NK-6 programme eventually pushed the '106's development timetable onto the beginning of the 1960s. When the bureau finally got back to full speed on the aircraft in 1960, in the hands of S M Yeger's Technical Projects Department, it had accumulated a vast amount of knowledge from the modifications made to the Tu-22. As a result several versions were now suggested – the '106R' reconnaissance and '106K' cruise missile carrier based on the

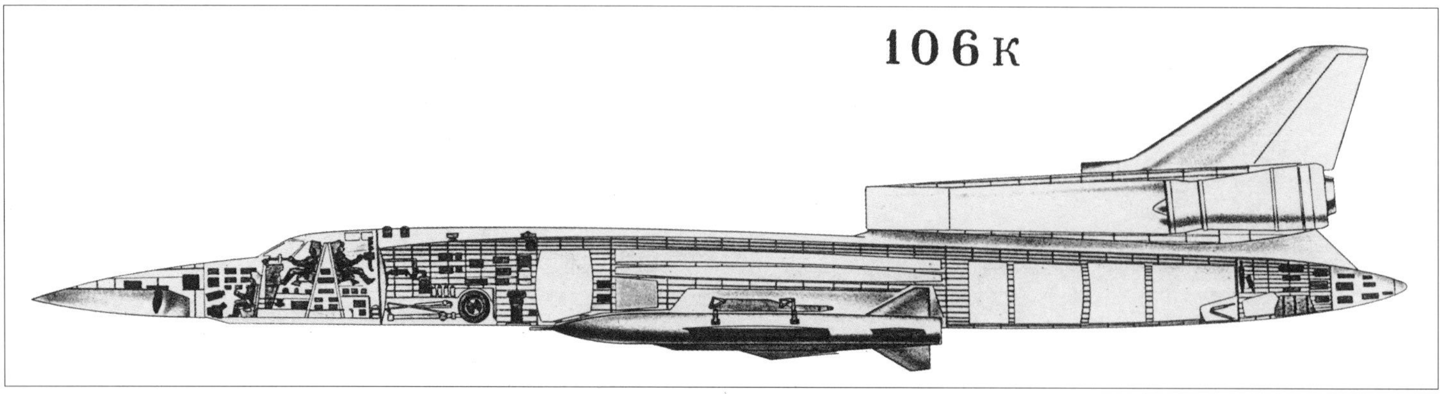

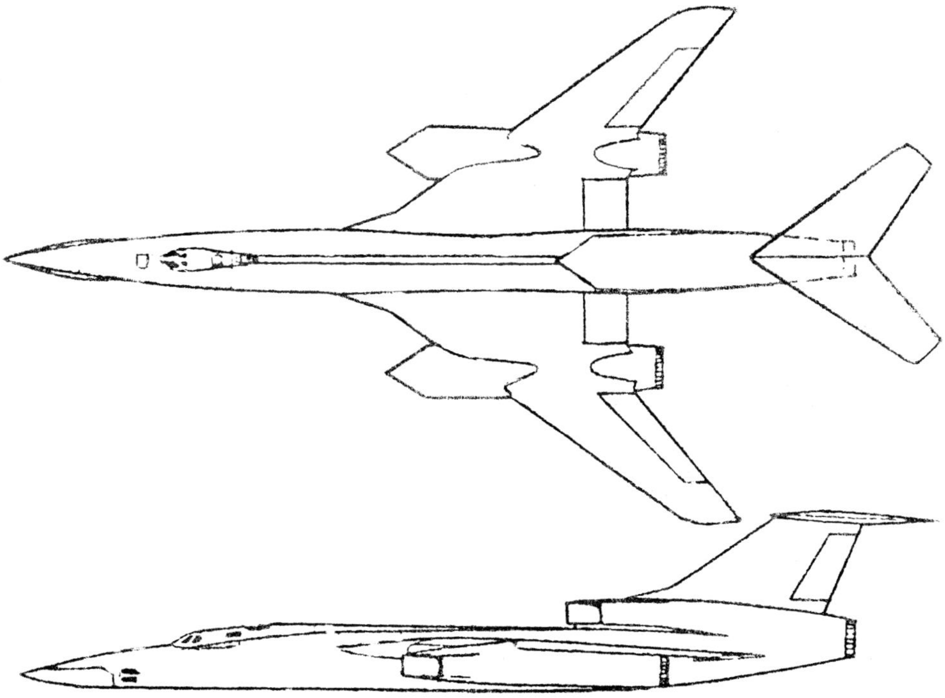

Variant of the Tupolev 'Aircraft 106A' powered by four Tumansky R-15B-300 or Dobrynin VD-19R2 afterburning turbojets in paired wing nacelles. George Cox

Tupolev '106A' with three Tumansky R-15B-300 engines, one under each wing and one at the fin root (1960). Russian Aviation Research Trust

Model of the R-15B-300-powered Tupolev '106A'. The design had a tricycle undercarriage with the main gears retracting into the nacelles. George Cox

'106' with NK-6s and standard Tu-22 wings (the 'K' was armed with a single Kh-22 air-to-surface missile and lacked defensive armament), the '106B' with a modified fuselage and the '106A'. The '106B's armament and systems were unchanged; span was 77ft 7½in (23.666m) and length 132ft 6½in (40.4m).

'Aircraft 106A' was a much more radical redesign of the Tu-22 and featured, in addition to new wings and engines, a T-tail. Four alternative powerplants were considered – two Kuznetsov NK-6s in underwing nacelles, four Tumansky R-15B-300 or Dobrynin VD-19R2 afterburning turbojets paired in wing nacelles, or three R-15B-300s. The latter configuration was certainly the most unusual since two of the engines were fixed under the wings and the third was placed at the base of the fin in a 'straight-through' mounting, along the lines of the American McDonnell Douglas DC-10 airliner. Offensive armament consisted of one Kh-22 carried semi-recessed in the fuselage and there were no defensive guns.

Altogether, with other versions introducing boundary layer blowing plus lift engines in the main landing gear nacelles to help reduce the take-off run, no less than two dozen different '106' configurations were eventually evaluated during the years spent on the project. Work on the original '106' plus the '106B' reached an advanced stage and a decision was finally made to adopt the thin 60° sweep wing. In due course the design of the NK-6 installation and the main elements of the airframe layout were completed, the choice and position for much of the equipment was selected and the drawings for the prototype were nearly finished. In addition the construction of the new wing and rear fuselage and tail had commenced and the NK-6 development programme progressed to an advanced stage with flight test articles being evaluated in a Tu-95LL flying test-bed aircraft.

However, in 1963 both industry leaders and the VVS doubted the wisdom of continuing the project and suggestions were made to cancel both the aircraft and its engine. In response Andrei Tupolev accelerated the pace of the programme as much as he could and, with the intention of rolling out a prototype in 1963, also decided to take a production Tu-22 airframe and install two NK-6s with minimal modifications. By doing this, and getting a flight test article into the air, Tupolev felt that it would show clearly the significant improvements in performance and flight characteristics that would now be available over the basic Tu-22. Armed with this information he could then push hard to get the more important NK-6-powered '106' cleared to continue, but in the end this 'stopgap' prototype was never completed because the Kuznetsov OKB failed to bring any of its NK-6 engines to a suitably airworthy condition. By this time the Tupolev OKB's experimental production factory had built the new rear fuselage with its huge NK-6 nacelles.

The NK-6-powered '106' was expected to cruise supersonically at 1,119mph (1,800km/h), have a subsonic range of 4,195 miles (6,750km), reducing to 2,486 miles (4,000km) when flying supersonically, and a service ceiling of 59,055ft to 65,617ft (18,000m to 20,000m). There were three crew. For the earlier '106K', range without external stores was 2,175 miles (3,500km) at 590mph (950km/h) or 1,865 miles (3,000km) in supersonic cruise, and service ceiling 52,493ft to 59,055ft (16,000m to 18,000m). The '106' programme was finally cancelled in 1965, when the bureau began its first studies into the variable sweep 'Aircraft 145' that eventually led to the Tu-22M *Backfire* (Chapter 11).

Myasishchev *Bounder* Family

Myasishchev M-50

To quote Bill Gunston in one of his many books on Soviet aircraft, this huge aeroplane was 'one of the most breathtaking aircraft of its day'. The seeds for the M-50's relatively brief moment of glory were first planted in 1954. Soon after Myasishchev's M-32 was dropped in 1953, the Soviet Union received its first news of the forthcoming American Convair B-58 Hustler bomber, which essentially comprised a relatively small aircraft with all of the disposable load, fuel and bombs, housed in an external pod underneath the fuselage; this was key because the pod helped to keep down the bomber's overall size. In response, the Myasishchev OKB, backed up by a new SovMin resolution dated 30th July 1954, began work on a 'composite long-range bomber', of its own (composed of a strike aircraft and a launch aircraft) which it designated the M-50. Other OKB's also looked closely at this idea and the concept as a whole is detailed in Chapter 5.

The aircraft, powered by four Dobrynin or Mikulin jet engines, was to be capable of 1,119mph (1,800km/h) and achieve a range approaching 8,080 miles (13,000km) when carrying a 11,023 lb (5,000kg) load. Early studies included a twin-boom carrier aircraft that looked more like a contemporary airliner with the supersonic bomber attached underneath, or the attachment of a large float to allow the bomber to take-off from water. Another design had a droop nose (somewhat similar to the Concorde supersonic airliner) and another used ten engines – four in pairs in two underwing nacelles, four stacked in pairs in two wingtip nacelles plus one either side of the fin root. Other preliminary projects looked at using between two and ten engines with the NK-6, VD-9 and AL-9 as the possible alternative engines.

However, some of the basic problems to appear during the first quarter of 1955 included a lack of sufficient knowledge for the stability and controllability afforded by the canard over such a wide speed range, that it was impossible to obtain a higher relative weight/load for the attack component of the composite bomber, the difficulties in flight testing this type of combined aeroplane (especially after separation since the attack element would not be recovered) and the difficulties of creating a launch system that could reach a speed of 404mph (650km/h) for a launch weight of at least 551,146 lb (250,000kg). TsAGI wind tunnel test also showed that the aerodynamic efficiency of a combined aeroplane was inferior to a conventional aircraft. In March 1955 the M-50 'composite bomber' was halted and replaced by a more conventional aeroplane.

The first 'single aircraft' designs were actually completed in February 1955 and on 19th July a new SovMin resolution upgraded the M-50 to a conventional long-range type with a powerplant of four VD-9 turbojets or NK-6 bypass turbofans, giving a maximum speed of 1,243mph (2,000km/h) and a service ceiling approaching 52,493ft (16,000m). I P Tolstykh was nominated as the chief designer and on 28th March 1956 another resolution was passed authorising the installation of 46,295 lb (205.8kN) Zubets RD16-17 engines or VD-9As. The RD16-17 could sustain supersonic flight without afterburning and it also offered a fuel consumption that did not exceed the requirements. Designed by OKB-16 led by P F Zubets, the engine offered 40,785 lb (181.3kN) of dry thrust and 46,295 lb (205.8kN) in reheat. The M-50 was to begin its state acceptance trials in the first three months of 1958.

Despite Myasishchev's heavy commitments on the M-4/3M series of bombers, the huge programme that this new aircraft represented meant there was no possibility of a competition between prototypes from different OKBs. One of the key factors in achieving the required performance was a big reduction in the weight of equipment carried, plus a crew of only two. Myasishchev and TsAGI conducted a joint research programme which analyzed no less than 39 possible con-

An early variant of the Myasishchev M-50.
George Cox

Another early 'composite bomber' variant of the Myasishchev M-50, partially inspired by the American Convair B-58 Hustler. On these designs the weapons were housed in the underwing or underfuselage pods.

Compared to the original OKB 'composite M-50' model this version shows slight differences to the fuselage behind the canopy. However, it gives a good idea of the general layout. George Cox

figurations, each tunnel tested in model form, and these embraced tailless and canard formats and tandem wing designs, but the final choice centred on a more conventional delta plus swept tail layout.

The primary elements were the thin delta wing, all-flying swept elevators and rudder, plus each engine mounted in its own nacelle and fitted with supersonic intakes and controlled by its own fly-by-wire system. With these features it was considered that sustained supersonic flight could be secured with cruise at Mach 1.7. The decision to use engine nacelles was brought about by the immense volume of installed power that was needed to push the aeroplane into the supersonic regime which, it was felt, could create a lot of problems if trying to fit the powerplant into the fuselage or wings. As one might imagine, the final result was achieved after a great deal of debate and the research work itself lasted for about a year; there was actually much support for some of the alternative layouts.

The position of the nacelles was vital in ensuring that the drag they created was kept to a minimum. There were four alternatives:

1. Two engines in nacelles on underwing pylons plus two at the wingtips
2. All four on underwing pylons
3. Two nacelles on each wing with one above and one below
4. Two on underwing pylons with the second pair mounted alongside the tail end of the fuselage.

Early results suggested that number 3 (the best from a structural point of view) gave the highest drag with the other three pretty similar, but 2 and 4 could not offer a suitably high level of structural rigidity without giving extra weight; consequently the solution was to use number 1 which was relatively easy to build and gave good aerodynamics.

The design and development of the M-50, both through its advanced shape and capability but also its sheer size, presented many new problems for which no existing data was available to help in solving them. For example, new calculations and formulae had to be devised by the Myasishchev design bureau to determine the resistance to deformation exhibited by this type of wing. In fact this was the first time that such logarithmic structural stress calculations were used in the USSR to determine the properties of a new aircraft's wing, but the task was still so complex that it took four years to complete, from 1955 to 1959. Other features that needed to be designed from scratch included the nacelles and their intakes and the need to ensure optimum stability and controllability at all speeds, for which the all-flying tailplane and tailfin were essential. Several fatal crashes of early supersonic aircraft types had occurred because the centre of lift forces moved rearwards when an aircraft passed from subsonic into supersonic flight; for the M-50 longitudinal stability and controllability were vital.

New methods of construction were also required including the employment of large stringers and milled skin panels, the structural materials used in the greatest quantities being V-95 light alloy and 30KhGSNA steel. Another substantial task was to match these features with the huge amount of fuel needed to achieve a satisfactory range because current jet engines were still heavy consumers of fuel. Since there were only two crew members, to keep down the workload many of the flight systems had to be made to operate automatically which meant much better and lighter electronics had to be created to prevent this new equipment from being too heavy, thus cancelling out the progress in weight reduction already achieved. Nevertheless, despite these measures, the final structure weight was 39,683 lb (18,000kg) higher than required; the empty weight was estimated to be 131,393 lb (59,600kg). Thanks to the extreme performance demands coupled with the size of the aircraft, the M-50 did have a new control system that was capable of transmitting signals electronically to the control surfaces. In addition an automatic system was used to adjust the balance of the CofG by transferring fuel between the various internal tanks.

The airborne requirements included the ability to fly at between 168mph and 1,243mph (270km/h and 2,000km/h) while the unrefuelled range had to be 7,458 miles (12,000km) maximum. Normal cruise speed would fall within the range 1,056mph to 1,119mph (1,700km/h to 1,800km/h) but the M-50 had to be capable of a dash at 1,180mph to 1,243mph (1,900km/h to 2,000km/h) over the target. A long-range mission would require at least two in-flight refuelling operations (the first after 1,243 miles [2,000km]) but the aircraft itself could carry 385,802 lb (175,000kg) of fuel at take-off (in an all-up-weight of 557,760 lb [253,000kg]), although it would need RATOG to help get it off the ground.

Although first designed as a bomber, there were soon plans to adapt the type as a cruise missile carrier as well, with Myasishchev's own '45B' missile the favoured choice. The M-50's preliminary project was concluded in December 1955 and work began on a mock-up early in the new year; when complete an official inspection by a Commission led by Marshal of Aviation V A Sudets was held in July 1956. It appeared that the aircraft should meet most of its requirements except for its unrefuelled range (which was eventually reduced to 6,215 miles [10,000km]) and a take-off run which was too long if RATOG was not used. The lower range meant IFR was essential which, because it was carried out at subsonic speed, also made the M-50 more vulnerable to enemy defences. Due to these weaknesses the Commission was unable to approve the M-50 mock-up but its findings were a surprise to the designers.

In response MAP arranged a meeting with various NII specialists and P V Dementyev wrote a letter to the Commander-in-Chief of the VVS, P F Zhigarev, to explain that, 'As a result of the consultations made, the MAP cannot accept the rejection of this preliminary project and of the mock-up of this aircraft'. Dementyev explained that the development of a supersonic long-range bomber with a cruising speed of 1,056mph to 1,119mph (1,700km/h to 1,800km/h) was a new and massive task to undertake and to achieve this with an aircraft weighing 584,215 lb (265,000kg) meant a longer take-off run was essential; thus, to keep the take-off run within the required 9,843ft (3,000m), RATOG had to be used. Two in-flight refuellings would give the specified range but adding further defensive weaponry would increase the weight yet again, so the M-50 would have to rely on electronic countermeasures to reduce the effectiveness of enemy defences (thanks to its high speed, the only defensive gun to be carried was a single 23mm in the rear fuselage). Dementyev concluded by saying that 'MAP has no other way of attaining this goal within the set deadlines'.

A modified full-scale mock-up was finally approved and the decision was also made to build two prototypes plus a static test airframe. The first machine, now called the M-50A and essentially a research aircraft, would carry out the preliminary flight test programme to clear the aerodynamics (including sustained supersonic flight) while the second, the M-50B, would get the full set of equipment, the correct engines and in-flight refuelling capability. There were further delays in producing some parts for the M-50 airframe, with numerous delivery deadlines missed, but it was the RD16-17 engines that fell badly behind schedule. As a result and to allow flight testing to get moving, the M-50A was eventually fitted with 24,250 lb (107.8kN) VD-7A power units.

The M-50A made its maiden flight on 27th October 1959 and performed well. Initial flights were made without afterburning on any of the engines, but this facility was added when the inner VD-7s were replaced by VD-7MAs, giving 35,275 lb (156.8kN) of thrust with reheat, in April 1961. It was hoped that these would deliver a top speed of Mach 1.35 but, in the event, reheat was only ever used on take-off. In 1958 the Myasishchev OKB was released from its obligation to present the M-50 for state testing because a decision had now been taken to use the two M-50s as part of the development programme for a derived follow-on design called the M-52 (below). The limited power of the substitute flight test engines meant that the M-50 was never flown at weights above 253,527 lb (115,000kg); reports also suggest that flying the beast made massive demands on the crew.

A total of eleven sorties had been completed when the decision was taken in 1960 to abandon both the M-50 and M-52. A number of factors influenced this move but the shortfall in performance, despite the achievement of completing such a massive aeroplane to such a tight schedule, was especially important. However, there was also plenty of politics behind the issue because, by this time, the VVS was now considered to be less important for the Soviet Union's planning for future warfare than in the past, strategic missiles were the way forward.

In late May 1961, after the Myasishchev OKB had been closed, it was decided that the M-50 should be displayed at that year's Tushino Show. The aircraft had been lying idle for about a year and quite a bit of preparation was needed to make it airworthy again. Seven practice flights were completed before the M-50 made its final trip over Tushino on 9th July 1961, although Western observers did not know this and the aircraft received considerable attention from the world's newspapers (and was eventually codenamed *Bounder*). The M-50 performed a roll over Tushino's spectators, leaving behind its escorting MiG-21 fighters, but it was never to be seen in the air again. Despite making such a big technical advance and introducing many new features, the M-50 actually proved to be relatively free of major problems. Its weakness were the engines which could not combine the large thrust the aircraft needed together with an acceptably low fuel consumption.

Myasishchev M-51

This was a proposed unmanned supersonic guided cruise missile development of the M-50 designed for a zero-length launch. The airframe was to be raised off the ground on a huge retractable pylon and then a massive battery of rocket motors (one group under each wing and another under the rear fuselage), plus the usual turbojets, would blast the machine into the air; a large volume of rocket fuel would have been carried within the centre fuselage and the total thrust from the rockets was to be 637,100 lb (2,832kN). The idea received official support through a SovMin resolution of 3rd August 1956 and a preliminary project was presented to MAP in October, but this project never did get off the ground. Soon afterwards, with the mounting level of work on Myasishchev's 3M and M-50 bombers and the successful launch of the first prototype Soviet ICBM, the M-51 was abandoned.

Myasishchev M-52

As the M-50's design and development proceeded, the design bureau looked as ways of improving the basic concept. Apart from a proposed airliner variant, the first fruit of this work was the M-52, also briefly known as the '50V', which was intended to serve as a supersonic launcher for a heavy stand-off missile. During this period bomber design worldwide was heavily influenced by the rapid progress in developing more advanced defensive systems, both fighters and surface-to-air missiles, which in America included the arrival of the McDonnell F-101 Voodoo and Convair F-102 Delta Dagger fighters armed with the latest air-to-air missiles. These presented quite a problem to Myasishchev's subsonic 3M bomber, and no little difficulty to the M-50, so another alternative strategic system was desirable.

The idea of using an advanced aircraft designed to get close enough to a target to release an unmanned cruise missile was a solution and the review report for the M-50 mock-up also stated that MAP should ask Myasishchev to 'create a weapon system comprising a supersonic launcher, based on the M-50, with a cruise missile'. Myasishchev himself supported the concept which would have a superior performance to the M-50. The result was the M-52K weapon system made up by the Kh-22 cruise missile (NATO codename AS-4 *Kitchen*), K-22U guidance system and the aircraft itself.

A draft project was submitted on 19th November 1957 that initially comprised the M-52 aircraft plus Myasishchev's own 'Article 44' Mach 3 cruise missile, carried externally, that weighed 24,250 lb (11,000kg) and was capable of a ceiling of 75,459ft (23,000m). The M-52's range would also be increased by using Myasishchev's M-70 supersonic flying boat IFR tanker (Chapter 9) to obtain more fuel.

Model of the Myasishchev M-52 before small winglets were introduced outside the tip-mounted nacelles. John Hall

The M-52K was authorised by a SovMin resolution of 31st July 1958 (this document also authorised the M-56K strategic bomber below). Myasishchev's OKB worked hard at preparing the M-52's design, based on the M-50, while the Mikoyan OKB was given the job of developing the Kh-22 cruise missile, which was initially proposed in rocket and ramjet-powered versions. After much tension between the two OKBs, with Myasishchev again working on its own cruise missile design, Mikoyan eventually agreed to a liquid-fuelled rocket engine. The M-52 powerplant would comprise four RD16-17 engines and the specified performance figures were to include take-off weights of up to 440,917 lb (200,000kg) and, with a warload approaching 22,046 lb (10,000kg), a range of between 4,972 and 6,215 miles (8,000km and 10,000km) at a speed of 1,243mph to 1,367mph (2,000km/h to 2,200km/h).

As first proposed the M-52 featured a pure delta wing of 89ft 10½in (27.39m) span, with the four power units in underwing nacelles, plus a conventional tail arrangement with the horizontal surfaces in the low position; its length was 187ft 4in (57.1m). However, during the design process the engine arrangement used by the M-50 was introduced to go with a tapered version of the wing. A small T-tail was also added solely to supply additional balancing during the aircraft's climb to altitude only – it could only be moved downwards to compensate for any excess pitch movements. Due to over-sensitivity the M-50's all-flying tailfin was also replaced by a conventional fin and rudder arrangement. The engine nacelles had central shock cones to give them supersonic capability.

Big changes were also made to the fuselage, which reduced the M-52's length over the M-50, but the bicycle undercarriage was retained (although revised) and the final fuselage weight was similar to the M-50. However, thanks to a redistribution of the structure mass between the main spars and skins, the wing weight was cut by some 15%, the torsion box-style of construction used in the M-50 having proved to be relatively unsuccessful. There were still two crew, who wore pressurised suits, but their seating had been rearranged so that they now sat side-by-side in seats which could eject upwards at all flight speeds; a proposed trainer version would have three crew. The KSB-1 flight navigation system was operated with a PN radar while modifications to the SAU-52 autopilot brought a big drop in the weight of this item of equipment.

The primary offensive armament of two Kh-22 cruise missiles was carried on the outside of the lower centre fuselage and these were to be replaced eventually by the more advanced Kh-44. There was also an internal bomb bay for bombs of any size up to 11,023 lb (5,000kg), or airborne torpedoes or mines, but when the cruise missiles were being carried this bay would be filled with another fuel tank. With two of the older cruise missiles aboard, the M-52's radius of action was expected to be 1,429miles (2,300km), rising to 2,331 miles (3,750km) with in-flight refuelling. Provisionally, for rear defence, a DB-52 barbette could be fitted which housed an AO-9M cannon controlled by a Krypton radar.

The M-52's take-off would be assisted by a re-usable jettisonable trolley, which would 'carry' a maximum of 187,390 lb (85,000kg) of the bomber's weight and be recovered by parachute, and a pair of 37,480 lb (166.5kN) Sevruk S3-43M RATOG motors; this extra power would be available for a period of one minute. By using the trolley the aircraft could take-off at a higher weight (546,737 lb [248,000kg]) and achieve an angle-of-attack of 13.5° at airspeeds of around 267mph (430km/h).

The whole story of the development of the M-50 and its derivatives appears to have been a constant saga of ever improving performance and capability with ever more changes to the design, which perhaps cannot be a surprise for such a large aeroplane involving so many different and difficult areas of research and development. Consequently, it will also be no surprise to learn that the second M-52 prototype (the M-52D or '50D') was to introduce several changes over the first machine. (The 'D' stood for 'doubler', a common Russian phrase used to denote an alternative version of an airframe which was intended to meet the same requirement).

The M-52's detailed preliminary project and mock-up was examined in June 1959 by a State Commission led by General Colonel Ye N Preobrazhensky, (this was also the point from where the M-50 prototypes were regraded as pure research aircraft) and once again it was rejected. The Commission's concluding report noted that the M-52K missile launcher, developed as 'a minor modification of the M-50 strategic bomber' had retained the latter's configuration with relatively little change to its size and weight, but the possibility of the type operating from Class 1 airfields near to areas of armed conflict had been ruled out and this would lead to 'extraordinary operating costs escalation as well as increased operating complexity'. In addition the aircraft's performance did not meet the SovMin resolution and VVS specification requirements. Range was the biggest weakness and at 363,757 lb (165,000kg) take-off weight the aircraft was incapable of achieving a radius of action of 2,548 miles (4,100km). This take-off weight was a limit set by the strength of the undercarriage and to get the maximum possible range, in-flight refuelling was essential. However, there was another problem.

Chapter 3 has noted how the Myasishchev OKB developed the M-4 *Bison* into an IFR tanker but the final 3MT tanker variant could only perform a refuelling operation at a maximum speed of 373mph (600km/h). Unfortunately the minimum airspeed at which the M-52 could safely receive fuel was between

348mph and 354mph (560km/h and 570km/h) and there was clearly little margin for error (this small margin was considered to be insufficient for manoeuvring during the IFR operation). As a result there were proposals to develop IFR versions of both the M-50 and M-52, the former offering the potential to refuel in flight at heights above 32,808ft (10,000m) and at supersonic speeds. Finally, a lack of power suggested that the lift against drag ratio would be reduced which would prevent the M-52 from becoming supersonic when above 374,780 lb to 396,825 lb (170,000kg to 180,000kg) weight.

Myasishchev's reaction to the review report was to reject the findings, saying that the estimated data met the requirements, but the OKB had to introduce further changes such as geometric camber on the wing together with additional wing area, which included raked leading edge extensions applied to the wingtips. A small additional surface was also fitted outside each outer nacelle, the objective being to improve the efficiency of the aerodynamics when the bomber was travelling at Mach 1.7. To back this up the Kh-22 missile itself was cleaned up. Revised brochure documents were prepared in November 1959 and during December the VVS specialists reviewing them confirmed that the M-52 design was indeed viable. In addition, an M-52R reconnaissance version was to have been proposed in October 1960.

It was initially intended to build five M-52 prototypes armed with Kh-22 missiles and to have the first pair finished in 1960 and the last three in 1961. Soon afterwards, however, both the Soviet Defence Ministry and GKAT (the State Committee for Aviation Equipment) asked the SovMin for a reduction to three prototypes only, which would be used to support to yet another new design, the M-56K. This new design was, in many respects, intended to be a response to and a match for the spectacular North American XB-70 Valkyrie bomber. Afterwards the Ministry, GKAT and GKRE (the State Committee for Radio-Electronics) proposed the complete cancellation of the M-52 but new resolutions, issued by the Party Central Committee and the SovMin on 5th February and 30th May 1960 respectively, declared that the three prototypes should be completed to give the Soviet Air Force and aircraft industry vital experience in handling such large and complex aeroplanes.

Despite being moved to Zhukovsky in readiness for flight testing, the first M-52's maiden flight was held up by delays in delivering the RD16-17 engines. In late 1959 the Myasishchev OKB was closed and then the whole M-50/M-52 programme was abandoned during the following year – the poor prospects for the M-52 were finally acknowledged although this decision did generate some protests. Both M-50 and M-52 prototypes languished in long-term store at Zhukovsky before, in 1968, the M-50 was taken to join the collection of aeroplanes at the Monino Museum; the M-52, however, was scrapped in the 1970s. The first M-52 was not too far off from reaching completion, lacking just engines, some control systems, ejection seats and other minor assemblies. VVS service evaluation of the aircraft had been expected to get going in 1963.

In 1962 Vladimir Myasishchev, after he had been placed in charge of TsAGI, tried to resurrect the M-52 by suggesting that the bomber should be refitted with 48,500 lb (215.6kN) thrust VD-7M or NK-6 engines. As a result a moderate development programme could be implemented but his proposals were felt to be unacceptable and were rejected. Such a task could take as long as four years and the result, an aircraft that no longer met contemporary requirements, was very unlikely to be put into production. By this time the Sukhoi OKB was working on what was to become the T-4 and Tupolev had begun the similar 'Aircraft 135' project (Chapter 10).

Myasishchev M-54

This designation referred to a proposed variant of the M-50/M-52; the main version had a large delta wing and four 40,785 lb (181.3kN) RD16-17 engines on pylon-mounted underwing pods (which were more suited for sustained supersonic flight), plus a single central fin; the M-52's horizontal tail surfaces, however, were dispensed with, which reduced the movement of the centre of lift to 10% when passing into the supersonic regime, thus reducing the level of CofG movement and, therefore, the need for fuel to be used as ballast. The weight saved approached 33,069 lb (15,000kg) while losing the tail also improved the aerodynamic efficiency. The M-54 also introduced a tricycle undercarriage and a single X-22 cruise missile was to be carried beneath the centre fuselage. In plan view it looked fairly similar to the American Convair B-58 Hustler but was an altogether larger aeroplane; span was 88ft 3in (26.9m) and length 144ft 3in (44.0m). Work on the M-54 did not get very far but did include in-depth wind tunnel testing to explore the design's flutter and dynamic load characteristics.

Myasishchev M-56

The end of the M-52 was still far from the end of the Myasishchev heavy bomber story because this further design was begun in 1957 as a private venture and a radical modification of the M-52K with an even better performance – 1,989mph (3,200km/h) top speed, 65,617ft (20,000m) ceiling and 9,944 miles (16,000km) range. Prior to the design of this new advanced cruise missile launcher (the M-56K), and its later reconnaissance variant (M-56R), which were intended to enter service in the late 1960s, it was clear that the M-50/M-52 layout could not be developed any further because of the substantial shift in the

Model of the original Myasishchev M-56 layout (1957). A cruise missile would have been carried beneath the fuselage.

The 1959 version of Myasishchev's M-56. Russian Aviation Research Trust

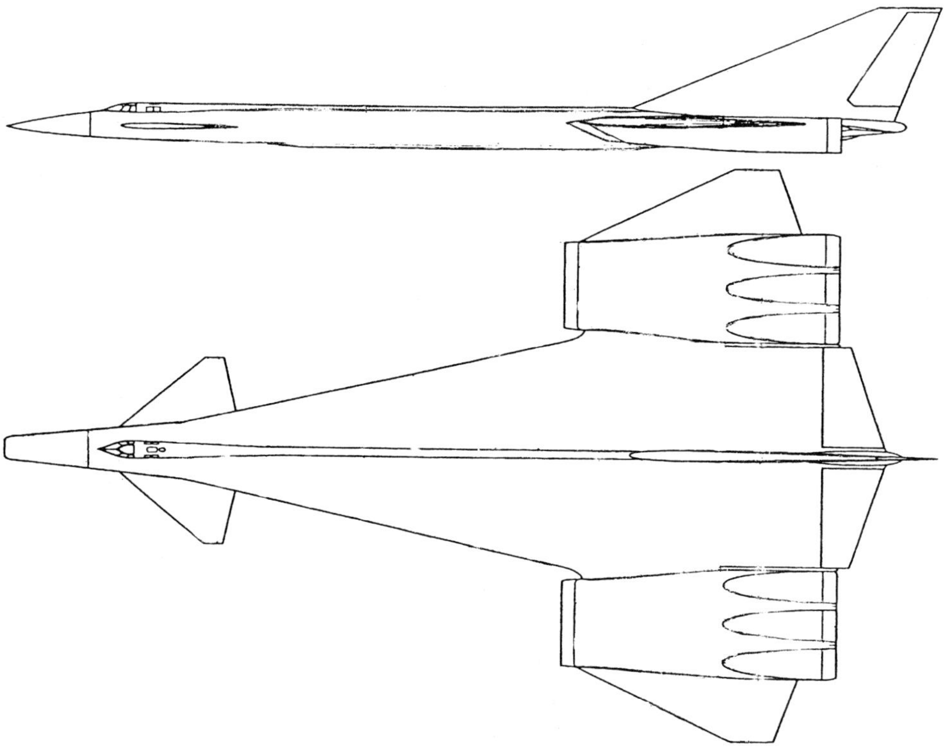

centre of lift and CofG suffered by their configurations when moving into supersonic flight mode (a tailless delta reduced the centre of lift change from 28% to 15%); there was also substantial airflow separation on the tailplanes and adverse ground effect during landing, and the angle-of-attack needed to be reduced because the bicycle undercarriage and long rear fuselage did not fit together well. Studies looked at tailless deltas, canard deltas and conventional straight wings and in 1958 it was the canard delta, with pylon mounted engines, that was adopted. The eventual layout was designed by L L Selyakov, one of the OKB's designers.

As noted, the transition of an aircraft from subsonic to supersonic speeds was accompanied by a rearward shift of the centre of lift, which also brought severe changes to the aircraft's longitudinal stability, and this new wing arrangement would reduce the problem. However, a further solution would be to move the internal fuel around inside the aeroplane to match the movements in centre of lift, and to use an *all-flying* canard foreplane with no tailplane, a configuration which at this date had rarely been employed in aircraft design. When flying subsonically the canard would be capable of full movement but this would not create any extra force that acted on the whole aeroplane. However, after passing into the supersonic regime, the canard would be locked solid and this would push the centre of lift forwards. The aircraft's elevons would maintain longitudinal control; at the time, the whole idea represented a very original way of maintaining longitudinal stability.

The combination of wing and engine intake position also ensured a sufficient increase in aerodynamic efficiency, to 6.2 compared to 5.0 to 5.5 for the M-50 and M-52, and that would allow the bomber to achieve the required range – at last! Compared to the efficiency figure for the later Concorde and Tupolev Tu-144 supersonic airliners, which was 7.8 and 8.2 respectively, this increase was relatively small, but it would prove sufficient provided that the engines themselves supplied enough power. For small targets the M-56 was intended to carry, beneath the fuselage, two Mikoyan Kh-22s (total weight 25,573 lb [11,600kg]) while one of Tsybin's large 28,660 lb (13,000kg), 1,865 mile (3,000km) range X-44 cruise missiles would be used for larger objectives such as air or naval bases. The M-56K go-ahead was given by a SovMin resolution dated 31st May 1958 and the aircraft was to be powered by four Kuznetsov NK-10B engines.

Trying to find a suitable solution to all of these requirements caused plenty of problems for the designers and the initial work suggested that a maximum speed of 1,554mph (2,500km/h) at about 50,853ft (15,500m) altitude was possible, when the specification called for something in excess of 1,865mph (3,000km/h). However, the M-56 moved forwards very quickly and several variations of the original layout were tried with the basic canard delta retained but the engines in different positions; there were also other layouts with 'straight' tapered wings or even more advanced delta wing versions.

An official NTS GKAT meeting was held on 4th July 1959 to discuss in general the nation's strategic weapon system projects, including the M-56K which was by now a hot topic. One of the department chiefs from TsAGI, V V Studminsky, together with the head of applied mathematics at the Academy of Sciences, M V Keldysh, and Sergei Ilyushin and Pavel Sukhoi, strongly opposed the M-56 as a strategic weapon system saying that it was too complex when compared to ICBMs and Burya type intercontinental cruise missiles; it should be used as a medium-range aircraft instead. The chief of TsIAM, G P Svishchev, gave strong support to Lavochkin's Burya cruise missile and wished to see other projects, including the M-56 cancelled, but the chief of TsAGI, A I Makarevsky, felt that the M-56 was still needed because unmanned intercontinental missiles, although filling a vital role, were as yet too unreliable and expensive: manned systems still had a role to play.

However, the M-56 was allowed to continue but the NK-10B engine was abandoned, the engine design bureau having other commitments, and the draft M-56 project was not reviewed by GKAT and VVS Command. Now the OKB's efforts were switched to the M-56R which was classed as a strategic reconnaissance and bombing system to meet new VVS requirements. This aircraft would undertake photographic or electronic reconnaissance, for which it would carry a reconnaissance drone or a detachable equipment pod, or attack fixed or mobile targets using the Kh-22.

The alternative M-56, proposed during 1959, was powered by six 22,050 lb (98.0kN) Klimov VK-15 engines. This was essentially a very highly swept flying wing with the engines placed in groups of three in each outer wing and most of the huge 'lifting body' fuselage/wing in between housing fuel. The canard was retained, there was a single central fin and also large elevons on the inner wing trailing edges; span was now 89ft 1½in (27.166m) and length 145ft 2in (44.25m). This was to be an incredibly advanced machine, with only two crew, which was intended to cruise at 1,678mph to 1,989mph (2,700km/h

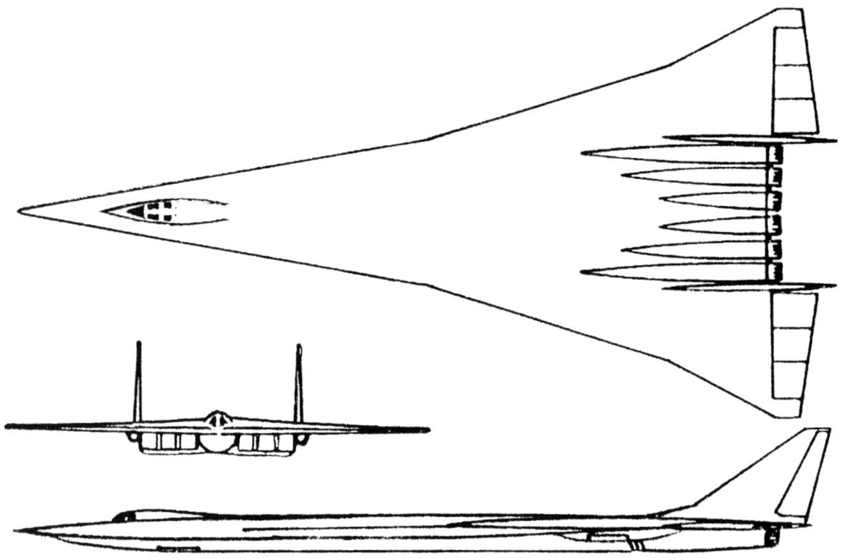

Myasishchev M-57 (1959).

to 3,200km/h), have a top speed of 2,175mph (3,500km/h) and offer a maximum range of at least 6,837 miles (11,000km) and a ceiling approaching 82,020ft (25,000m); the missile itself would have a range of 1,243 miles (2,000km).

A preliminary project for the six-engine design was put together during the spring of 1960 and the final powerplant was to comprise RD16-16 or RD17-117F engines of 27,557 lb (122.5kN) thrust dry and 38,580 lb (171.5kN) with reheat. One problem was that the take-off weight had to be limited to 396,825 lb (180,000kg) to allow runways built with the thickest surfaces to handle the aircraft but, in due course, this figure was exceeded and reached 462,963 lb (210,000kg).

A new SovMin resolution, dated 30th May 1960, authorised the M-56's development with a total of four prototypes to be built powered by Tumansky RD-17-117F engines and a further series of studies made by Myasishchev into the ideal configuration confirmed that the canard, plus separate elevons on the wing, offered the greatest advantages; in addition the 'slab-form' air intakes for the 'engine packs' was indeed the best solution. Myasishchev built two 1/25th scale flying models, one flying as a glider and the second powered by a piston engine, to record the design's flight characteristics on camera and it was concluded that the 'power-pack' layout blended well with the contours of the airframe. Some sources say that an M-56 mock-up was built and that the aircraft's first flight was expected to take place during the summer of 1963, but by now there were heavy restrictions on the VVS's budget for acquiring aircraft and, as a result, the development of the service's strategic bomber arm was brought to a standstill. After the Myasishchev Bureau was closed, a SovMin resolution terminated all versions of the M-56 on 3rd October 1960.

Myasishchev M-57

One of the Tumansky VK-15-powered M-56 projects was eventually redesignated M-57, its primary objective being to improve the lift/drag ratio against the figures obtained for the M-50 to M-56 series. This was believed to be a canard design with six Klimov VK-15s mounted in two 'three-engine' nacelles under the wings, and fitted with an 'all-flying' fin, but a display model shows a 'cranked delta flying-wing' with the delta extending from the nose, twin fins and the six engines mounted side-by-side in a box between the fins with their intakes beneath the wing. The M-57 was to have had a tricycle undercarriage and carry two Kh-22 missiles in a central weapon bay. Three crew were carried, maximum weight 485,009 lb (220,000kg), maximum speed 1,865mph (3,000km/h), service ceiling 78,740ft (24,000m) and range with 302,028 lb (137,000kg) of fuel aboard 10,068 miles (16,200km). The M-57's research was later merged into the M-56 project.

–

A great deal has been written and published over the years describing the cancellations of many advanced British combat aircraft, together with their politics and the effects on industry that these decisions created. Such cancellations also feature in the histories of other countries who have possessed their own aircraft industries. Between 1945 and 2000 the USA abandoned several important fighter and bomber programmes for a variety of reasons while Canada, Sweden and even Israel lost major aircraft projects before they entered service. As more and more information comes out of the former Soviet Union, it is becoming clear that 'cancellation disease' was also a frequent visitor to this nation's aircraft industry and in many cases the reasons were purely political.

During the late 1950s and early 1960s the apparent superiority of guided missiles over manned aircraft for both defensive and strategic offensive purposes, which in 1957 became a decisive factor in Britain military planning, also affected the USSR. This powerful theory, based on the considerable advances achieved in missile systems at that stage, seems to have concentrated the minds of the Communist Party leadership; Nikita Khrushchev for example was much more interested in missiles than aircraft. The missile lobby became a powerful force within the Soviet Union and influenced both the political and military leadership in bringing to an end aviation's function as the supplier of much of the strategic deterrent. The country's strategic air force (even its tactical aircraft) and the aviation industry as a whole was badly affected, the slogan 'the missile will replace the aeroplane' became popular and in due course major changes were introduced to the structure of the USSR's military industry.

Several aircraft OKBs and MAP factories were switched to other tasks and their aircraft manufacturing equipment chopped up. In addition a substantial number of Soviet Air Force units were disbanded and their airfields became aviation graveyards. The British aviation press has regularly highlighted the widespread losses in future fighter and bomber projects that occurred following the 1957 Defence White Paper (which declared that manned fighter aircraft were obsolete and guided missiles would be the way to go), and the cutbacks in RAF strength that followed, but this same theory seems to have hit the Soviet Union's Air Forces equally as hard. The Lavochkin OKB for example, a specialist in the design of fighter aircraft, was closed and the members of its design team moved on to creating missiles and satellites.

The arrival of the missile age, and Khrushchev who became the Soviet leader in 1956/57, were the prime reasons for the end of Myasishchev's work on strategic bombers. Khrushchev concluded that the money and energy pumped into types like the M-50 and

M-52 would be better spent on intercontinental ballistic missiles (ICBMs) and in 1960 he made the decision that closed down the Myasishchev OKB and moved many of its design staff onto missile and space programmes as well. Khrushchev was also very opposed to the policies of his predecessor Stalin, who was a strong supporter of aviation, and this did not help many of his judgements and decisions. Vladimir Myasishchev himself became the head of TsAGI before, in 1967, returning to aircraft design to work on the very secret M-17 reconnaissance aircraft.

Of course the huge costs of developing types like the M-50 and M-52 were a factor, and the relationships between the various personalities who led the competing design bureaux sometimes did not help either; for example Andrei Tupolev seems, at times, to have been very difficult to work with. That said, Aleksandr Yakovlev was very politically motivated and more in alignment with Communism and the Soviet Union's leaders than many of his contemporaries, and his criticisms of Tupolev's leadership of the aircraft industry in the 1930s assisted in bringing about Tupolev's imprisonment in 1937. As a result Yakovlev remained less than popular with much of the Soviet Union's aircraft industry and, after Tupolev was released, there was a bitter and continuing hostility between the two men and their supporters.

Fifteen years after the end of the Cold War the factors and facts behind these cancellations are in many cases only just coming to light as Russian historians work their way through the mountains of archives now available to them. It may be some time yet before a definitive history is written which gives a full account of the 'politics' of Soviet aviation. In the meantime the reader will have to settle for the rather basic coverage given here and make his or her own judgements as to how wise it was to abandon types like the Myasishchev M-50 and M-52.

Supersonic Bombers – Data / Estimated Data

Project	*Span ft in (m)*	*Length ft in (m)*	*Gross Wing Area ft² (m²)*	*Max Weight lb (kg)*	*Powerplant Thrust lb (kN)*	*Max Speed / Height mph (km/h) / ft (m)*	*Armament*
Ilyushin Il-54 (project 23.3.53)	57 11 (17.65)	84 3 (25.69)	?	79,365 (36,000) (normal)	2 x TRD-1 18,960 (84.3)	780 (1,255) at 16,404 (5,000)	1 x 1 (fixed) + 1 x 2 23mm cannon, 6,614lb (3,000kg) bombs
Ilyushin Il-54 (flown)	57 11 (17.65)	95 0 (28.963)	910 (84.6)	91,711 (41,600)	2nd prot: 2 x AL-7F 22,045 (98.0)	2nd prot: 718 (1,155) at S/L, 777 (1,250) at 16,404 (5,000)	1 x 1 (fixed) + 1 x 2 23mm cannon, 11,023lb (5,000kg) bombs
Myasishchev M-31	113 8 (34.64)	160 9 (49.0)	4,301 (400.0)	401,235 (182,000)	4 x VD-5 28,660 (127.4)	684 (1,100) (?) at 36,089 (11,000)	2 or 4 x 23mm cannon, up to 26,455lb (12,000kg) bombs
Myasishchev M-32	84 4 (25.7)	158 10.5 (48.43)	?	370,370 (168,000)	4 turbojets	1,150 (1,850) at 36,089 (11,000)	No defensive cannon fitted, up to 26,455lb (12,000kg) bombs
Tupolev '98'	56 8 (17.274)	105 1.5 (32.055)	941 (87.5)	85,979 (39,000)	2 x AL-7F 14,330 (63.6) dry, 20,945 (93.1) reheat	848 (1,365) at 36,089 (11,000)	1 x 1 + 1 x 2 23mm cannon,
Tupolev '105' (flown)	77 11 (23.745)	137 5.5 (41.921)	1,791 (166.6)	198,413 (90,000)	2 x VD-7M 35,275 (156.8) reheat	est 901 (1,450) at 36,089 (11,000)	2 x 23mm cannon, max 19,841lb (9,000kg) bombs
Tupolev '105A'/Tu-22 (flown)	77 5 (23.6)	138 5 (42.2)	1,745 (162.25) with LERX	187,390 (85,000)	2 x VD-7M	938 (1,510) at 32,808 (10,000)	1 x 23mm cannon, max 19,841lb (9,000kg) bombs
Tupolev '106K'	77 7 (23.646)	131 10.5 (40.195)	?	?	2 x NK-6 up to 48,500 (215.6)	1,367 (2,200)	No defensive cannon fitted, 1 x Kh-22 cruise missile.
Tupolev '106'	77 7 (23.646)	131 10.5 (40.195)	?	up to 233,686 (106,000)	2 x NK-6 up to 48,500 (215.6)	1,367 (2,200)	Various bombs or 1 x Kh-22 air-to-ground missile
Myasishchev M-50A (flown)	82 4 (25.1)	188 7 (57.48)	3,125 (290.63)	319,665 (145,000)	2 x VD-7A, 2 x VD-7MA 24,250 (107.8), 35,275 (156.8)	653 (1,050) max achieved	None carried
Myasishchev M-50 (estimated)	80 9 (24.62)	172 0 (52.42)	3,153 (293.24)	546,737 (248,000)	4 x M16-17M 46,295 (205.8) reheat	1,243 (2,000)	1 x 23mm cannon, 11,023lb (5,000kg) bombs
Myasishchev M-52K	86 0 (26.22)	178 9 (54.49)	3,247 (302)	c546,737 (248,000)	4 x M16-17F 44,090 (196.0)	1,243 (2,000)	Provision for defensive cannon, 2 x Kh-22 cruise missiles
Myasishchev M-56K 1959	80 8.5 (24.6)	153 3 (46.7)	3,806 (354)	462,963 (210,000)	4 x NK-10B 35,275 (156.8)	1,647 (2,650)	1 x Kh-44 or 2 x Kh-22 cruise missiles + nuclear or conventional bombs up to a max of 19,841lb (9,000kg)
Myasishchev M-57	105 0 (32.0)	204 11 (62.45)	?	529,101 (240,000)	6 x VK-15M 22,050 (98.0) dry, 34,830 (154.8) reheat	2,300 (3,700)	Kh-22 or Kh-44 cruise missiles.

Chapter Five

Composite Bombers and Reconnaissance Aircraft

The Tsybin NM-1 model test aircraft.

The previous chapter, during its description of the early stages of the Myasishchev M-50 series, touched on the concept of the 'Composite Bomber', an idea stimulated in part by the existence of the American B-58 Hustler supersonic bomber. The idea was to have a two part strategic weapon system, a strike aircraft and a launch aircraft, the latter carrying the former to a point which brought the strike aircraft within range of its target. The bomber would then be released to make its attack and it was thought that this method should bring any target on the earth's surface within reach of the Soviet Union's strategic weapons. Myasishchev's studies were soon overtaken by the aeroplane that eventually flew as the M-50, but there were other design bureaux who were interested in the 'Composite' concept.

Tsybin RS Series

By the early 1950s almost all of the experimental design bureaux (OKBs) were well established. In fact their chief designers had formed a strong monopoly with individual OKBs usually specialising in certain types of aircraft, so it was quite difficult for a new organisation to compete and enter into the arena of aircraft design. The Myasishchev OKB was one that managed to break this barrier and another was OKB 256 led by Pavel Tsybin. Tsybin's background was in gliders and sports aircraft but in 1945 to 1947 a team led by him also worked on and flew the LL series of rocket-powered transonic research aeroplanes. In the 1950s Tsybin moved on to high-speed nuclear delivery aircraft which were initially intended to be launched from a larger carrier aeroplane. This research eventually progressed into pure reconnaissance types which, to be accurate, fall outside the parameters of this book; however, they are far too interesting to leave out. Tsybin was allowed to form his own OKB but this was absorbed by Myasishchev in October 1959.

Tsybin RS

The arrival of thermonuclear weapons (that is, hydrogen or H-bombs) meant that new types of systems would be required to deliver them. In truth, until the arrival of ballistic missiles, the only solution was a supersonic bomber and on 4th March 1954 Pavel Tsybin took his ideas for such a type to the Kremlin.

Tsybin RS (1.56).

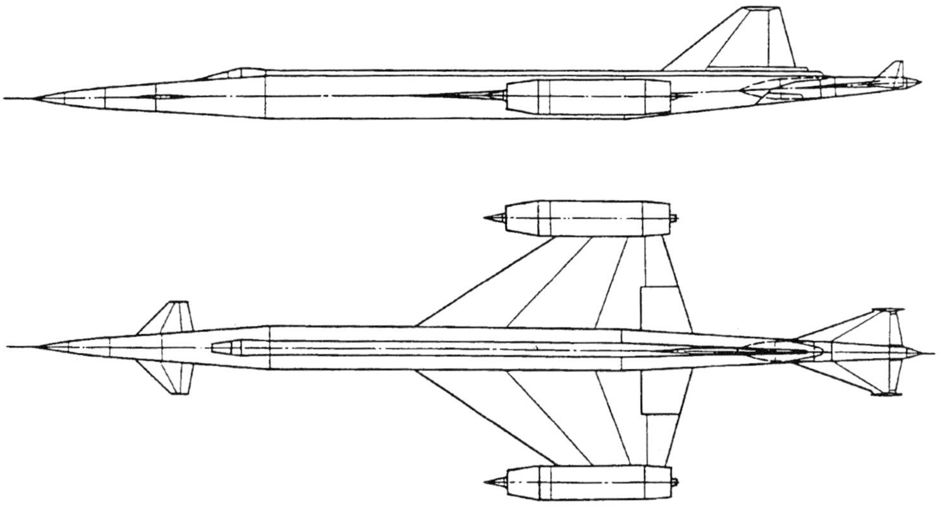

Model of the Tsybin RS complete with its '244N' nuclear weapon housed in a delta flight vehicle fitted to the back of the fuselage. George Cox

His proposals for a jet aeroplane (a *Reaktivnyi Samolyot* or RS) had come together after a good deal of consideration and actually described a machine capable of 1,865mph (3,000km/h) top speed, 8,701 miles (14,000km) range and a ceiling of 98,425ft (30,000m); 36,376 lb (16,500kg) of fuel would be carried.

To achieve these objectives such an aeroplane would need very thin wings, as light a structure as possible and extremely powerful engines, the latter giving a high thrust/weight ratio; however, no such engines were, as yet, under development. TsAGI research also showed that a long slim fuselage would help together with a canard foreplane and placing the engines at the wingtips. A winged bomb based on the '244N' thermonuclear device could also be specifically designed for the aircraft. (Curiously, similar research carried out in 1952 at the Royal Aircraft Establishment at Farnborough in England had established that a similar type of twin-engined aircraft, but with a conventional tail and mid-wing engines, would be ideal for research into to speeds approaching Mach 2.5. This eventually led to the Bristol 188.)

The Kremlin was very enthusiastic and took the proposals most seriously, despite some who felt that bringing such a concept to reality might prove to be very difficult. As a result a more detailed proposal was prepared and this was assessed very closely by MAP, TsAGI and a commission made up of industry and VVS members. They decided that the idea was indeed feasible, although yet more assessments were to delay the start of work by about a year. Finally, on 5th May 1955 Tsybin completed a presentation to Communist Party leaders and MAP which resulted in a SovMin resolution, dated 23rd May, allowing him to establish his own design bureau to bring the RS to fruition. The first prototype was to be flying by 1st February 1957, the second two months later. (Ivnamin Sultanov quotes Andrei Tupolev telling Tsybin at this time 'If you succeed with it, I will learn from you and will then do the same, only faster and better than you. But if you make a mess of it [the terms he used were rather stronger] you will only prove that it was a waste of time establishing a new OKB'.)

The setting up of the Tsybin design bureau to produce such an advanced aeroplane in such a short time was a near impossible feat. Many high quality staff were needed, who were specialists in their fields, and numerous other problems also needed to be sorted out which often pushed to one side the primary difficulties of the RS itself, leaving them unsolved. In retrospect, the task as a whole was beyond the capability of a brand new design bureau. By late 1955 it had been established that the RS's range would only reach a maximum of 4,661 miles (7,500km) and it was at this time that a proposal was made to carry the RS for part of the journey on an adapted Tupolev Tu-95, perhaps as much as 2,486 miles (4,000km), before releasing it to continue alone (for Tu-95N, N = *Nositel* or 'carrier'). In the meantime the Bondaryuk engine OKB had begun work on a supersonic ramjet, the RD-013, to power the RS.

A preliminary project for the RS was issued on 31st January 1956 (a supplementary proposal described the 2RS reconnaissance variant [below]). Aerodynamically the design had a good deal in common with the British Avro 730 supersonic bomber and reconnaissance aircraft (although it was a much smaller aeroplane) with a long narrow fuselage, small trapezoidal wing and wingtip-mounted engines, plus a canard foreplane of 10ft 6in (3.2m) span. The RS would be loaded aboard the Tu-95N which would then take it to an altitude in the region of 29,528ft (9,000m) and, at the required point, release it to make its own way to the target (at a speed approaching 1,865mph [3,000km/h]). The RS would first accelerate using two rocket motors (these would eventually be jettisoned) and then continue on power supplied by the wingtip ramjets. These were configured for Mach 2.8 operation with fixed-geometry multi-shock inlets and convergent/divergent nozzles. A total of 23,082 lb (10,470kg) of fuel was to be carried and the '244N' nuclear weapon, weighing 2,425 lb (1,100kg), would (according to one drawing) be carried to the target in a small tailless delta

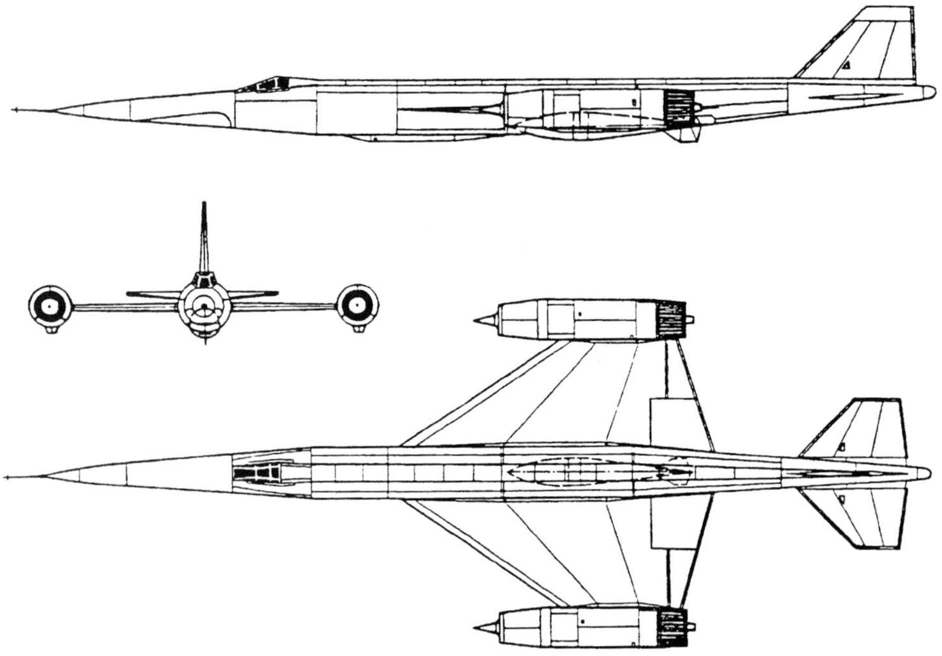

Tsybin 2RS (1956).

vehicle mounted in the back end of the RS's fuselage; this would then be released when required.

At the same time as Tsybin's RS was taking shape, the Korolyov OKB was working on an intercontinental ballistic missile (ICBM) called the R-7 and during 1957 this made both its initial launch and a flight which achieved the weapon's full design range. Such rapid progress rather put the RS concept in the shade and, as a result, the jet bomber was abandoned, although work did continue on the 2RS. In addition, both the Lavochkin and Myasishchev design bureaux had been working on ground-launched pilotless intercontinental cruise missiles, called 'Article 350 Burya' and 'Article 40 Buran', which were to be powered by Bondaryuk RD-021U and RD-018A ramjets respectively and intended to cruise at speeds in the region of 1,865mph (3,000km/h).

Tsybin 2RS

The 2RS, first proposed in the January 1956 brochure, showed few changes from the RS and was intended to use the same ramjet powerplant and Tu-95N launch aircraft. However, the deletion of the nuclear delivery vehicle meant that the canard foreplanes were replaced by slab tailplanes while large cameras were housed in the fuselage ahead of the wing. One drawing also shows that a free-fall bomb containing a 244N device could be carried recessed in the lower fuselage, level with the aft portion of the wing. Dimensionally the 2RS was the same as the RS except that its length had been reduced to 89ft 11in (27.4m). Maximum weight (with cameras only) was 46,186 lb (20,950kg) and maximum speed 1,678mph (2,700km/h) Mach 2.54 at 65,617ft (20,000m); service ceiling was 88,583ft (27,000m) and range at high altitude 4,351 miles (7,000km).

By this time the concept of a bomber/reconnaissance aircraft air-launched from a carrier aeroplane was seen as a less attractive idea while the use of ramjets would also be restrictive. Consequently, the 2RS was soon redesigned as the RSR with conventional jet engines and reheat and from August 1956 the Tsybin OKB's main effort was switched to this alternative. The RSR was also slightly longer which added to the problems of loading it beneath the carrier. Nevertheless the 2RS was not cancelled until early 1957, at which point Tupolev also abandoned converting a Tu-95 into the prototype Tu-95N (progress on this at the Tupolev OKB had been very slow, and made with some reluctance, because Tupolev had become a strong opponent to the whole project).

Tsybin RSR and 3RS

The RSR received its official go-ahead on 31st August 1956. The preliminary project was completed on 26th June 1957 and described an aeroplane that was now intended to take-off from a runway. A much stronger bicycle undercarriage was fitted with twin-wheel main and nose gears and the ramjets were replaced by low ratio (0.6%) Solovyov D-21 turbofan engines, which in cruise flight would perform almost like ramjets anyway. Almost all of the RS's steel and titanium structure had been replaced by lighter materials because the lower power engines meant that there would be less stress on the airframe and fewer problems with kinetic heating, which meant that the overall structure could be lighter (although skin temperatures in excess of 200°C were expected in certain areas).

A growing threat of enemy surface-to-air missiles needed to be countered and so, to reduce the chances of being detected by their radars, the RSR's lower fuselage and wing surfaces were coated in a porous material that was designed to absorb the radar signal's electronic energy. In addition the entire airframe was restressed to allow it to make a barrel roll to a height of 137,795ft (42,000m) or to perform a climb and turn together with a rapid change in altitude, both manoeuvres being intended to assist in avoiding surface-to-air missiles although the aircraft was not to exceed 2.5G.

A total of 26,455 lb (12,000kg) of fuel was carried in two fuselage tanks while two long slim external 'supersonic' tanks, 37ft 5in (11.4m) in length and holding another 4,850 lb (2,200kg) of fuel each, would be carried beneath the inner wings. An automatic trim control system was fitted to pump the internal fuel backwards and forwards so that the CofG could be maintained at the position required by any particular flight condition. The RSR's estimated range was 2,339 miles (3,760km) and the covering SovMin resolution stated that the prototype had to be rolled out during the first quarter of 1958.

The Tsybin 3RS was a version of the RSR intended to be capable of both a normal runway take-off or an air launch. The Kremlin was still interested in the huge ranges (up to 8,080 miles [13,000km]) that could be achieved by launching from a carrier aircraft and a SovMin resolution dated 20th March 1958 authorised that the Tu-95N conversion should be completed; however, Tupolev's response to this was a strong refusal. During a meeting held at the Kremlin on 15th May 1958 to discuss the programme as a whole, Andrei Tupolev suggested that Myasishchev should provide the launch aircraft instead. Tupolev considered both the Myasishchev and Tsybin OKBs to be undesirable competitors and if they were merged to work together on one programme, he felt that it would be easier to get rid of them in one go. Setting up a joint development programme between Tsybin and Myasishchev would, of course,

Model of the Tsybin RSR.
Russian Aviation Research Trust

The Tsybin NM-1 model test aircraft.

create further delays which would not help the new OKBs.

Tsybin NM-1

To assess the RSR's overall layout and to obtain concrete data very quickly, Tsybin also designed a scale model aircraft with simple engines called the NM-1 (NM = *Naturnaya Model* or 'life-like' model), and in the autumn of 1956 funding was made available for this aircraft. Two Mikulin (Tumansky) AM-5 engines were fitted in simple nacelles which, due to their lower power rating, meant that the NM-1 could only fly at subsonic speeds. Retractable skids were fitted on the fuselage and under the nacelles and there was a small tailwheel; for take-off a two-wheel trolley was fitted on the main skid which was jettisoned once the aircraft was in the air. The nose section was shorter than the RSR, which meant 1,543 lb (700kg) of ballast had to be carried to restore the CofG, and a simplified fuel system was used, but the structure and its materials were similar to the RSR.

The NM-1's taxi trials began on 1st October 1958 but bad weather and a series of minor defects brought much delay, flight clearance only being received in the early spring of 1959. The first of 32 test flights was made on 7th April (the last sortie took place in 1960) and the results showed that sustained flight with an RSR-type wing was impossible. During landing the NM-1 was easier to fly than fighters like the Mikoyan MiG-21 and Sukhoi Su-7 and Su-9 but its take-off characteristics gave some difficulties and during flight there was a tendency to roll. These factors resulted in considerable changes being made to the RSR itself, which meant that during the prototype's construction the airframe had to be modified. Tsybin worked desperately to keep his project alive and eventually, after appealing to the TsAGI, the Politburo and VVS Command, received new support and permission to postpone the completion date to late December 1960. In addition S K Tumansky and Artyom Mikoyan were ordered to give their assistance on the installation of alternative Tumansky R-11F engines.

Tsybin RSR 'Mk.2'

Despite having to be redesigned to a quite considerable degree, the manufacture of the first machine was not delayed for too long. However, the next problem was the non-availability of the D-21 turbofan engines which, in fact, were never to be delivered, primarily because Andrei Tupolev put severe pressure onto Factory No 19, the manufacturing facility at Perm, to give priority and extra production facilities to the D-20 engine (a non-afterburning D-21) used by his Tu-124 airliner. As a result the RSR's D-21s were replaced by an alternative pure jet engine with reheat, the Tumansky R-11F, which required slimmer and longer nacelles and a central shock cone in each intake along the lines of that used in the MiG-21F fighter; with these power units the aircraft was known as the RSR R-020.

A great deal of effort went into getting the RSR's weight down. The volume of internal ribbing in the wing was increased with the original five spars, present in Tsybin's designs for three years, being replaced by sixteen thinner spar webs welded to the skin panels. Much thinner sections were applied throughout and welding replaced many riveted joints – a reduction in the airframe's fatigue life to just 200+ hours, and possibly only three flights, was considered acceptable because of the RSR's operational task. Dural was now the primary structural material but an attempt to introduce some beryllium alloys caused problems, not least in their toxicity to the workforce.

The resulting airframe was exceptionally light and the RSR's earlier large supersonic external fuel tanks were replaced by smaller 2,866 lb (1,300kg) capacity alternatives that would need to be jettisoned at a maximum of 528mph (850km/h). More taper was applied to the wing trailing edge and the tailplane and fin were reduced in size, while the main undercarriage units were changed from two large to four smaller wheels. Total fuel load was 23,589 lb (10,700kg) and estimated service ceiling 73,819ft (22,500m); the aircraft's range was 2,486 miles (4,000km), calculations suggesting that the required range could still be achieved because of the lower weight.

Five pre-series aircraft were ordered and drawings were issued to Factory No 99 at Ulan-Ude in early 1959. However, yet again the tide was turning towards ballistic missile systems and on 1st October of that year Tsybin's design bureau was closed down by an order from Khrushchev. The responsibility for completing the RSR was passed on to the Myasishchev OKB at Khrunichev who completed a feasibility study on the project and made little or no changes to it; in fact, being very preoccupied with his own bomber designs, Myasishchev left Tsybin to continue without interference. The first RSR was moved to Zhukovsky on 29th September 1960 but the Myasishchev Bureau itself was now transferred to missile and space work and, as a consequence, the RSR programme found itself in the hands of the Chelomey design bureau, who also continued with it. However, in October 1960 Vladimir Myasishchev was made head of the TsAGI and lost control of all

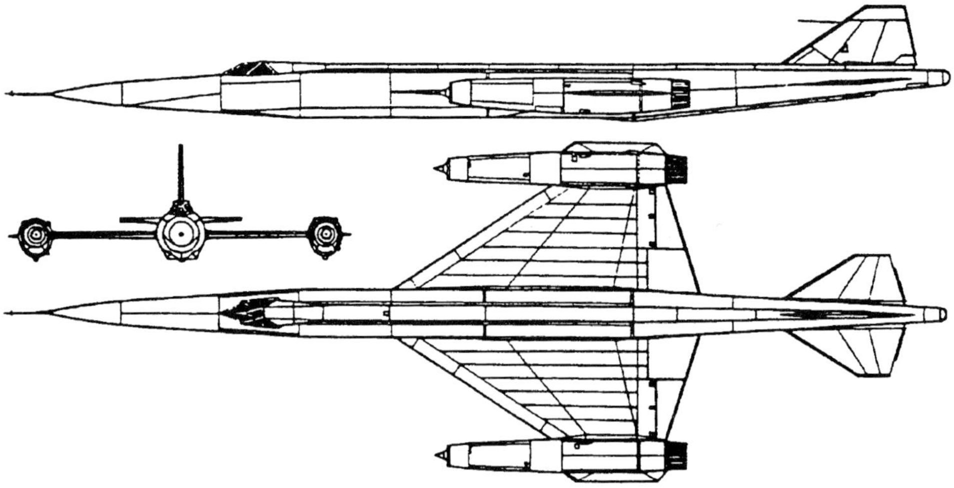

Tsybin RSR R-020 (1959).

Tupolev 'Aircraft 100' (1953).

MiG-25). This could, of course, use all of the cameras and other equipment which, by now, had been designed for the RSR types. Mikoyan's aircraft was eventually built but, as the MiG-25, it struggled to reach Mach 3 and so the reconnaissance version was only used for tactical purposes.

The Bartini design bureau also worked with Tsybin in designing advanced strike systems and the former's enormous A-57 flying boat was intended to carry one Tsybin RSS, an unmanned development of the RSR, on its back. This proposal is described in Chapter 9.

of his design staff, and at the same time the RSR was officially cancelled and Tsybin went to work on Soyuz space modules.

The available documents are not clear in revealing at what stage the construction of the R-11F-powered R-020s had reached when the cancellation arrived. Some state that Factory No 99 had completed three aircraft and ten more were being built, but there are strong doubts about this and it may be that no example was ever actually completed. Certainly their manufacture did get under way, but no aircraft was flown and all of the assembled material was eventually scrapped; some pieces, however, did make their way to the Moscow Aviation Institute to serve as study tools.

After the abandonment, Vladimir Myasishchev wrote a letter to Tsybin that contained a drawing of a supersonic RSR-type layout based on TsAGI's ideas. The engines were still housed in mid-wing nacelles but the inner wing had been given much greater sweepback, a change that was intended to reduce the movements in the centre of lift during the transition from subsonic to supersonic flight. This design, which had a span of 47ft 7in (14.5m), had been drawn in 1958 but there had been delays in sending a copy to Tsybin (or in fact it may deliberately have not been sent); Myasishchev had only recently discovered the drawing and one can only imagine Tsybin's thoughts on seeing it. With the abandonment of the British Avro 730 in 1957, it was left to the Americans to put into service the only type of this class of aircraft to actually make it into the air, the Lockheed SR-71 Blackbird.

The article written by Ivnamin Sultanov shows just how much Tsybin's project was blocked or tampered with by prominent people, with Andrei Tupolev playing the lead. Yet his reconnaissance aircraft was brought to the hardware stage and the NM-1 test specimen was flown. Sultanov also states that Artyom Mikoyan, head of the Mikoyan design bureau, did his best to prevent the programme from succeeding. He wanted the machine's R-11F engines to be used by his MiG-21 fighters and also offered to produce a Mach 3 research aircraft of his own called 'Article 155' (which later became the

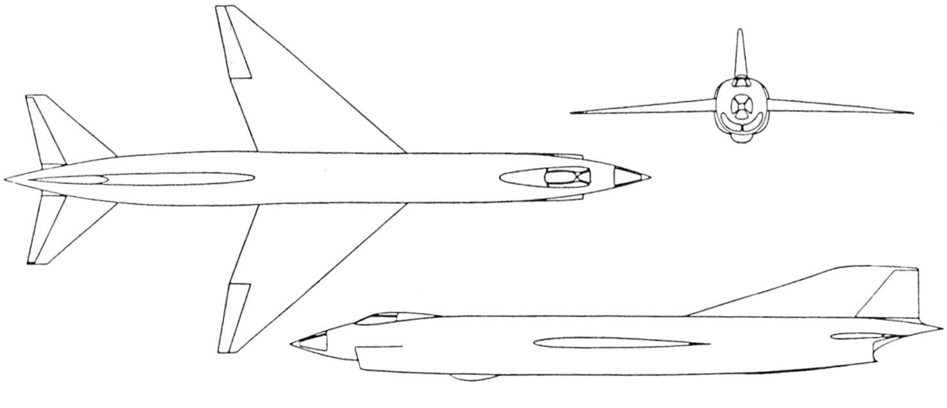

Tupolev Composite Bomber Systems

Besides working on a variant of its Tu-95 to serve as a carrier for Tsybin's aircraft, the Tupolev design bureau also proposed some 'Composite Bomber' systems of its own. The Tu-108 long-range strategic system ('Aircraft 100' and '108' combined) envisaged the design of two new aeroplanes; 'Aircraft 113' on the other hand was a proposed cruise missile intended to be carried and launched by a Tu-95. The Bureau's objective was to create a heavy intercontinental carrier aircraft (preferably supersonic) that was capable of reaching United States territory itself, or at least getting to a specified launch area for either a piloted nuclear bomb carrier or a pilotless flying bomb, suspended beneath this parent aeroplane, to continue to a specified target.

Tupolev 'Aircraft 100' Suspended Bomber

The first part of the design bureau's studies into a composite bomber was this project, begun in 1953 as a component of a strategic strike system. It was intended to be suspended under either a turboprop-powered Tu-95 or Tu-96 (Chapter 3) or the all-new Tu-108. When mounted under these aircraft in flight the '100' would be carried over a distance of 3,729 to 4,040 miles (6,000km to 6,500km) before being released to fly by itself for another 497 to 622 miles (800km to 1,000km) to the target at an airspeed of up to 932mph (1,500km/h); then the '100' would fly back to the nearest friendly base at about 622mph (1,000km/h). A pilotless version of 'Aircraft 100' was also considered which, after release from its carrier, would cover around another 622 miles (1,000km) to a target at a speed of 932mph (1,500km/h).

By March 1954 the bureau had completed sufficient studies to be able to select what seemed to be the optimum layout for 'Aircraft 100', powered by two 12,125 lb to 13,230 lb (53.9kN to 58.8kN) thrust Tumansky AM-11F engines. On 30th July the USSR Council of Ministers issued a Decree which officially set the OKB the task of developing and building the '100' suspended bomber as a part of Tu-108 long-range strategic system also currently under development (see below). The stipulated range of the entire system was now set at 4,350 miles (7,000km) and 'Aircraft 100' would carry two crew and a nuclear weapon weighing between 2,756 lb and 3,307 lb (1,250kg and 1,500kg). In order to test and refine 'Aircraft 100', a special test-bed aircraft was required which was to be converted from a production Tu-95.

The composite system as a whole was redesigned many times during the development process and many of the changes affected 'Aircraft 100'. Several alternative layouts were evaluated for the suspended bomber with either a diamond-shaped wing, a delta wing or a tailless arrangement. One of the final versions was a mid-wing monoplane which had both a swept wing and a tail. Its fuselage employed a semi-monocoque structure and had a pressurised cabin in the nose which housed two crew, pilot and navigator/bomb-aimer, both of whom had ejection seats. Fuel tanks were located in the centre and rear fuselage and the two AM-11M turbojets were placed at the back of the fuselage. Fixed air intakes were located under the crew cabin and a relatively spacious bomb bay for nuclear weapons was positioned in the middle fuselage beneath the wing centre plane. The bomb aiming system was based on the Initsiativa radar while the landing gear comprised a nose leg and landing skid.

In 1955 the '100's powerplant was switched to the new D-20 turbofan engine which offered a maximum thrust of 13,230 lb (58.8kN) and was more fuel-efficient than the AM-11s; it also increased the estimated maximum speed to 1,119mph (1,800km/h). Development work on 'Aircraft 100' lasted until the middle of 1958, at which point the programme was cancelled.

Tupolev 'Aircraft 108' Suspended Bomber Carrier

Work on this aircraft, designed to go with 'Aircraft 100', began in 1952. At the end of that year the OKB's 'Brigade of Projects' (or Advanced Projects Department) led by B M Kondorski began to assess the problems related to the development and manufacture of a heavy supersonic aeroplane. Different wings (delta, diamond, swept and tailless) were analysed while the experience of the Soviet Union's foreign adversaries was also studied and consideration was given to alternative powerplants, either AL-7 and VD-5 turbojets with or without afterburners; the number of engines also varied from three to twelve.

The USSR Council of Ministers Decree issued on 30th July 1954 to cover 'Aircraft 100' also ordered the bureau to develop and build this second element of the composite supersonic strategic strike system. The document officially described the overall project as a 'composite long-range bomber' and the carrier element was to be an aircraft powered by six VD-5M non-afterburning turbojets giving 33,860 lb (150.5kN) of thrust at take-off; it was designated 'Aircraft 108' by Tupolev. As noted under 'Aircraft 100', the two aircraft together would complete around 3,729 miles (6,000km) of the distance to the target at 932mph (1,500km/h) before the suspended 'Aircraft 100' was dropped in the target area; both carrier and suspended aircraft would then return to their bases after the mission was over.

Also proposed was a separate strategic bomber version of the '108' which, when carrying a 11,023 lb (5,000kg) bomb load and flying at 590mph to 622mph (950km/h to 1,000km/h) at 52,493ft to 55,774ft (16,000m to 17,000m) altitude, had a range of 7,769 to 9,323

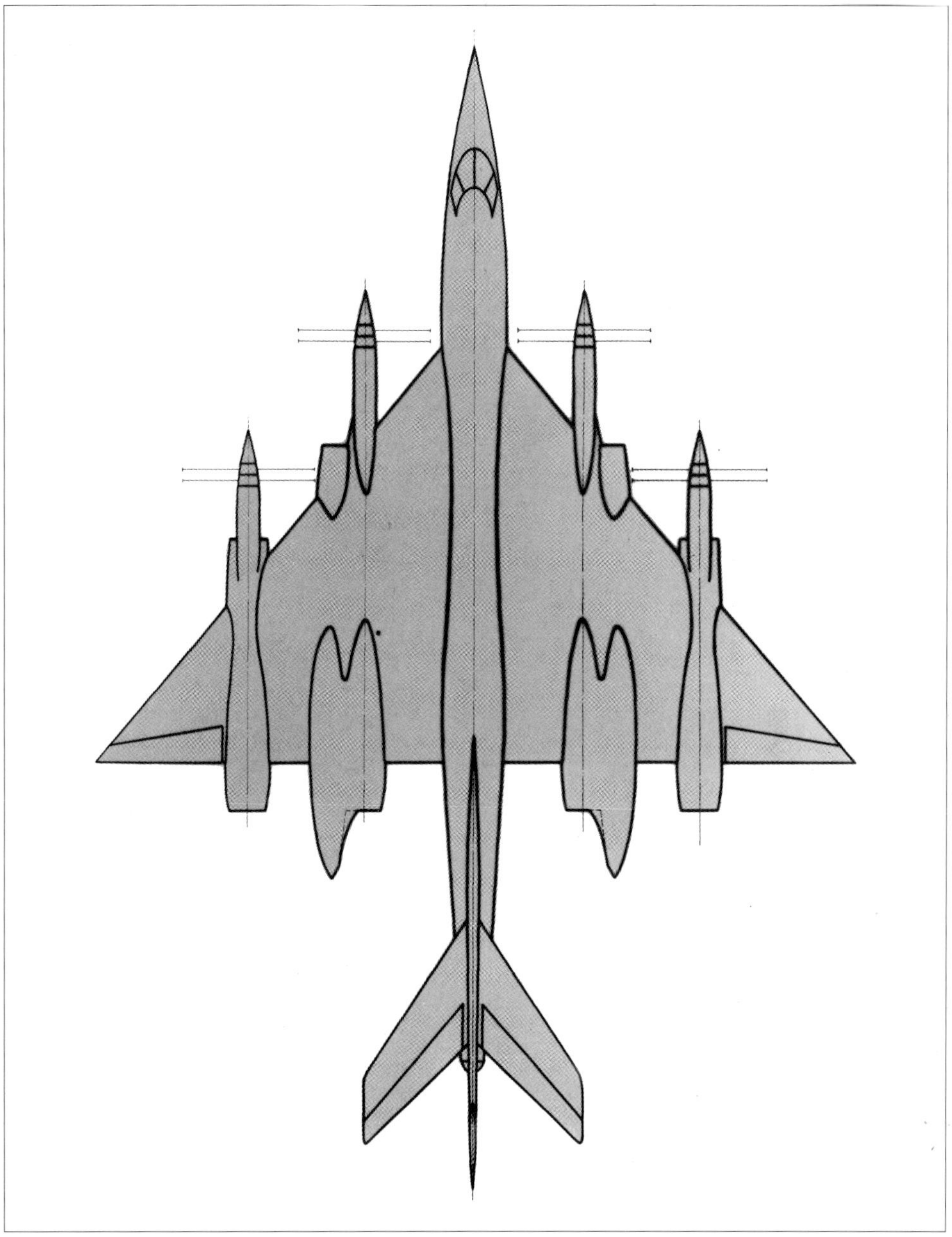

This version of Tupolev's 'Aircraft 108' was powered by P-8 turboprop engines (1952/53).

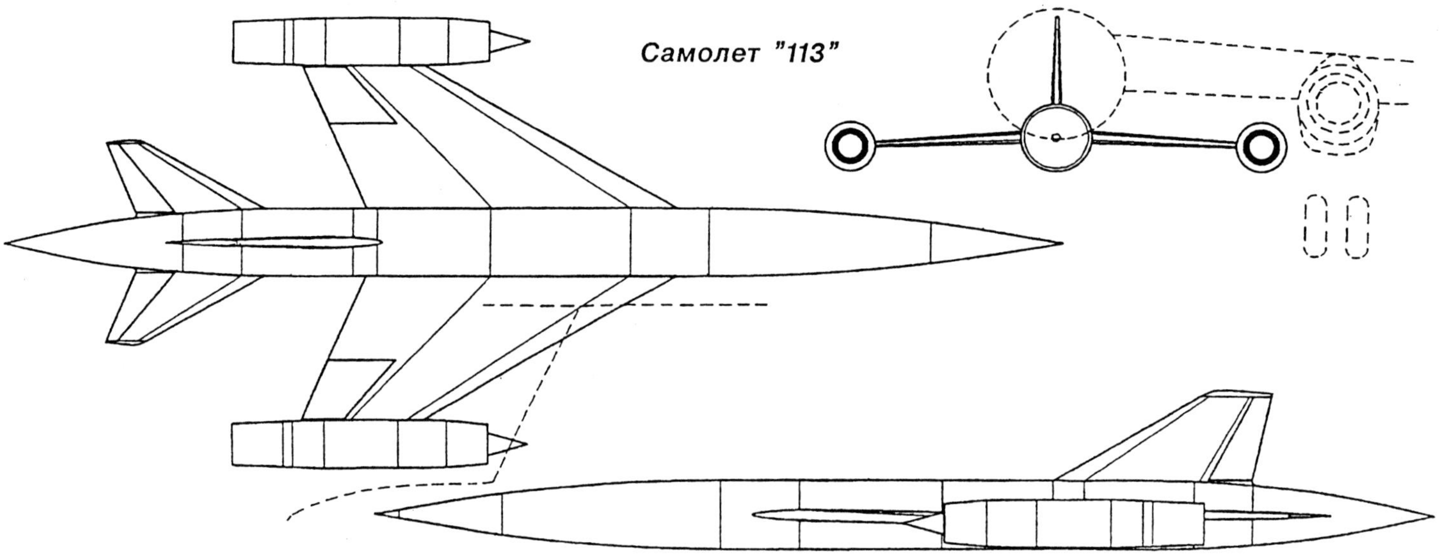

Tupolev 'Aircraft 113' (1955). The weapon's semi-recessed position beneath its carrier aircraft is also shown.

miles (12,500km to 15,000km). If flown at supersonic speed (870mph to 932mph [1,400km/h to 1,500km/h]) and at 45,932ft (14,000m) altitude, the range would fall to 2,486 miles (4,000km). Service ceiling over the target was stipulated to be 42,651ft to 45,932ft (13,000m to 14,000m), normal bomb load 11,023lb (5,000kg) and maximum 26,455lb (12,000kg). The '108's defensive armament would consist of one twin 30mm cannon mount positioned in the fuselage tail and three or four crew would be carried. Three aircraft were to be manufactured by 1957.

In the course of further evaluations the OKB considered heavy airplanes with delta and swept wings (with sweep angles from 40° to 60°) and powered by different engine types (AM-17, VD-5M, VD-7M and VK-9F turbojets with and without afterburning, P-8 turboprop engines or P-4 turbofan non-afterburning engines). Now the most suitable arrangement was found to be four 55,115lb (245.0kN) P-4 turbofans on an aircraft having a 35° to 40° swept wing of 3,978ft² to 4,301ft² (370m² to 400m²) area and a thickness/chord ratio of 6%; its take-off weight was estimated to be 595,238lb to 705,467lb (270,000kg to 320,000kg). In due course the preliminary project was redesigned with P-6 (NK-6) engines. A carrier aircraft was also evaluated with ramjet engines which would offer an even higher top speed of up to Mach 2.85; however, this was seen to be too much for what the overall system required and was also beyond the technology available at the time. In the summer of 1955 the '108' was redesigned again to be fitted with six 22,050lb (98.0kN) VD-7M engines or the four NK-6s.

During the '108's preliminary development a vast amount of experimental and theoretical research work was completed that looked into the specific structure and strength of delta wings. Subsequently, this experience was employed in the development of Tupolev's pilotless 'Aircraft 121' and '123' strategic ground-launched missiles and also its Tu-144 supersonic airliner. By February 1956 it was possible to draw some conclusions as to what was needed to make the overall 'composite' system work. An analysis of the possible flight routes needed to ensure that an attack could be made against American territory showed that, to guarantee that the main targets could be reached from bases in the USSR (when considering that the carrier had to return to its base), two in-flight refuellings would be required en route. 'Aircraft 108' had a lift/drag ratio of 13 to 15 at subsonic airspeeds and 5.5 to 6 at supersonic airspeeds and, in respect of the fuel, an unusually high load ratio of 72 to 75%.

In fact, based on the results obtained, 'Aircraft 109', a version of the '108' fitted with four non-afterburning P-4 engines, was the most suitable solution and the switch to afterburning NK-6 power units aggravated the problem. Moreover, it would be necessary to develop techniques and methods for flight refuelling to be carried out at great distances from home bases and also to create effective systems for navigation and communication.

Each of these problems on its own was a separate and a complex task and, in truth, the development of these multi-mode strategic supersonic striking carrier aircraft was considerably in advance of the technology available at this time. Such types were only finally realised in the 1970s and 1980s when the American Rockwell B-1 and Soviet Tu-160 *Blackjack* (Chapter 11) multi-role strike aircraft were brought to fruition, each of which employed a variable geometry wing. However, in spite of these problems, suggestions were made on how to overcome the crisis that had been reached in the '108's development. Among them were improvements to the aircraft's lift/drag ratio in the cruise, take-off and landing regimes, the introduction of new wing profiles, and the application of 'area rule' and a boundary layer blowing system. In addition, improvements to the layout's general aerodynamics were suggested in the form of a canard and a movable nose plane.

In the meantime, while these problems were being dealt with, on 28th March 1956 the USSR Council of Ministers issued a new Decree that requested the continuation of work on the '108' but with four NK-6 engines now in place, the start date for the flight test programme being shifted to 1959. Gradually, the development programme was transferred to the Technical Projects Department controlled by S M Yeger, which introduced some considerable modifications to the original design. The layout was changed back to a conventional low-wing monoplane with a tail unit, with the wing swept 45° at the leading edge, and the engines were placed in pairs adjacent to the fuselage over the wing centre section.

The main part of fuselage and wing was filled with fuel tanks and the fuselage nose housed a pressurised cabin for the pilot and navigator, while the tail cannon gunner/radio operator was positioned in the rear fuselage inside a second pressurised cabin. The lower nose housed a PN radar and the piloted strike aircraft or pilotless flying bomb was suspended on a special lowering carrier fitted inside the cargo compartment of the centre fuselage.

The preliminary development work on 'Aircraft 108' was to continue for another two years but, eventually, the project shared the fate of Myasishchev's M-50 and M-52 bombers, which possessed similar characteristics to the '108'. Then the 'rocket era' began and the Soviet political leadership put its money on the development of strategic intercontinental ballistic missiles. On 31st July 1958 the USSR Council of Ministers issued a Decree which ordered the cancellation of all work on 'Aircraft 108', and indeed the entire 'composite bomber' programme.

Tupolev 'Aircraft 109'

This was a version of 'Aircraft 108' fitted with 22,050 lb (98.0kN) thrust non-afterburning Kuznetsov P-4 turbofans. It had a span of 123ft 0in (37.5m), wing area 3,763ft² (350m²), maximum speed 1,119mph to 1,243mph (1,800km/h to 2,000km/h), cruise speed 466mph to 497mph (750km/h to 800km/h) and, when a 932 to 1,119mile (1,500km to 1,800km) portion of the distance was flown at supersonic speed, a range of 6,215 miles (10,000km).

Tupolev 'Aircraft 113'

The development of this 'suspended' air-to-ground missile project began in 1955. In many ways the weapon was similar to the work undertaken by Tsybin on its suspended RS manned bomber, for which Tupolev had expressly designed the Tu-95N version of its turboprop-powered bomber as its carrier aircraft. It was expected that a cruise missile of this class should have a range of about 1,865 to 2,486 miles (3,000 to 4,000km) and a top speed corresponding to Mach 2.5 at 36,089ft (11,000m) plus.

The preliminary proposals were completed in May 1955 and the OKB suggested a low aspect ratio 60° sweep wing and a D-20 or VK-11 turbojet housed in a nacelle on each wingtip (there was the possibility of replacing these with ramjets). The mid-fuselage tailplane was also highly swept and all of the weapon's fuel and equipment would be housed in the fuselage. A target aiming system was placed in the nose and then, moving backwards, came a fuel tank, the nuclear warhead, another fuel tank, and then in the tail the flight control system and some radio equipment. A total of 39,683 lb (18,000kg) of fuel would be carried.

It was intended that a suitably modified version of the Tu-95 or Tu-96 would serve as the carrier aircraft. The '113' would be conveyed in a semi-recessed housing beneath the parent's fuselage and it was the choice of these aeroplanes as the carrier that governed the size of the '113' itself, because the maximum possible size of the semi-recessed compartment and the span between the Tu-95's inner engine nacelles had to be taken into account. 'Aircraft 113' never progressed beyond the preliminary design stage.

Tupolev 'Aircraft 135'

The Tu-95 was used again in another composite bomber 'system' project produced during 1958. On 31st June of that year the USSR Council of Ministers issued a Decree requesting the development of a new strategic system which would employ a carrier aircraft based on the Tu-95 to deliver a pilotless long-range winged cruise missile version of the Tsybin's RS (designated S-30) or one of Tupolev's '100' or '113' missiles. The carrier would cruise at a speed between 435mph and 497mph (700km/h and 800km/h) at a height approaching 39,370ft (12,000m), after release the missile would fly at Mach 2.5+ at over 65,617ft (20,000m) altitude and the range for the whole system would be as much as 2,486 miles (4,000km). The preliminary project had to be ready for official assessment in the second quarter of 1959 with flight testing to begin during the first half of 1961. When completed, the preliminary project made use of different versions of the Tu-95, or the Tu-96, but the work did not progress further. This project represented the first use of the 'Aircraft 135' designation – two years later it was re-used for a long-range supersonic bomber described in Chapter 10.

Composite Bombers – Data / Estimated Data

Project	*Span ft in (m)*	*Length ft in (m)*	*Gross Wing Area ft² (m²)*	*Max Weight lb (kg)*	*Powerplant Thrust lb (kN)*	*Max Speed / Height mph (km/h) / ft (m)*	*Armament*
Tsybin RS	29 6 (9.0)	90 3 (27.5)	688 (64.0)	46,649 (21,160)	2 x RD-013 ramjets 9,700 (43.1) + 2 rocket motors	1,865 (3,000), Mach 2.82 at 91,864 (28,000)	1 x 2,425lb (1,100kg) '244N' nuclear store carried in its own delivery vehicle
Tsybin RSR 1957	33 6 (10.23)	89 11 (27.4)	688 (64.0)	46,296 (21,000)	2 x D-21 4,850 (21.6) dry, 10,470 (46.5) reheat	Cruise 1,740 (2,800), Mach 2.64 at 87,598 (26,700)	None carried
Tsybin NM-1 (flown)	34 4 (10.48)	87 2 (26.57)	688 (64.0)	20,282 (9,200)	2 x AM-5 4,410 (19.6)	311 (500) max achieved	None carried
Tsybin RSR (R-020) 1960	35 0 (10.66) (small tip extns)	91 10 (28.0)	688 (64.0)	43,805 (19,870)	2 x R-11F 8,685 (38.6) dry, 12,675 (56.3) reheat	Cruise 1,616 (2,600), Mach 2.44 at 39,370 (12,000)	None carried
Tupolev 'Aircraft 100'	41 0 (12.5)	77 9 (23.7)	495 (46.0)	68,342 (31,000)	2 x AM-11M	932 (1,500)	max 2,756lb (1,250kg) bombs
Tupolev 'Aircraft 113' (cruise missile)	26 3 (8.0)	75 6 (23.0)	452 (42.0)	67,681 (30,700) (at separation from carrier)	2 x D-20 or VK-11	Mach 2.5	1 x 8,157lb (3,700kg) nuclear warhead

Chapter Six

Nuclear Power and Flying Wings

Model of the extraordinary Moskalyov DSB-LK missile carrier.

Nuclear-powered Bombers

At the end of the 1940s and into the 1950s the Soviet Union began research into the development of nuclear reactors as a source of power to drive ships. The work was carried out by an Institute headed by academician I V Kurchatov and, quite soon, aviation was added to the list of possible recipients for nuclear powerplants. Within the Institute the possibilities and problems of nuclear-powered aviation was examined by a group led by another academician, A P Alexandrov, and on 12th August 1955 the USSR Council of Ministers issued a Decree which ordered certain groups within the aviation industry to join in with this research.

As a result the design bureaux led by Andrei Tupolev and Vladimir Myasishchev became involved with the development and manufacture of several aircraft designs intended to be powered by nuclear energy, while the design bureaux of N D Kuznetsov and A M Lyulka were tasked with developing engines for these aircraft. These powerplants would use the energy supplied by the nuclear reactor to replace the jet engine's combustion chambers. If such an aeroplane could be created it would open up a possibility that the Air Force could put into service a combat system whose duration and flight range would be limited only by the stamina of its aircrew.

Several alternative nuclear powerplants were considered, based on either ramjet, turbojet or turboprop engines, with different schemes for transferring the nuclear-generated thermal energy across to them. Several types of reactors and coolant systems were also evaluated, plus various types of protective shielding for the aircraft's crew and equipment to keep them away from the radiation produced by the reactor. Aviation applications, of course, brought extra problems when compared to those of ships, not least the need to keep down weight, so the task of making nuclear power acceptable for use in aeroplanes would not be easy.

The Americans also looked closely at the idea of bombers of unlimited range powered by nuclear engines and one key discovery was that early nuclear powerplants did offer a very long range but would not have sufficient power to push a large aeroplane through the sound barrier. A solution suggested by Boeing in 1954, which would give more speed if only for a short period, was to

The Tupolev Tu-95LAL nuclear-powered flying laboratory.

Close ups of the incredible nuclear powerplant carried by the Tu-95LAL test aircraft.

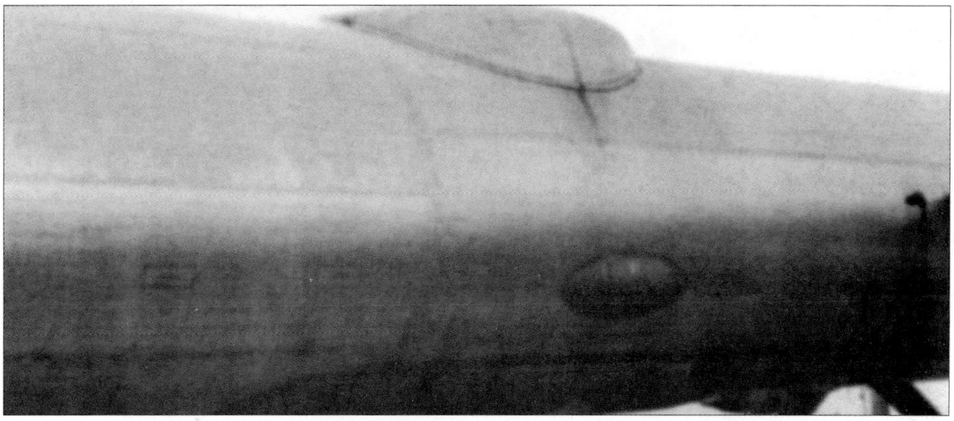

add a pair of conventional jet engines, known at the time as 'chemical engines'. To back up the research, it was intended to convert two Convair B-36H bombers into X-6 research aircraft powered by a nuclear reactor while a third B-36H was selected to serve as the NB-36H flying laboratory. The latter aircraft, although conventionally powered, would take a test reactor into the air to find out just how much shielding would be needed to make the concept a success. In the event the NB-36H was the only element of the programme to fly (between 1955 and 1957) and the difficulties of making the concept work, and the massive costs involved, saw the entire programme cancelled in 1961.

The Tupolev design bureau, working in co-operation with other establishments and enterprises who were specialists in the nuclear field, worked out a long-term development programme (estimated to cover two decades) for the design, development and evolution of heavy combat aircraft with nuclear powerplants. Eventually this programme was expected to lead to the construction of subsonic and supersonic military aircraft during the 1970s and 1980s that would cover the full spectrum of duties undertaken by the Soviet Union's air forces. To begin it was intended to build a ground test rig specifically for testing nuclear aviation powerplants. Afterwards a similar powerplant would be installed in a flying test-bed which would be used to look into the specific features and problems associated with the operation of a nuclear reactor on board an aeroplane, and also work on perfecting the radiation shielding for its crew and equipment. On 28th March 1956 the USSR Council of Ministers issued a new Decree which ordered Tupolev to start work on producing this flying test-bed, which was to be a converted Tu-95M bomber.

The engineering work needed to put together the ground test rig and to install the reactor on the aircraft was carried out by a branch of the Tupolev design bureau based in Tomilino and led by I F Nyezval. Radiation shielding for both the test rig and the flying test-bed, the latter designated Tu-95LAL (LAL = *Letaiouchaia Atomiaia Laboratoriia* or nuclear flying laboratory), was manufactured using materials that were completely new to the aircraft industry. To go with them came new manufacturing processes and their associated technology but the task was successfully undertaken by the bureau's non-metallic materials department under A S Fainshtein. New shielding materials, and airframe components made in these materials, were produced with the co-operation of specialists from the Soviet Union's chemical industry and then checked by nuclear engineers; in due course parts to an acceptable standard were produced.

In 1958 the ground test rig was completed and transported to a test area near Semipalatinsk while, at the same time, the flying-test bed's nuclear powerplant was prepared for its airborne trials. A trial start up of the ground-based reactor was made during the first half of 1958 and the required level of reactor power was achieved, thus opening the way to go ahead with the flight test programme. Following the installation of its experimental power system the Tu-95LAL completed a total of 34 research flights between May and August 1961 and these were made both with the reactor in operation and shut down. The main task was to examine the effectiveness of the radiation shielding, the aircraft's crew and scientists being placed in the front pressurised cabin, while a sensor, installed in this cabin registered the levels of radiation produced. The results showed that the level of shielding was sufficient, opening the way for the design bureaux to get moving with proposals for all-new aircraft with nuclear powerplants.

Tupolev 'Aircraft 119'

The next stage was to produce a new research aircraft designed specifically for a nuclear powerplant and Tupolev's proposal was designated 'Aircraft 119'. This project was based on the Tu-95 but two of this type's four standard NK-12M turboprops were to be replaced by two NK-14A turboprop engines with heat exchangers – the two outboard engines would still be NK-12Ms and their fuel would be carried in conventional tanks mounted in the wing torsion box. The N K Kuznetsov design bureau began its development of the NK-14A 'nuclear' turboprop while Tupolev was preparing its '119' preliminary project. The bomb bay location of the reactor was the same as that used by the Tu-95LAL and the connections leading from the reactor to the engines passed through the fuselage, up through the wings and then out to the heat exchangers attached to the two inboard engines. The nuclear reactor was to be encased in lead but some radiation was to be absorbed by the conventional fuel tanks.

The completion of the first 'Aircraft 119' was scheduled for 1965 but in addition, during 1964, two experimental engines were scheduled to be installed in a modified Tu-95, which would then serve as a '119' test-bed. Once the '119' had been evaluated, the next step would be to produce a military service aircraft powered by four NK-14As that would be capable of very long flights (in particular, there were plans to develop a nuclear-powered anti-submarine warfare aircraft based on Tupolev's Tu-114 airliner). However, these plans were never realised and all of the research into nuclear-powered aircraft was cancelled in the first half of the 1960s. Work on 'Aircraft 119' was halted at the preliminary design stage.

Tupolev 'Aircraft 120'

Almost simultaneously with the start of work on 'Aircraft 119', the design bureau began research into the feasibility of a nuclear-powered long-range supersonic bomber. A large volume of research was completed in the areas of both turbojet development and defining the optimal layout for a nuclear powerplant on an aircraft of this class (taking into account again the need for an adequate level of radiation shielding for the crew and sensitive equipment). According to the design bureau's estimates, if the time schedule could be met a full-scale supersonic aircraft could enter flight test during the second half of the 1970s. The programme itself actually covered the evaluation of an entire series of heavy supersonic military aircraft – a long-range bomber, low-altitude bomber and a strategic intercontinental aircraft, all of them nuclear-powered and embraced by the designated 'Aircraft 120'.

The first in the series was to be the long-range bomber which would have much in common with, and have a similar specification to, the Tupolev Tu-22 (Chapter 4). It was planned to equip the '120' with two turbojet engines developed by the Kuznetsov design bureau and these, and the nuclear reactor and shielding, were to be installed around the rear part of the fuselage as far from the crew cabin as possible. The crew, pilot and navigator, were housed in a pressurised cabin in the forward fuselage that was surrounded with even more heavy radiation shielding. The aircraft itself would have a conventional aerodynamic configuration with a high-mounted 45° swept wing, swept empennage and a tricycle landing gear, but without Tupolev's trademark main gear wing nacelles.

The next type was to be the low-altitude strike aircraft, with capabilities similar to Tupolev's 'Aircraft 132' described in Chapter 8; in the target area this aeroplane was expected to fly at heights in the region of 492ft to 1,640ft (150m to 500m). It would have the nuclear reactor mounted in front of two turbojet engines, the entire arrangement being packed into the rear fuselage, and the jets were to be capable of operating either from the reactor or from conventional kerosene fuel. The kerosene would be used for take-off and landing only and was to be stored in a tank installed in the rear fuselage in front of the reactor; this tank would also serve as part of the radiation shielding. The forward fuselage again housed a pressurised and heavily shielded cabin for the two-man crew. Once more a conventional aerodynamic configuration was employed, this time using a low-position 'delta wing' with variable sweep angles over the leading edge and a small extension in the wing root. The empennage was also swept back and the horizontal stabiliser was mounted on top of the fin.

The third nuclear-powered supersonic project was a long-range strategic bomber similar in concept and capability to Tupolev's '108' and '135' projects described in Chapters 5 and 10 respectively. This aircraft was designed to have six turbojet engines, two of them equipped with heat exchangers which were to operate from the reactor (these were again designed by the Kuznetsov OKB). In general layout the aircraft was close to the American Convair B-58 Hustler supersonic medium bomber; there was no tailplane but the area rule had been applied to the airframe and the project's delta wing had a sweep angle of 52.5° at the leading edge and a thickness/chord ratio of 4.5%. The four engines not

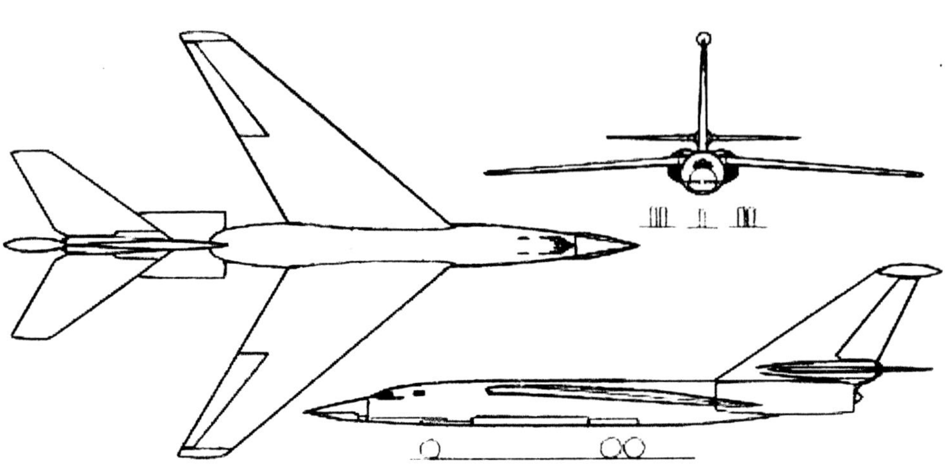

Tupolev 'Aircraft 120' (mid-1950s).
Russian Aviation Research Trust

Myasishchev M-60 (c1955). This was one of the basic designs to be drawn within this project number and was studied in some depth. The crew were housed towards the nose in an internal cabin which had no canopy. Russian Aviation Research Trust

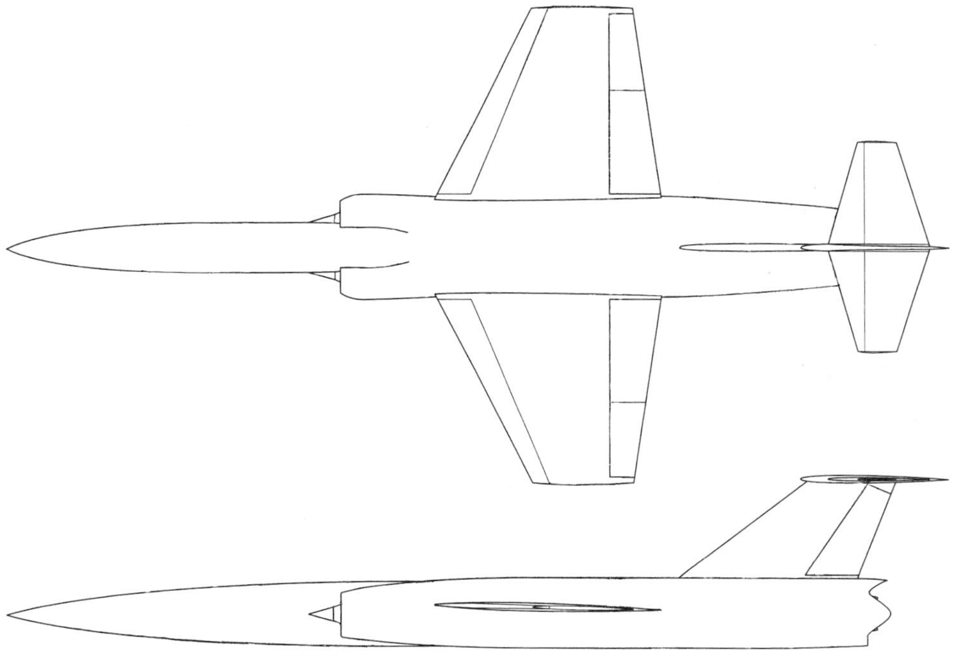

fitted with heat exchangers were mounted on pylons under the wing while the two that had heat exchangers were placed in the fuselage tail. Again the reactor and the crew's pressurised cabin were positioned in a similar way to the two other projects.

This line up of heavy nuclear-powered aeroplane designs gives a general picture of the Tupolev design bureau's research into the nuclear theme. However, several other projects for this type of engine were considered: for example, in the course of its studies into the 'Aircraft 135' supersonic strategic carrier, a proposal was also made for a version powered by a nuclear reactor. All of the OKB's work into nuclear-powered supersonic aircraft was cancelled in the first half of the 1960s. The main reasons for abandoning what had become a large-scale scientific and technical programme were the estimated lower costs of introducing strategic missile systems aboard nuclear-powered submarines and also the possible ecological problems that could have arisen during the operation of nuclear-powered bombers. Should one of these machines have crashed then the whole of the surrounding area could have been polluted with radiation. (For every test flight made by the American NB-36H, the aircraft was escorted by a C-97 transport carrying a group of paratroops. These would have sealed off and protected the NB-36H had this aircraft ever had to make a forced landing.)

Myasishchev M-60

The Myasishchev design bureau undertook considerable research and completed many studies into nuclear-powered bombers. Experience with the M-50 (Chapter 4) had demonstrated that a range in excess of 9,323 miles (15,000km), even with in-flight refuelling, was not possible with this type of aircraft; an altogether different power source was needed and nuclear energy was seen as one possible solution. As a result, on 19th May 1955 a SovMin resolution was passed instructing Myasishchev to produce a supersonic bomber powered by nuclear engines, known officially as 'special engines by A M Lyulka', and the work was designated 'Project 60' or M-60. A range approaching 15,538 miles (25,000km) would be possible which, in other words, meant that range was no longer a key factor and the crew could select whichever route to the target it preferred. However, they would need considerable protection from the powerplant's radiation while the effects made by the radiation upon the aircraft's structure also needed to be investigated; several other establishments, including the NII of Academy of Sciences, became involved.

The draft project was ready in July 1956 and Lyulka's powerplant comprised a combined nuclear/turbojet engine and the heat that the nuclear element generated, transferred through air to the jet, was expected to give a thrust of 49,600 lb (220.4kN) (a powerplant using air for heat transfer was known as an 'open' system). The engine would be placed at the rear of the fuselage, as far as possible from the crew, and the M-60 itself was expected to travel at Mach 2 (conventional engines would be used for the initial flight tests). However, the crew would sit in an enclosed cabin right in the centre of the forward fuselage that was protected on all sides by anti-radiation shielding to ensure that they were not exposed to intensive levels of radiation. This would make visual observation impossible and so the cockpit canopy was replaced by other visual aids; the crew cabin, complete with its protective shielding, weighed 30% of the gross take-off weight and so only two crew would be carried. The draft acknowledged that many new problems needed to be solved before the nuclear concept could be proved; nevertheless a detailed preliminary project was completed in March 1957.

Initially the M-60 showed a long slim aircraft with trapezoid wings, a trapezoid T-tail, two large nuclear-powered jet engines mounted side-by-side in the fuselage and a cruise missile was carried beneath the centre fuselage. Span was 86ft 11in (26.5m) and length 169ft 3½in (51.6m). This was soon enlarged to have four engines stacked in pairs at the rear of the fuselage (side intakes were still used) and a touch more sweep angle was added to the leading edge. As before, a tricycle undercarriage was employed with four wheels on all three legs and with a nose leg of greater length than the main units. The nuclear cruise missile was now semi-recessed in the lower middle fuselage but an alternative proposal showed this aircraft with a large nuclear bomb housed in a central bomb bay. This basic design appears to have been studied in some depth but many other peripheral skills would also need to be learnt, such as the ability to undertake servicing and maintenance in safety.

In due course other alternative designs were considered, all of which were embraced by designation M-60. The above layout was modified with a nose intake (and later had the tailplane moved to a mid-fuselage position) while several swept wing projects were also tried, most of which had an internal bomb bay. The swept wing efforts used different engine arrangements, all packed into the back end of the fuselage and fed by a large dorsal intake, and the biggest example had a single massive engine, a span of 103ft 8in (31.6m) and length 200ft 1½in (61.0m). Both crew were still in their enclosed cabin in the nose and, seated side-by-side, could eject through the bottom of the aircraft. Eventually some unmanned versions were also studied.

Finally, designs were suggested showing a delta wing with podded engines both on underwing pylons and in tip nacelles and these had much in common with the M-50 *Bounder*. As noted, the M-60's engines were to be designed by OKB Lyulka, in conjunction with A P Alexandrov's Nuclear Reactor Establishment, and for take-off and landing this would use a chemical fuel with the reactor acting purely as a booster. On reaching the required height, however, the nuclear element would be fully activated for sustained cruise with the engine operating in ramjet mode. Provided a reliable powerplant along these lines could be produced, Myasishchev stated that a strategic bomber capable of 1,989mph (3,200km/h) speed, at least 15,538 miles (25,000km) range and 65,617ft (20,000m) ceiling was feasible. The M-60 series of projects was eventually continued as part of the M-30 programme described shortly.

Myasishchev M-60M

A corresponding series of studies was undertaken for a nuclear-powered flying boat derivative of the M-60, called the '60M' or M-60M. Preliminary work was started in April 1956 against a Government authorisation and a SovMin resolution of 15th August gave support to the project. The M-60M was to take full advantage of the research made into the M-60, and also the M-70 supersonic flying boat described in Chapter 9.

Three of the earliest layouts showed versions of the same basic design with wing, tail and fin all highly swept; the horizontal surfaces were placed high on the fuselage to keep them out of the water. Combinations of either four, five or six engines were crammed into the back end of the fuselage and two crew were again secured in a sealed and radiation shielded cabin in or near the nose – it was the four-engined arrangement that was the most favoured. One design had side intakes level with the wing roots, the other pair had nose intakes, and the largest had a span of 111ft 6½in (34.0m) and length 199ft 4in (60.75m), although all were of roughly the same size. Other layouts introduced a trapezoid wing and T-tail, one having a dorsal intake (the drawing also showed a cruise missile in a centre fuselage weapon bay) the others a nose intake, and overall these had a longer fuselage but a reduced span. Water ingestion through the intakes was just one problem to be considered which resulted in their being placed at least 4ft (1.2m) above still water level.

All of these projects were to have employed the hydroski principle (the early designs also had retractable tip floats). This allowed a high-speed aircraft to take off from water by having a ski or set of skis extended under the surface like seaplane floats which, as the aircraft's speed increased, pushed the fuselage upwards clear of the water so enabling it to skim the surface until take-off. The bonus was that once retracted into a relatively small space as an integral part of the body, the skis offered much less drag than either buoyant floats or a flying boat hull, which was of great value to designs such as these that were intended to fly at high supersonic speeds. In each drawing there was a single large central nose ski placed under the forward fuselage, a tail damping unit and two outboard float skis. After take-off all of these would retract flush with the boat-shaped lower fuselage or hull.

Research confirmed that the water ski and hydrofoil alighting gear gave a 10-15% reduction in cross section area, superior hull streamlining, a reduced structure weight and also ensured that the engine intakes, wings and tail were kept well clear of the water. A modified Beriev Be-8 piston-engined flying boat was used for some full-scale experiments to test the ski and hydrofoil arrangement and the improved water characteristics that these demonstrated confirmed that the M-60M would not need a conventional flying boat hull. The Myasishchev Bureau's documents also show drawings for a launch pad coming out of a rocky coastline with the M-60M 'hangared' in a 'dry dock' as part of a huge network of underground facilities and tunnels supplying engines, weapons, fuel, and the like.

The draft design was completed in December 1956 and Myasishchev's studies established that such an aeroplane, with performance parameters and armament close to the land-based M-60, was feasible provided, once again, that the problems of nuclear propulsion could be solved. However, in 1958 Vladimir Myasishchev himself completed a report which stated that the introduction of nuclear power in aircraft could be successfully achieved, but the work required to perfect it, and the necessary reorganisation and re-equipment of the supporting ground forces, would ensure that its completion would take many years and even decades.

Myasishchev M-62

Little information has been traced for this project, a heavy nuclear-powered supersonic bomber designed in 1958-59. Lyulka engines were to have been fitted.

Myasishchev M-30

The first M-30 designation covered a supersonic high-altitude reconnaissance bomber project designed in 1953 which is described in Chapter 3. Curiously the designation was re-used in 1959 to cover another set of supersonic bomber designs with a 'closed circuit' nuclear power unit. The 'closed' system used a sodium/lithium liquid crystal method of heat transfer where the reactor, heat transfer medium and heat exchanger unit provided all of the energy for the engines (that is, there was no chemical fuel-powered back-up); experience with Soviet nuclear submarines

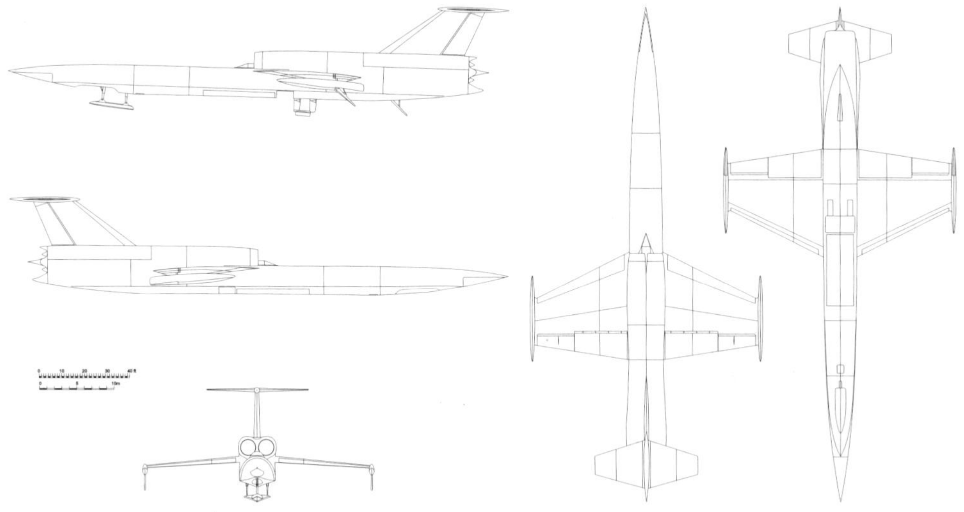

Drawing of one version of the Myasishchev M-60M nuclear-powered flying boat (1956). Jens Baganz

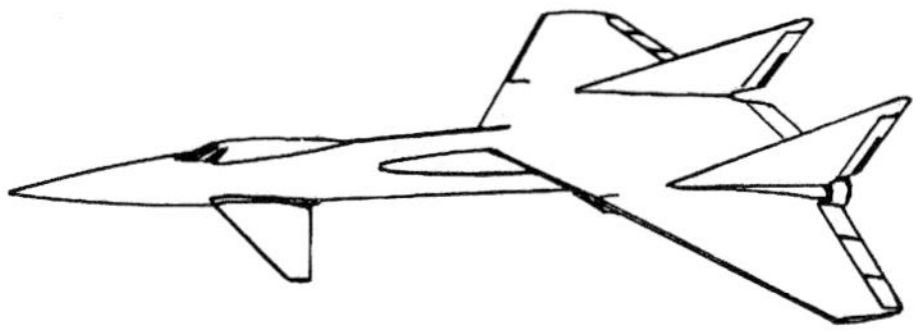

Impression of an early version of the Myasishchev M.30 (1959 onwards). Russian Aviation Research Trust

Model of a later version of the Myasishchev M-30.

had shown that the 'closed circuit' gave a much reduced level of radiation, which made it a good alternative to the 'open circuit' that used air from the atmosphere to transfer its heat. A SovMin resolution dated 19th June 1959 requested that the OKB should go-ahead with the development of this aeroplane and, to back it up, Myasishchev began working on a nuclear-powered variant of the M-50 (Chapter 4) as an experimental research aircraft, calling it the M-50LL.

The M-30's requirements stated an aircraft capable of carrying a 6,614 lb (3,000kg) cruise missile at a cruise speed of 1,678mph (2,700km/h) for a range of 7,769 miles (12,500km); the missile would be launched at about 78,740ft (24,000m) height and cruise to a target 622 miles (1,000km) away at 2,175mph (3,500km/h). One surprising discovery was that, unlike conventional kerosene-fuelled jet engines which lose thrust with increased height due to the reduction in air density, nuclear jets suffered a loss in thrust when flying below their 'optimum operating altitude' of 55,774ft (17,000m). The nuclear engines also needed an aerodynamic layout that matched their ideal operating height and the M-50's delta-style wing was considered unsatisfactory. Eventually the chosen solution was a compound sweep wing plus a canard and a 'pack-mounted' powerplant with six engines, an arrangement already examined in the conventional M-56 (Chapter 4) – the earliest M-30 'canard' arrangement had actually used four engines placed beneath a high-mounted tapered wing.

This aircraft was intended primarily to attack land and maritime targets in the European theatre (for instance, ICBM sites and aircraft carrier task groups) and was to be powered by Kuznetsov NK-5 engines. The nuclear power unit itself would weigh in the region of 66,138 lb (30,000kg) and another 83,774 lb (38,000kg) of protective shielding was afforded to the two-man crew. The design would have had an estimated service ceiling approaching 62,336ft (19,000m) and a range of 15,538 miles (25,000km). A preliminary project was finished on 30th December 1959 with 'detailed definition' completed exactly one year later, but the biggest weakness was the very heavy nose, containing the crew and its radioactive shielding, and six engines placed very much to the rear, which suggested that the airframe would suffer substantial strain, possibly as much as 12G. However, this problem was eased when it was learnt that neutron radiation would not impair the structural strength of conventional airframe materials.

The M-30 was designed with an all-metal structure, including titanium airframe skins plus steel for components carrying heavy loads, and a tricycle undercarriage. Its development programme envisaged a mock-up inspection in mid-1961, the construction of a '30' test aircraft with NK-5K engines by mid-1965 and the completion of the first full M-30 at the end of 1966. However, during 1959 official interest in nuclear power began to wane although work on the M-30 did continue, the original 'load' of one Kh-43 cruise missile being replaced by two Kh-22s. In due course the M-30's design was changed again by the introduction of an ogival fuselage to replace the earlier conical form plus the replacement of the two fins with a single central tailfin. Then, at the start of 1961, Soviet research into nuclear powerplants and nuclear-powered aeroplanes was abandoned, signalling the end of the M-30 and indeed Myasishchev's work on large heavy bombers.

Flying Wings

Chetverikov RK-1

This rather attractive design is something of a one-off, being the only known all-new jet bomber design to be produced by the design bureau of I V Chetverikov. It was to have

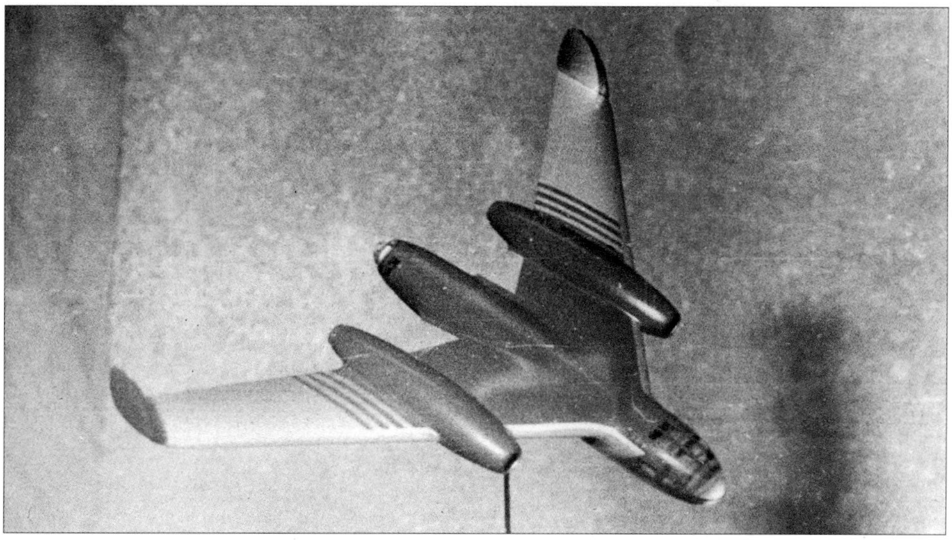

Model of the Chetverikov RK-1. Russian Aviation Research Trust

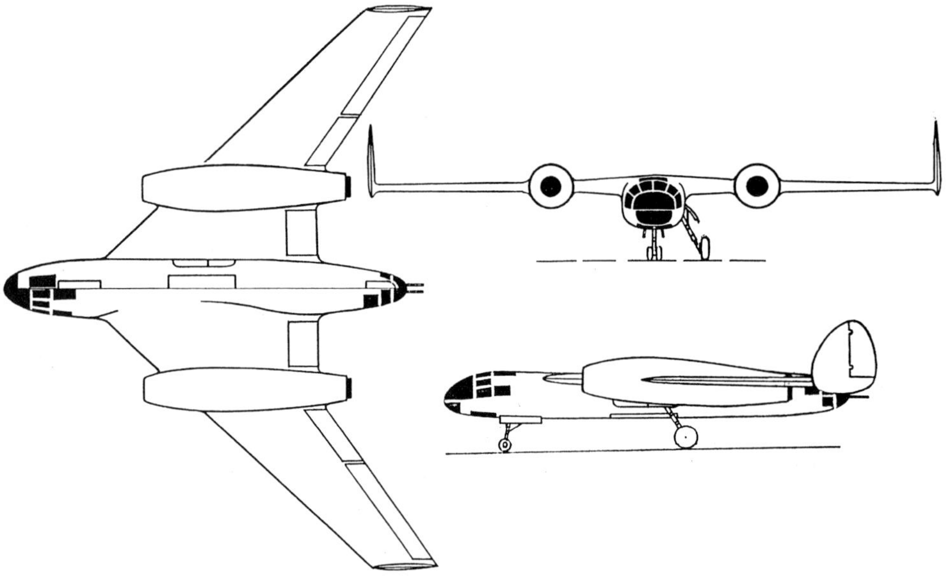

Chetverikov RK-1 (Drawing dated 1948). Russian Aviation Research Trust

been powered by two Rolls-Royce Nene engines mounted in wing nacelles and had tip fins (a feature of several British medium bomber designs of the period), a tricycle undercarriage and three crew. Two ShVak guns were mounted in a defensive tail turret and the maximum warload was 4,409 lb (2,000kg), comprising either four 1,102 lb (500kg) or two 2,205 lb (1,000kg) bombs. The VVS ordered the start of the RK-1's development in 1947 and the preliminary project and mock-up were approved in December 1948, but the aircraft was never built.

Moskalyov DSB-LK

An altogether more advanced flying wing, which makes a fascinating contrast to the RK-1, this cruise missile carrier project was first proposed in 1957. Between 1957 and 1960 a great deal of research was undertaken into the possible strategic weapon systems that could be developed for future use by the Air Force and these embraced all manner of possible solutions – long-range aircraft, flying boats, cruise missiles or air launched missiles. This work was undertaken on VVS Staff orders and much of it was done by a research team at the LKVVIA (the Leningrad Red Banner Engineering Academy) led by Aleksandr Moskalyov; one of the most interesting projects was the DSB-LK long-range strategic bomber.

The design parameters for possible long-range strategic bombers were 330,688 lb to 1,102,290 lb (150,000kg to 500,000kg) maximum weight, 11,023 lb to 33,069 lb (5,000kg to 15,000kg) warload, speeds in the region of Mach 2 to Mach 4 and an altitude of 65,617ft to 114,830ft (20,000m to 35,000m) over the target area. The configurations evaluated included conventional wings, tailless, flying wing, canard with a swept wing, crescent or delta wings or a straight or tapered planform while the aircraft were to be powered by afterburning turbojets or a combined power unit made up of turbojet and ramjet engines.

Once the preliminary analysis was complete it was possible to establish a more specific requirement for land-based bombers including a take-off weight which was not to exceed 661,376 lb (300,000kg). The bombers were to fly at Mach 4.4 to 4.6 at 98,425ft to 114,830ft (30,000m to 35,000m), although if the usual D-23 alloy airframe parts were replaced by alternatives manufactured in titanium alloys, these speeds and the aircraft's range could be increased. Between six and ten turbojets could be fitted, giving an estimated speed of Mach 2.0 to 3.2, while the ramjet combinations offered Mach 3.8 or even higher. Defensive armament would include four air-to-air missiles (with a range of 6.2 miles [10km]) plus two cannon barbettes and ECM equipment; a Rubin-1 radar was also carried.

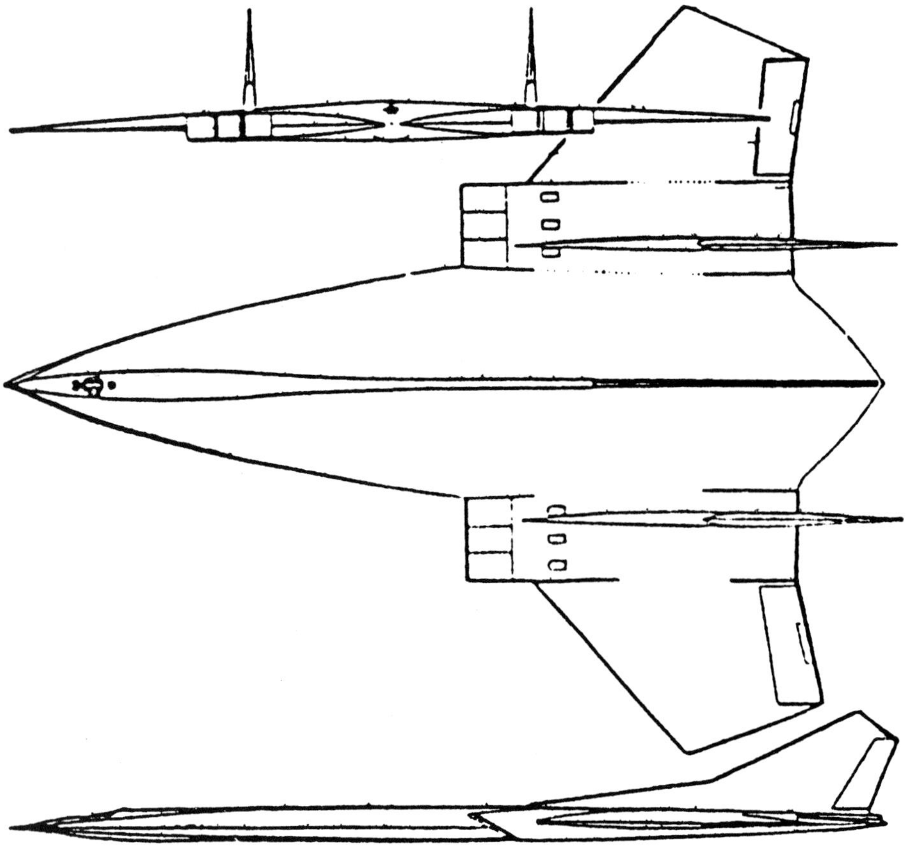

Moskalyov DSB-LK (1957). Russian Aviation Research Trust

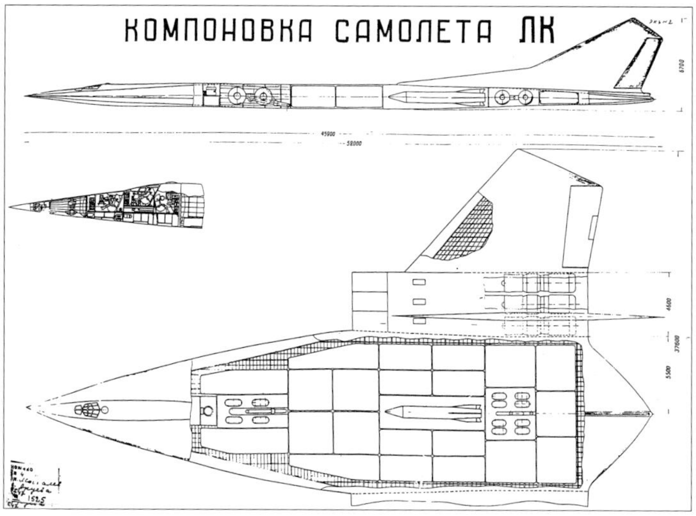

Cutaway drawing of the DSB-LK.

Model of the extraordinary Moskalyov DSB-LK missile carrier.

This research indicated that, using the technology available at this time, the development of such an aircraft was feasible. A preliminary project for the DSB-LK flying wing bomber was completed with the assistance of TsAGI, the Myasishchev OKB and others. The DSB-LK itself had a 'cranked delta' wing swept 72° inboard and 42° outboard, a bicycle undercarriage with four wheels on the front leg and eight on the double rear leg, six VK-15M engines in two boxed sets of three with the twin fins mounted over the inner units, and three crew housed in a small cabin in the nose. Its flying bomb, cruise missile or 11,023 lb (5,000kg) bomb would be carried in a bay placed right in the centre of the aircraft and loads between 11,023 lb and 33,069 lb (5,000kg and 15,000kg) could be carried. Range was 10,441 miles (16,800km) and service ceiling 114,830ft (35,000m). Work on the project ended in 1960.

Nuclear-Powered Bombers – Data / Estimated Data

Project	*Span ft in (m)*	*Length ft in (m)*	*Gross Wing Area ft² (m²)*	*Max Weight lb (kg)*	*Powerplant Thrust lb (kN)*	*Max Speed / Height mph (km/h) / ft (m)*	*Armament*
Tupolev 'Aircraft 120' (long-range bomber)	80 0.5 (24.4)	100 9 (30.7)	1,828 (170)	187,390 (85,000)	2 turbojets	839 to 901 (1,350 to 1,450) at 26,247 (8,000)	max 19,841lb (9,000kg) bombs
Tupolev 'Aircraft 120' (low-altitude strike)	62 4 (19.0)	100 9 (30.7)	1,849 (172)	224,868 (102,000)	2 turbojets	777 to 870 (1,250 to 1,400)	max 11,023lb (5,000kg) bombs
Tupolev 'Aircraft 120' (long-range strategic bomber)	100 4.5 (30.6)	132 10.5 (40.5)	3,441 (320)	337,302 (153,000)	6 turbojets	Supersonic	max 11,023lb (5,000kg) bombs
Myasishchev M-60 (early design c1955)	100 4.5 (30.6)	190 11.5 (58.2)	?	?	4 turbojets	cMach 3	2 x 23mm defensive cannon, 1 x cruise missile or nuclear store
Myasishchev M-60M (preliminary project)	101 6 (31.0)	217 8 (66.35)	3,753 (349)	493,827 (224,000)	4 turbojets	c1,367 (2,200)	2 x 23mm defensive cannon, 1 x cruise missile or nuclear store
Myasishchev M-30 (preliminary project)	88 3 (26.9)	150 5 (45.85)	?	335,097 (152,000)	6 x NK-5	1,989 (3,200) at 6,562 (2,000)	1 x 13,228lb (6,000kg) Kh-43 cruise missile

Flying-Wing Bomber – Estimated Data

Project	*Span ft in (m)*	*Length ft in (m)*	*Gross Wing Area ft² (m²)*	*Max Weight lb (kg)*	*Powerplant Thrust lb (kN)*	*Max Speed / Height mph (km/h) / ft (m)*	*Armament*
Moskalyov DSB-LK	123 4 (37.6)	170 7 (52.0)	?	617,284 (280,000)	6 x VK-15M 34,830 (154.8)	Mach 2.8	Ballistic or cruise missile or bombs up to maximum 33,069lb (15,000kg)

Chapter Seven

Ground Attack Aircraft

The Sukhoi T8-1 (Su-25) prototype led to the Su-25 family.

The massive and spectacular machines covered by several of the chapters in this book do not represent all of the bomber story. A less obvious area, but one just as important, is ground attack and in fact the Soviet Union's salvation during the second World War rested a good deal on an example of this type of aircraft. Ilyushin's Il-2, first flown on 2nd October 1939, was a key element in the USSR's wartime defences and was built in greater numbers than any other aircraft in history, the final tally approaching 36,000 examples. The Soviets called this category of aircraft *Shturmovik* and persevered with such types after the war had ended – today it is represented by the Sukhoi Su-25. This chapter examines the Soviet Union's post-war attempts to put close-support ground-attack aircraft into service (an area which also brings some the OKB fighter specialists into the picture). The next chapter looks at their larger cousins, the tactical strike aircraft.

Projects for Air Force and Navy

Alekseyev I-218

Semyon Alekseyev controlled his own OKB for only a short period, from 1946 until August 1948, and in that time he designed several jet fighters. There was also an attempt to produce an armoured ground attack *Shturmovik* called the I-218 which actually embraced several versions and went into competition with Ilyushin's Il-20. The original I-218-I was a large twin-boom aircraft powered by a single Dobrynin VM-251 piston pusher engine and fitted with a contra-rotating propeller using two three-blade units. Built with an all-metal stressed-skin structure, the I-218-I had its two crew housed in a forward compartment protected by armour manufactured in a new chrome-nickel steel alloy. There were four forward-firing cannon (four 23mm, or two 57mm plus two 37mm) and two rearward-firing 23mm, one on the outside of each tail-

Alekseyev I-218 (1947).

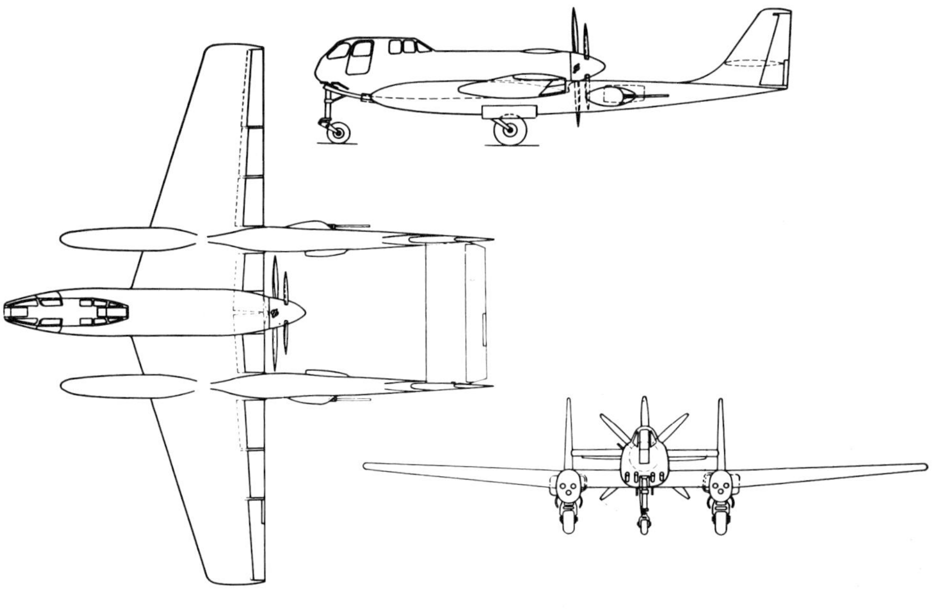

Model of the Alekseyev I-218.

boom; the latter were remotely controlled by the rear crewman. For offence a total of 3,307 lb (1,500kg) of weapons were carried, made up of bombs or rocket projectiles loaded mostly beneath the fuselage but with six of the rockets under the wings. The aircraft had a tricycle undercarriage.

There were three proposed developments. The I-218-Ib or I-219 shared the same twin-boom layout but had swept fins, a tailwheel undercarriage and a new cockpit and canopy arrangement. Next came the bigger I-218-II (I-221) with a tricycle undercarriage and the twin booms replaced by a conventional empennage with swept tailplanes and fin. This was to be powered by a 10,140 lb (45.1kN) Lyulka TR-3 turbojet fed by wing root intakes and carried its defensive armament in the back of the cockpit. Finally, the I-218-III or I-220 was a derivative of the I-221 which was intended to receive a 4,000hp (5,364kW) Dobrynin VD-4 piston engine. All of this aircraft's flying surfaces were straight and quite a large ventral fin was added to help directional stability.

A key capability for these designs was to be long endurance at low altitude (the I-218's predicted range was 746 miles [1,200km]) but by 1948 the Soviet Union had decided that such aircraft, certainly with piston powerplants, were becoming obsolete (the success of the American piston-powered Douglas AD Skyraider during the Korean War was still a few years away). The eventual closure of Alekseyev's design bureau, plus criticism from Aleksandr Yakovlev stating that Alekseyev's fighter designs were all copies of the wartime German Me 262, meant that the prototype I-218-1 attack aircraft was never built (it was probably never started – there no available evidence to confirm that construction did begin). The follow-on designs stayed on the drawing board and, two months after the closure, Alekseyev moved to the Baade OKB to help with the development of that organisation's Type 140 and 150 bombers (Chapter 3).

Ilyushin Il-20

Having achieved great success with the Il-2 family, Ilyushin continued with his attack aircraft ideas after the war had ended. On 11th March 1947 a Government directive tasked his OKB with developing a new attack aircraft that would be superior to the Il-10 in performance and fire power (the Il-10 was a 1944 development of the Il-2). Ilyushin's response was a heavy armoured attack aircraft designated Il-20, a single-engined all-metal machine using a tailwheel landing gear. It also had a very unusual fuselage, not used anywhere in the world before, in that its cockpit was placed above the engine. Coupled with a very large windshield extending right down to the propeller hub, this afforded the pilot an exceptionally good view both forwards and downwards. Theoretically this cured a problem suffered by wartime *Shturmoviks* in that during a low-level bombing run the pilot lost the target beneath his aircraft's nose before it was time to drop the bombs. Here, in horizontal flight the pilot had a 37° downward angle of vision and in a 40° to 45° dive he could see a target right under the aircraft. The Il-20 was powered by a Mikulin M-47 liquid-cooled piston engine (also known as the MF-47), although documents for early projected versions show the MF-45Sh which offered the same power.

Several variations in gun armament were considered. The first option (with the MF-45Sh fitted) comprised two 23mm wing-mounted cannon for firing in a dive and two more obliquely mounted in the fuselage (at an angle of 23° to the flight line) for strafing large area targets in a level flight. The normal bomb load was 882 lb (400kg), stretching to 1,543 lb (700kg) in the overload configuration, and there were launching rails for four 132mm rocket projectiles. A second layout

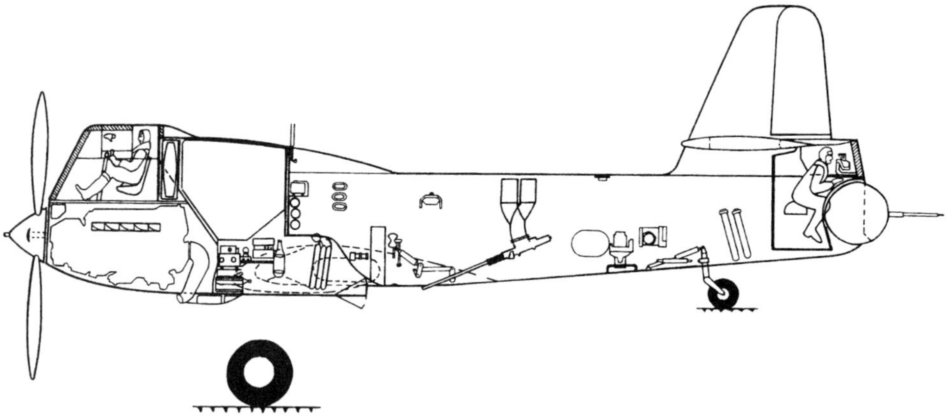

Early proposed layout for an M-45Sh-powered Ilyushin Il-20 fitted with an Il-K8 tail turret (1947). Note the fuselage cannon mounted at an oblique angle.

Mock-up of the Il-20 showing the aircraft's massive forward fuselage.

envisaged the use of one 45mm cannon, two 23mm cannon and six rockets. Finally, another M-45Sh-powered version featured an Il-K8 type tail turret for defence against rear attacks, which made the fuselage longer and moved the wing further aft to compensate for changes in the CofG.

These layouts were abandoned in favour of the design that was eventually built, which was fitted with four wing-mounted 23mm cannon and a rearward-firing 23mm dorsal turret behind the cockpit; it could also carry bombs in wing cells plus eight small or four larger rocket projectiles under the wings. The obliquely mounted fuselage cannon, although at one time considered to be an important feature, were discarded as being of limited use because they created difficulties with sighting. However, the wing-mounted cannon could be deflected downwards to an angle of 23° for use against targets such as enemy troop columns, which could now be strafed by the aircraft when flying horizontally. The offensive load rose to 2,623 lb (1,190kg) with small bombs accommodated in four bays in the wing centre section and two 1,102 lb (500kg) bombs on external racks under the wings. There were also guide rails for four RS-132 rocket projectiles or eight smaller weapons.

The Il-20 mock-up was officially reviewed in July 1948 and the prototype, nicknamed *Gorboon* or Hunchback, was completed by 27th November. It made its maiden flight in early December but during trials only achieved a maximum speed of 320mph (515km/h) at 9,186ft (2,800m), when production Il-10's had achieved 342mph (551km/h) at the same height. Thanks to the decision to place the pilot above the engine, the Il-20 had a larger cross-sectional area which had increased its weight and drag. There were also problems with the M-47, an engine as yet not fully developed, which gave powerful vibrations during flight, and eventually these weaknesses prevented the Il-20 from being submitted for its state acceptance trials. There were also difficulties with maintenance (having the cockpit placed so high made this, to say the least, a tricky affair).

As a consequence on 14th May 1949 the USSR Council of Ministers issued a directive terminating all work on the Il-20. Another factor contributing to this decision may have been the then current view that the general transition to turbojet-powered aircraft had dictated a need for an attack aircraft giving

Two views of the only Il-20 to be built, the side angle making a good contrast to the earlier drawing.

The Il-20's wing-mounted guns could be depressed to allow the aircraft to fire at ground targets without having to enter a dive.

higher-performance. Before the end Ilyushin studied two more versions which featured an Il-10-style layout that sacrificed the Il-20's 'better vision' for improved aerodynamics and, potentially, an improvement in performance. An anti-submarine version never left the drawing board and the basic ground-attack role was eventually filled by variants of the Mikoyan MiG-15 jet fighter.

Tupolev Tu-91

At the beginning of the 1950s the Tupolev design bureau started looking into carrier-based combat aircraft as part of the USSR's plans to build light and medium-sized aircraft carriers. In 1950 the Brigade led by B M Kondorsky evaluated a twin-engined torpedo-bomber called 'Project 509' which looked very similar to the bureau's 'Aircraft 81'/Tu-14 design (Chapter 2) and was essentially a Tu-14T torpedo-bomber variant fitted with folding wings, an arrester hook and other naval items to allow to operate off carriers, but this did not progress beyond the preliminary project stage. The Bureau's alternative proposal, 'Project 507', was eventually turned into the combined dive-bomber torpedo-bomber attack aircraft project designated 'Aircraft 91' ('Projects 507' and '509' actually refer to Projects '7' and '9' of 1950). Andrei Tupolev appointed V A Chizhevsky (previously one of the heads of the Bureau of Special Designs or BOK) as the '91's Chief Designer while Pavel Sukhoi, who had extensive experience in the design of attack aircraft but was currently without an OKB of his own, gave assistance during the initial development work.

However, by the spring of 1953, when work on 'Aircraft 91' was in full swing, the USSR's military and political leadership began to lose interest in building aircraft carriers and, as the result, the Command of Naval Forces Aviation changed its requirements. From now on 'Aircraft 91' was to continue as a land-based tactical aircraft. On 29th April the USSR Council of Ministers issued a Decree covering the development of 'Aircraft 91' with one

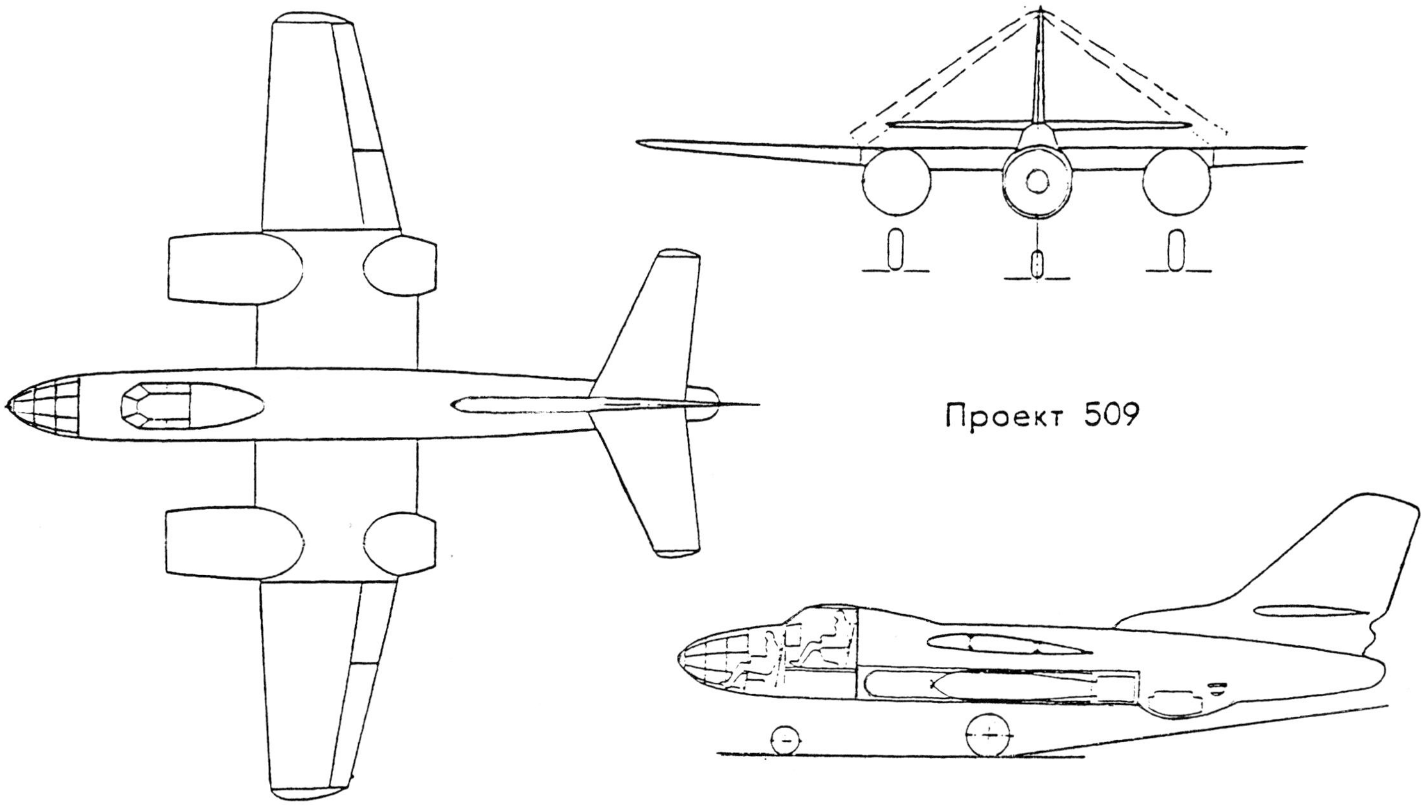

Tupolev 'Project 509' (1950). Russian Aviation Research Trust

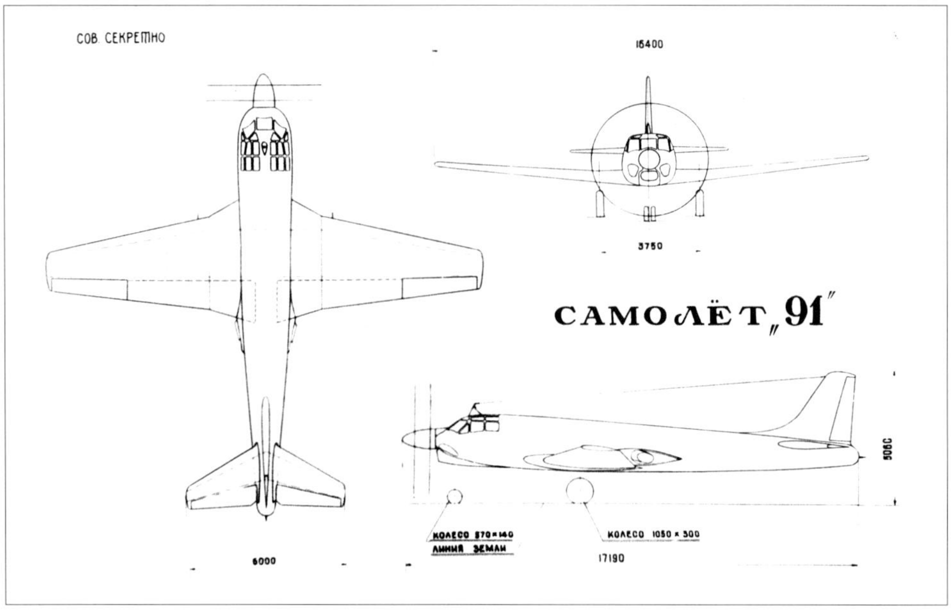

The sole Tupolev 'Aircraft 91' to be built was also known as the Tu-91.

Three-view general arrangement of the Tupolev 'Aircraft 91'.

Kuznetsov TV-2M turboprop engine and in June the Navy provided the design bureau with its Tactical and Technical Requirements for the type. This aircraft was now intended to attack enemy shipping and ground installations and to operate from concrete and grass runways.

A particularly interesting feature of the '91's layout was the position of the TV-2M turboprop behind the crew cabin. Initially the aircraft had a 6,250hp (4,661kW) TV-2F fitted but it was intended eventually to bring the maximum power up to between 7,000hp (5,220kW) and 7,650hp (5,705kW). The reduction gear, housed in the nose, was rotated with the help of a long shaft running through the crew cabin between the pilot and navigator's seats and this transferred the engine power to two counter-rotating airscrews. 'Aircraft 91' was to carry various weapons including torpedoes, bombs, mines, rocket projectiles in special containers with extendable mounts, two cannons in the wing panels and another in a remotely controlled tail mounting. In all the offensive capability of a single '91' was described as being equal to the broadside of a heavy cruiser's main armament.

This aircraft was an unusual task for the Tupolev design bureau and initially brought forth a multitude of different designs but, in spite of this, its development moved on without serious problems. The mock-up assessment was held in September 1953 and in April 1954 the construction of the prototype was completed. On 2nd September 1954 it achieved its maiden flight and the first stage of its manufacturer's testing, which showed much promise and brought forth many positive comments, lasted until 21st January 1955. Joint testing by manufacturer and customer was held from the end of January until the end of April and confirmed the aircraft's good qualities; as a result, 'Aircraft 91' was recommended for series production. During the test programme the rocket armament and powerplant were improved and in summer 1955 the aircraft undertook 'combat application' testing in the Black Sea. According to reports

Two views of the prototype Ilyushin Il-40 fitted with wing root intakes and a solid nose.

made by flight and ground crews, it was a success; in fact the aircraft was seen to be of great value to, and a vital need for, the Soviet Navy.

However the rest of 'Aircraft 91's story was a failure. Principally because of personal reasons (in other words Khruschev did not like the aircraft), its development was terminated. As a result, the USSR lost an excellent combat aircraft which had offered a unique combination of flight and tactical performance characteristics and would have provided a very effective air support system for both the Soviet Union's land and naval forces. Series production was to have taken place at Factory No 31 in Tbilisi (the capital of Georgia) for which the supply of a full set of documentation had been arranged. The second prototype was partially completed and, after some changes had been made based on the test results, this machine was to have served as a standard for series production (the crew cabin was to have been widened and there were alterations to the equipment).

The development of other versions – anti-submarine aircraft, trainer and an electronic countermeasures aircraft, had also begun, but all of these were stopped. The subsequent experience of relatively small conflicts, such as those in the Middle East and Vietnam, eventually forced many air forces to return to the concept of a well protected and powerfully armed subsonic close support aircraft. Aircraft developed in the 1970s and 1980s, such as the Sukhoi Su-25 (below) and American Fairchild-Republic A-10 Thunderbolt, employed many of the ideas that had evolved during the research and trials undertaken with 'Aircraft 91'. The '91's service ceiling was 38,058ft (11,600m) and range 1,461 miles (2,350km), and the machine was nicknamed 'Bychek' after the goby fish which has eyes on the top of its head.

Jet-Powered *Shturmoviks*

Sukhoi Type 'N'

During the war Pavel Sukhoi's design bureau had built several light bomber types but this project was its first foray into a jet-powered armoured attack aircraft. Work began in 1948 and the concept of fitting a good deal of armour to such aircraft, particularly around the cockpit area, was new and intended to increase the aeroplane's survivability during low-level operations. This design used a single Klimov VK-1 jet installed in the rear fuselage and had two crew, the pilot and his gunner who operated a rearward-firing turret fitted with a single 23mm cannon. Another six-cannon mounting fired forwards and the 'N' could carry bombs either in a fuselage bay or beneath the wings; it was given a tricycle undercarriage but remained a paper project. Sukhoi's OKB was closed in 1949 but re-opened in 1953.

Ilyushin Il-40

This aircraft, Ilyushin's first *Shturmovik* to be powered by a jet engine, was begun in 1950 as a private venture. The preliminary studies showed that a relatively small but powerful engine would be ideal to power such a machine and a full technical proposal was completed at the end of 1951 and passed to official circles for assessment in January 1952. It was to be fitted with two afterburning Mikulin AM-5F engines placed in the wing roots, swept wings, a tricycle undercarriage and would carry two crew. The concept of heavy armour protection was continued in this design and the crew, much of the aircraft's equipment plus six fuel tanks, were all housed in an armoured shell made in 3mm to 8mm thick steel plate, to which the wings, tail and engines were attached. In fact a total of 4,228 lb (1,918kg) of steel armour and toughened glass was worked into the Il-40's structure.

The maximum weapon load that could be carried internally was only 935 lb (424kg) but the external weapons pushed the total figure up to 3,038 lb (1,378kg); in total ten 220 lb (100kg) bombs could be carried. The aircraft had four retractable rocket launchers in the wings which could each carry either six TRS-82 82mm 13.2 lb (6kg) or four TRS-132 132mm 58.4 lb (26.5kg) rockets, and six more stores stations beneath the wings could take another 54 82mm or 36 132mm. A battery of six NR.23 cannon (three per side) was mounted in the nose while a single tail-mounted 37mm was operated remotely by the gunner.

The official response was quick and favourable and on 1st February a SovMin resolution was passed giving Ilyushin permission to proceed with a prototype. Work progressed at a good pace and the mock-up was

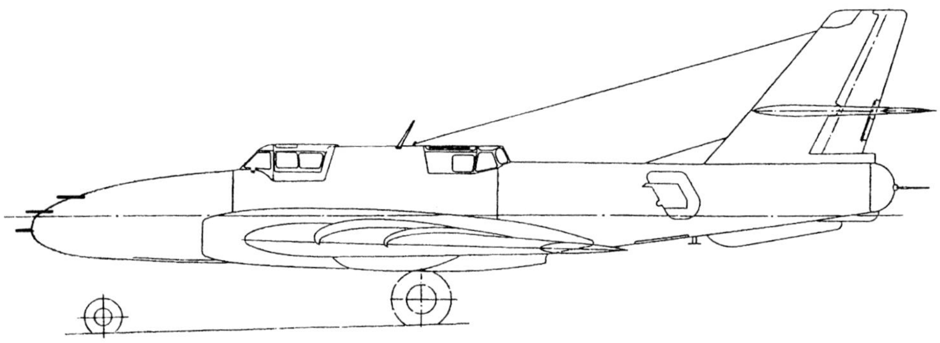

Side view of the Ilyushin Il-40 showing the original nose armament of three pairs of NR-23 cannon.

A production Il-40P shows the changes made to the intake position from the original prototype.

An impression of the Il-40 from the original project documents which shows a shorter nose and six NR-23 cannon.

officially inspected in mid-May. The Il-40 made its first flight on 7th March 1953 and, when flown, it was one of the first Soviet jet-powered aeroplanes to be painted in camouflage rather than being left in a natural aluminium finish. However, the heavy cannon armament created problems in that the guns made the engines surge when they were fired and it took two years for a solution to be found. In the end the intakes were lengthened and placed side-by-side in the nose while an alternative gun arrangement was adopted comprising four more powerful AM-23 23mm cannon in a line beneath the nose aft of the nosewheel bay; another AM-23 was now fitted in the tail.

The new version was labelled Il-40P and other changes included an increase in the normal bomb load to 2,205 lb (1,000kg) and the introduction of more powerful Tumansky RD-9V engines. A new Il-40P prototype made its first flight on 14th February 1955 and, during trials, this machine was able to fire its guns under every possible flight regime. Il-40P State trials began during the following October, the machine's pilots being confident that all of the major problems were now cured, and a Decree was issued ordering the production of forty aeroplanes. By the spring of 1956 no less than five production machines were ready for flight test and the programme was in full swing, but on 13th April another Decree abandoned the requirement for Il-40s and closed down all further work on the project. All of the assembled production jigging was scrapped.

The next step was another Decree, dated 20th April, which abolished ground-attack aviation altogether within the Soviet Air Force. A new military doctrine had been formulated where fighter-bomber aviation would take the place of ground attack and strikes on targets out of range of ground artillery were considered more important than sorties over the battlefield. The employment of tactical nuclear weapons was another factor and types like the Il-40 were felt to be no longer necessary; instead bombers such as Yakovlev's Yak-26 (Chapter 8) would be the alternative solution. Not until the 1970s was it realised that aeroplanes like the Il-40 could actually be key elements in the support of ground forces in a battle zone. In June 1956, after the programme had been cancelled, the Il-40 prototype was displayed at Kubinka, a move which brought the allocation of a Western codename, *Brawny*. There were several proposed Il-40 derivatives including the Il-40T which was adapted to carry a torpedo and had the navigator placed in a glazed nose to enable him to aim at his target.

Modern Equivalents – The Su-25 Story

Soviet military doctrine during the mid-1960s concentrated on nuclear conflict and the operation of military aircraft in a nuclear environment; ground-attack aircraft were still considered to be of very limited value. However, in March 1969 a competition was announced by the Soviet Air Force, the Ministry of Defence and by the Ministry of the Aviation Industry calling for designs for a new battlefield close-support aircraft. The competition placed great emphasis on the ability to provide the best possible combat effectiveness with a relatively cheap aircraft which would be simple both to manufacture and to operate and have very high survivability and reliability. Just as important the aircraft had to be capable of being mass-produced in the

The proposed Il-40K missile carrier and Il-40T torpedo carrier had alternative noses.

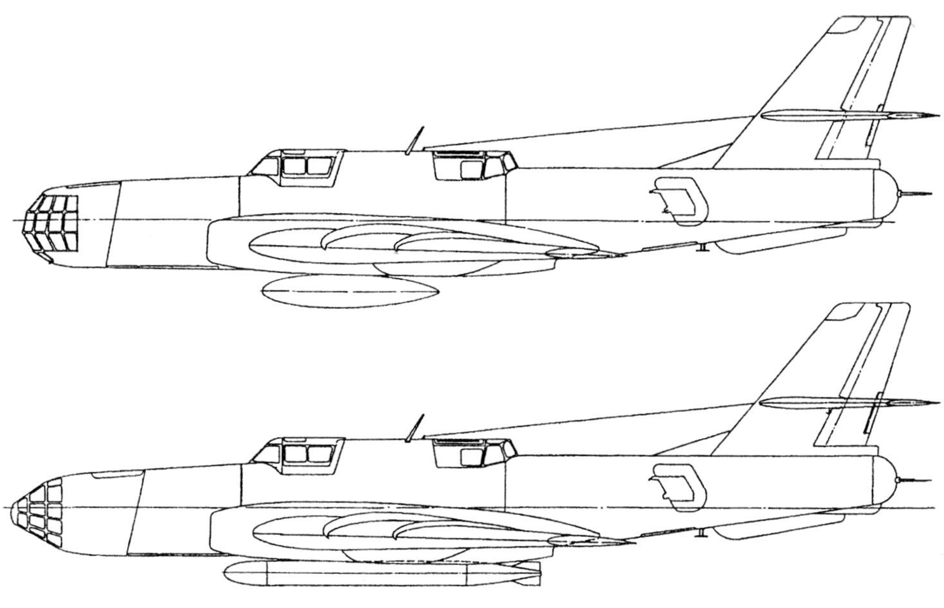

shortest possible time and, for this reason alone, most of the competing OKBs originally considered projects for a light close-support aircraft (*Lyogki Samolyot-Shturmovik* or *LSSh*) based either on existing operational types or on prototypes already test flown. Only a few designers actually considered the idea of creating a new aircraft from scratch that would fully meet the requirements.

The history of the Sukhoi Su-25, the winning design, actually goes back another year to March 1968 when the senior instructor in the Department of Tactics of Combat Application at the Yuri Gagarin Air Force Academy, Ivan V Savchenko, approached his friends in the Sukhoi design bureau with a suggestion for developing jointly the concept of a new close air-support aircraft. Soon afterwards an 'action committee' was formed, made up of members of the design bureau, and it was unofficially decided to draft an outline proposal for a project given the preliminary designation SPB (*Samolyot Polya Boya* or Battlefield Aircraft).

At this stage the Soviet Air Force had not yet formulated its own operational requirements and so the group also had to define the basic concept for the type, its performance and weapons, and its place in Soviet Air Force operational doctrine. The designers avoided the currently fashionable idea of a multi-role type that could undertake both strike operations and air combat, a concept embodied in the Sukhoi Su-17 and Mikoyan MiG-23B which were undergoing flight trials at this time (Chapter 8). It was concluded that the SPB should be a small, highly manoeuvrable aeroplane, designed to operate at low level, powered by two well proven and reliable engines and offering the pilot excellent visibility from the cockpit. The committee examined several alternative configurations, including those employed on the American Douglas A-4 Skyhawk, Fairchild Republic A-10 Thunderbolt and North American OV-10A, and the Swedish SAAB 105. One of the group's first designs showed a twin-boom, twin-engined layout but, in the end, the preference was for a conventional arrangement with a shoulder-mounted wing and two engines mounted in separate nacelles placed either side of the centre fuselage.

Sukhoi T-8

The committee's Oleg Samoilovitch and Yuri Ivashechkin presented the SPB to Pavel Sukhoi on 29th May 1968 and he readily approved the design. After introducing a few minor changes he ordered that work should commence on the aircraft's development under OKB designation T-8. After the aerodynamicists had been brought into the project, led by Deputy Chief Designer I E Baslavski, the type's overall shape could be defined more precisely. Studies had shown that the greatest probability for the survival of a ground-attack aircraft occurred at a speed not exceeding 528mph (850km/h) while, over a battlefield, a flight speed greater than 497mph to 528mph (800km/h to 850km/h) was of little practical benefit because it gave too little time for visual acquisition of the target. In addition the chosen subsonic speed range would permit the use of simple fixed air inlets and exhaust nozzles.

A month after work had started on the T-8, Pavel Sukhoi decided that the project was sufficiently advanced to report its progress to the official authorities. A full proposal was formulated in August 1968 and a summary of the design project was sent to the Scientific Technical Committee of the General Staff of the Soviet Defence Ministry, the Headquarters of the Soviet Air Force, the Ministry of Aviation Industry and the Headquarters of the Soviet Navy, plus TsAGI. The first to respond, in September, was the Scientific Technical Committee of the Ministry of Defence who had concluded that such an aircraft was not required. However, the Soviet Air Force's leading scientific research institute sent a more cautious reply, suggesting that work on the T-8 should be continued. The others never replied but many highly-placed military figures realised that the inclusion of a cheap and simple, mass-produced *Shturmovik*-type in the Soviet Air Force inventory would permit the renewal and replacement of the Frontal Aviation aircraft fleet in a comparatively short time. Just as important, it would enable the Air Force to substantially reduce the number of heavy tactical strike aircraft required by its front-line regiments.

Taking this factor into consideration the Minister of Defence, Marshal Andrei Grechko, wrote to the Minister of Aviation Industry, Pyotr Dementiev, suggesting that a competition should be held with the objective of creating a new light ground-attack aircraft. By 19th March 1969 the Soviet Air Force had outlined its own preliminary requirements for this type of aeroplane which specified a normal combat payload of 2,205 lb (1,000kg), a maximum of 6,614 lb (3,000kg) and a range at sea level when flying at 497mph (800km/h) of not less than 466 miles (750km). The competition for a subsonic *Shturmovik* could now be launched but it was already clear that the design put forward by Sukhoi was the only one which satisfied the main criteria.

Returning to the T-8, it was eventually decided to fit an existing production engine to the prototype during the early stages of its flight testing. The choice was the Klimov RD-9B with its afterburner removed, a version of the engine used on the MiG-19 fighter, and as the Tumansky R9-300 this offered 5,950 lb (26.4kN) of thrust. Later on this powerplant would be replaced by the Izotov TR7-117, Lyulka AL-29 or Tumansky R53B-300, all of them offering similar take-off thrust ratings in the region of 6,615 lb to 7,165 lb (29.4kN to 31.8kN) and little difference in terms of weight and size. The T-8's estimated service ceiling was 22,966ft (7,000m) and its range at sea level when carrying a 2,646 lb (1,200kg) of

Model of the Mikoyan 'Article 7-23' (1969).

Model of the MiG-27Sh (1969).

weapons, but no drop tanks, was 466 miles (750km). The Sukhoi OKB felt that a rear defensive gunner would be useless under contemporary conditions of aerial conflict. The other proposals submitted for the competition were as follows.

Ilyushin Il-42

From the very beginning Sergei Ilyushin had supported the idea of a subsonic *Shturmovik* and his competing Il-42 project retained the basic features of the old Il-40 but was fitted with new engines, new weapons, and much more advanced avionics; however, a big difference from the T-8 was the Il-42's second cockpit for a rearward facing gunner (who operated by remote control a GSh-23 23mm twin turret with super-imposed barrels). This feature was to have a negative impact on the ultimate fate of this aircraft because the Soviet Air Force requirement specifically called for a single-seat aeroplane.

However, Ilyushin had analyzed very closely the type of environment in which such an aircraft could find itself and realised that a key threat would be ground-based shoulder-launched surface-to-air missiles. Second World War experience had shown that a ground-attack aircraft protected from its most vulnerable position, the rear, with guns operated by a second crewman allowed the pilot to concentrate entirely on his mission and, in general, created more confidence. Ilyushin felt that the Il-42's rear gun would also prove valuable in preventing attacks by ground-based missiles launched by enemy troops. Experience in Afghanistan later on was to bear this out because Soviet single-seat aircraft were often hit by such weapons when the, by now, obsolescent Il-28 was able to frustrate such attacks through the use of its rear-mounted guns. No data is currently available for the Il-42.

Mikoyan LSSh Studies

The Mikoyan OKB produced several designs against the LSSh specification. The simplest solution, designated 'Izdyelie 7-23' or 'Article 7-23', was a MiG-21PFM *Fishbed-F* light tactical fighter altered to match the new requirements. This had a modified wing with leading edge root extensions that were intended to improve the aircraft's low-speed manoeuvring performance, while under each wing there were three pylons for bombs or rockets and the fuselage centreline pylon was retained to carry a supplementary fuel tank.

A second project, known as the 'MiG-27Sh', was a hybrid of a MiG-21 tail and a MiG-23B *Flogger-F/H* forward fuselage (the latter a variable geometry-wing type which later became the MiG-27 strike aircraft – Chapter 8) and used a fixed swept wing and lateral air intakes. Compared to the MiG-21 the new wing had a greater span and area and was equipped with slots and three weapons pylons on each side. Three more weapon stations were placed beneath the fuselage, making nine in all, and these could carry a maximum 6,614lb (3,000kg) of bombs. A fixed twin-barrel GSh-23 23mm cannon was also mounted under the fuselage and, for the first time in a Mikoyan aeroplane, the cockpit was fully armoured. This design would have been capable of transonic performance although the engine did not have an afterburner.

The most advanced of Mikoyan's proposals was the '27II', better known as the MiG-21LSh, which was effectively a hybrid of the MiG-21I *Analog* research aircraft and the MiG-23B. The first of two *Analogs* made its maiden flight on 18th April 1968 and this aircraft was originally built to test the wing shape and profile of the Tupolev Tu-144 supersonic transport. As a result the MiG-21LSh had a tailless delta configuration embracing the MiG-21I's ogival wing and the MiG-23B's forward fuselage, plus small side intakes level with the cockpit feeding two engines placed side-by-side in the rear fuselage. The adoption of the large ogival wing enabled the designers to offer a greater number of underwing hardpoints for weapons carriage, four beneath the fuselage and eight under the wings with each capable of taking a 551lb (250kg) bomb, while the nose held the fire control systems for the air-to-ground missiles plus some jamming equipment. The usual mix of offensive weapons could be carried, again to a maximum of 6,614lb (3,000kg). Mikoyan also stated that its designs would be capable of a speed of 746mph (1,200km/h) but relatively little data is currently available for any of these projects.

Two views of a model of the MiG-27II (or MiG-21LSh) (1969).

Yakovlev Yak-LSh

The Yakovlev OKB developed its own Yak-LSh light *Shturmovik*, details of which are, surprisingly, still classified. However, information published in the open press suggests that this two-seat aircraft was to have been based on a highly modified version of the Yak-28 *Brewer* light supersonic tactical bomber (Chapter 9) but, according to the Yakovlev OKB, the design actually showed little resemblance to the Yak-28. Following its elimination at the project stage of the competition, all further work on this particular design was terminated.

–

During this period, when the Soviet Air Force leadership was totally focused on creating a fleet made up exclusively of supersonic fighter-bombers, the results of some research conducted by GosNIIAS (the State Research Institute for Aviation Systems) into subsonic attack aircraft was, to certain extent, sensational. It was met with disbelief by the military and attempts to turn it into reality led to serious opposition within the Air Force. In fact, this opposition was so strong that, if it had not been for the decisive position adopted by the GosNIIAS leadership, the production line for any subsonic *Shturmovik* would have been shut down right from the start and not re-opened. The GosNIIAS's leaders, actively supported by Pavel Sukhoi and Sergei Ilyushin, had expressed their agreement with the conclusions of the institute's researchers and stated their position in letters to the appropriate State officials. The overwhelming conclusion was that there was a clear need for a subsonic battlefield close air-support aircraft for the Soviet Air Force, although the Air Force top brass fought against the idea for a long time.

In June 1969, three months after the announcement of the competition, all of the competing LSSh projects were submitted for scrutiny by the MAP's Scientific-Technical Commission. In due course, preference was given to the single-seat Sukhoi T-8 and Mikoyan MiG-21LSh designs, and both OKBs asked to continue working on their projects and build prototypes. The Il-42's rear protection was called into question and was expected to require extra crew and specialist training for gunners so both this, and the other two-seat design from Yakovlev, were rejected. For a number of reasons, the Mikoyan design bureau abandoned its participation in the competition, but the development of Sukhoi's T-8 design was continued. Work on the Il-42 was also continued by the Ilyushin design bureau, which had decided to develop the aircraft as a private venture and build a prototype for flight test (the Il-102 described later). However, in August 1971 the Soviet Air Force announced that the required maximum speed had been increased to 746mph (1,200km/h) at sea level. In view of this radical change, Pavel Sukhoi ordered that all work on the T-8 project should be suspended.

Sukhoi T-58Sh

At this point, largely in response to the above-mentioned high-speed performance requirement and undertaken on the instructions of Deputy Chief Designer Evgeny Ivanov, the Sukhoi Bureau began work on a highly modified variant of its Su-15 *Flagon* (T-58) interceptor to fulfil the amended specification. The assault version of the Su-15 was given the designation Su-15Sh, or T-58Sh (*Shturmovik*), and had its delta wing replaced by a trapezoid-shaped swept wing of greater span and area, while the shape of the nose as far back as frame ten was also changed. The cockpit and engine compartments were to be protected by armour plate to give improved survivability, with self-sealing protection also afforded to the fuel tanks, whilst the radar and other avionics associated with the old interceptor mission were replaced by an ASP-PF gunsight, a PBK-2 bomb sight optimised for lob-bombing and a Fon (Background) laser rangefinder.

The T-58Sh featured eight weapons hardpoints and a built-in 225P Gatling cannon. External stores options included bombs of up to 1,102 lb (500kg), assorted unguided rockets, Kh-23 air-to-surface missiles, UPK-23-250 cannon pods and SPPU-17 pods with movable cannons for strafing ground targets in level flight; for self-defence the aircraft would be armed with K-55 and K-60 air-to-air missiles. Maximum take-off weight with a combat load of 8,818 lb (4,000kg) was projected to be 38,580 lb (17,500kg), internal fuel capacity 9,921 lb (4,500kg), projected maximum speed at sea level 777mph (1,250km/h) and estimated range at sea level (without auxiliary tanks and carrying a normal combat payload of 4,409 lb [2,000kg]) 373 miles (600km). Billed as an 'in-depth upgrade', to all intents and purposes the T-58Sh was a new aircraft.

At the same time, Sukhoi's designers, who did not agree with the concept of a supersonic *Shturmovik*, tried to explain that an aircraft operating 19 to 31 miles (30 to 50km) behind the front line could not escape detection by enemy air defences but it should be able to operate permanently within his air defence protection zone. They therefore recommended that the maximum speed should be limited to 528mph (850km/h), or Mach 0.7 at sea level, thereby avoiding the unpleasant phenomenon of shock stall. In the end the maximum sea level speed, written into the tactical operating requirements, was 622mph (1,000km/h) or Mach 0.82. On the whole the more 'modest' objectives suggested for the

Sukhoi T-58Sh (c1971).

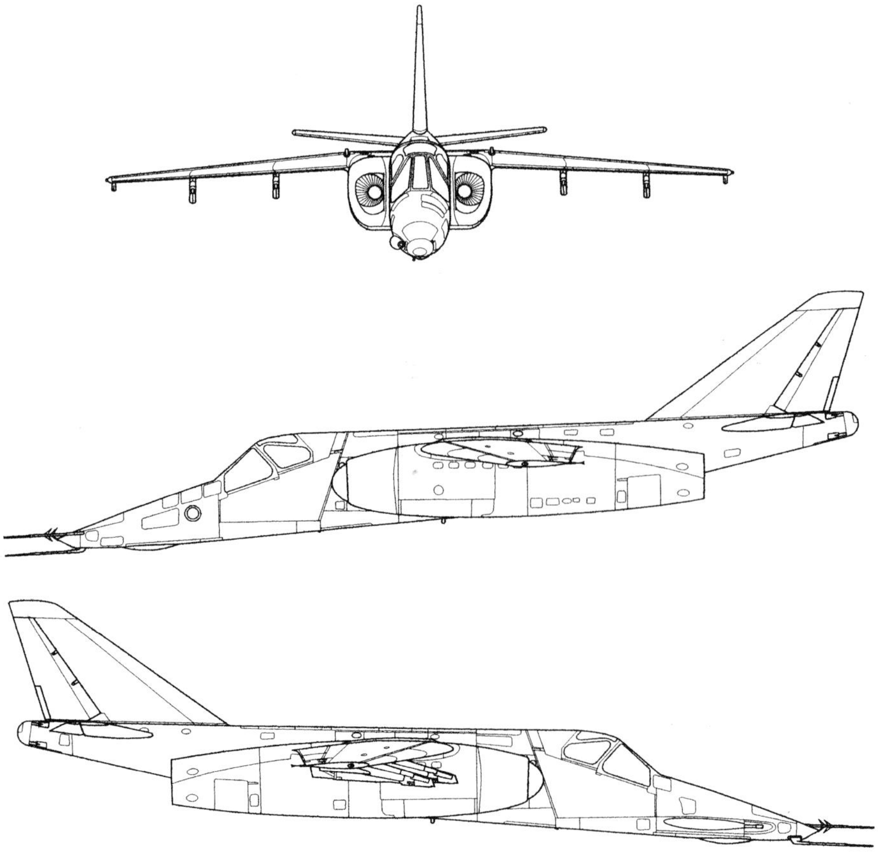

Drawing of the Sukhoi LVSSh in the form it had reached at around the end of 1971.

new *Shturmovik* were agreed with the Soviet Air Force very quickly. This was largely due to the influence of the Sukhoi design bureau's Zelika Yoffe, at the time the head of a section dealing with combat survivability, who used his old Air Force connections to expedite the matter, and in three days he was able to obtain an agreement with the military over the operational requirements proposed by Sukhoi. Before joining Sukhoi, Lieutenant General Yoffe had been in control of one of the Soviet Air Force's scientific research institutes and had initially been opposed to the T-8 project. As a result of these moves, Pavel Sukhoi ordered that the T-58Sh should be shelved.

–

At the end of November 1971, the Sukhoi design bureau received more precise operational requirements from the Soviet Air Force which now included a maximum combat load increased to 8,818 lb (4,000kg); the machine was also now being referred to as the LVSSh (*Lyogki Voiskovoi Samolyot-Shturmovik*) or light army support *Shturmovik*. Work on the T-8 was only re-started by the OKB at the beginning of 1972, when Pavel Sukhoi himself confirmed the overall layout and signed the order authorising the start of detail design, with Mikhail Simonov designated as project leader. The broad aerodynamic concept was retained but the systems and equipment layout were now completely changed. The LVSSh T-8's normal take-off weight was increased to 23,214 lb (10,530kg) which in turn led to an increase in the aircraft's overall dimensions – fuselage length rose from 41ft 2in (12.54m) to 44ft 11in (13.7m) whilst the wing area went up to 303ft^2 (28.2m^2), albeit retaining the original shape and aspect ratio.

Two full-scale LVSSh mock-ups were built, one 'fully equipped' and the other with simplified installations. Between 12th and 15th September 1972 a mock-up commission studied both these and the preliminary design documentation, the aircraft by this time having been given the military designation Su-25, and concluded that all of the Air Force's operational requirements had been achieved. It was decided that there should be two flight test prototypes, T8-1 and T8-2, plus an additional airframe, T8-0, for static fatigue testing. Pavel Sukhoi took this decision on his own, without the formal resolution of the Central Committee of the Soviet Communist

Two views showing the Sukhoi LVSSh mock-up, the second with its prospective weapons.

Party and the USSR Council of Ministers, and even without an appropriate order from the Ministry for Aviation Industry (MAP). Such a decision was extremely bold because, in the Soviet Union at that time, no prototype aircraft could be built without the customary instructions from the highest authority. Throughout 1973 the design bureau conducted work on the new aircraft on an unofficial basis, as a result of which it experienced great difficulties in financing the project.

In early May 1974 during a visit to the design bureau, the Minister for the Aviation Industry Pyotr Dementiev saw the assembled airframe in the assembly shop and gave his approval for the preparation of a joint resolution to cover the prototypes. This cleared the way, in the absence of a special decree from the Communist Party and the Council of Ministers, for Sukhoi to receive all of the components it required from relevant suppliers to finish manufacturing the two aeroplanes; T8-1 was finally completed at the end of October 1974. However, it had not been possible to commence the construction of T8-0 until T8-1 had vacated its jigs and so, although the static airframe was finished on 12th September 1974, the time needed to acquire sufficient fatigue data prevented the T8-1's maiden flight from taking place until 22nd February 1975. The prototype's take-off weight was now 26,896 lb (12,200kg), which considerably exceeded the 22,046 lb (10,000kg) originally specified. In due course the Su-25 entered full production although many changes had to be made from the original prototype.

Sukhoi Su-39

During 1981 the Sukhoi OKB began work on a tank-busting variant to be called the Su-25TK and this made its maiden flight on 17th August 1984. It was built in small numbers and this version, and the follow-on Su-25TM project flown in 1996 with many more improvements including a modern glass cockpit, was also designated Su-39. However, by then the airframe was obsolete and not really suitable for the introduction of advanced avionics.

Sukhoi T-12

In 1997 Sukhoi released details of a design for a most unusual 'twin-fuselage' attack aircraft

The sole Ilyushin Il-102 prototype seen just about to land at Zhukovsky after a test flight.

intended as a possible successor to the Su-25. Work was apparently under way on this project as early as 1980 and the design, called T-12 and Sh-90 (*Shturmovik* for the 1990s), had two crew in side-by-side 'fighter-type' cockpits, a large flat central fuselage, an intake between the cockpits and a single engine, two large weapon bays inside the central fuselage, twin outward-canted fins and wings that were swept slightly forward; its sensors were expected to be similar to or the same as the Su-25TM. Span was about 52ft 6in (16m), length 64ft 0in (19.5m) and take-off weight 44,092 lb (20,000kg) and stories were circulated that a lot of serious design work had been completed and plans were in place to put the type into production at Irkutsk. However, the general shortage of finance in Russia during the 1990s ensured that this was another advanced military aeroplane that failed to reach the hardware stage.

Ilyushin Il-102

Despite having the Il-42 rejected, the Ilyushin OKB continued working on its ground-attack ideas and the original design was completely reworked. Analysis of operations in the Arab-Israeli conflicts indicated that a two-seat design really could be the ideal while results from other countries operating single-seat types suggested these achieved a combat efficiency some 25% lower than expected. Advances in avionics would also allow the gunner to become more actively involved in observation over the battlefield. Sergei Ilyushin retired in 1970 to be replaced by Ghenrikh V Novozhilov, who became the OKB's general designer, and he continued with a new version of the Il-42 designated Il-102. This had improved visibility for the crew and was fitted with two Klimov RD-33I turbofan engines, a version of the unit used by the Mikoyan MiG-29 fighter but without reheat; they did, however, have thrust vectoring nozzles for improved take-off and over field performance and enhanced manoeuvrability.

The Il-102 had a fixed external forward-firing twin-barrel GSh-30/II 30mm cannon housed under the fuselage centreline plus a flexible 23mm twin barbette to the rear; in addition two gun pods could be carried externally beneath the wings. Fourteen hardpoints were available for weapons carriage which meant a big mix of stores could be taken on board – bombs of up to 1,102 lb (500kg) in weight, 57mm, 80mm and 130mm rocket pods, a KMG-U submunitions dispenser or R-60M or R-73 air-to-air missiles; up to 5,071 lb (2,300kg) of ordnance could be carried internally.

Prototype construction began in May 1980 without official authorisation but progress

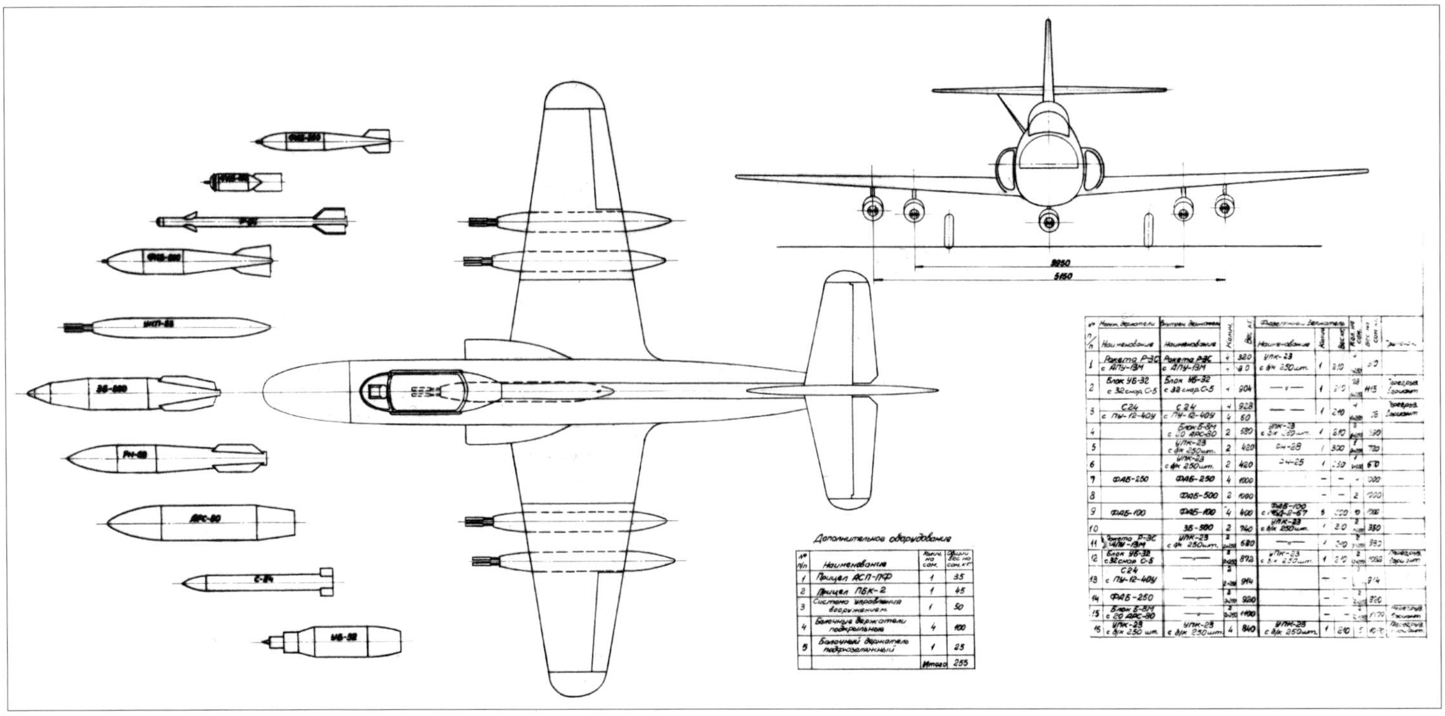

Yakovlev Yak-32Sh (1961).

was slow because certain items such as ejection seats had to be taken 'on loan'; there were plans for second prototype with enhanced performance. Support was forthcoming from Chief Air Marshal Pavel S Kutakhov, VVS Commander-in-Chief, but the Minister of Defence, Marshal Dmitri F Ustinov, mindful of the official support given to Sukhoi's aircraft, refused to even inspect the machine. Furthermore he ordered that all Il-102 testing was to stop and that the prototype had to be destroyed, showing clearly to Novozhilov that unofficial projects would not be tolerated. However, support was also forthcoming from the Minister of Aircraft Industry, Ivan S Silayev, and Novozhilov decided to quietly keep the project moving under a new in-house designation OES-1 or 'prototype experimental aircraft No 1'. A first flight was finally achieved on 25th September 1982 and performance and weapon testing showed that the Il-102 was a promising aircraft, but it was found to be slightly inferior to the Su-25 and so was not adopted for service. Il-102 was unveiled to the public at the Mosaero Show held in August 1992, which then generated some further interest with Russian Vice-President and former Su-25 pilot, Aleksandr Rutskoy, closely involved; however, his arrest following the failed 1993 coup signaled the end for the Il-102.

Yakovlev Yak-32Sh

In 1960 the Yakovlev OKB flew the prototype of a jet training aircraft called the Yak-104 which, in due course, entered service as the Yak-30, and shortly afterwards a single-seat trainer and civil aerobatic version designated Yak-32 (the Yak-30 designation was also used in 1948 for a jet fighter and both Yak-30 and Yak-32 were also applied to supersonic reconnaissance types described in the next chapter). In 1961 Yakovlev demonstrated to Defence Minister R Ya Malinovsky a version of the Yak-32 modified as a lightly-armed counter-insurgency aircraft called the Yak-32Sh, with two wing racks for a variety of possible weapon loads. According to a report dated 25th May 1961, the thrust of the aircraft's Tumansky RU19-300 engine was to be increased to 2,425 lb (10.8kN) and its fuel capacity to 1,601 lb (726kg). Maximum take-off weight was 6,393 lb (2,900kg) with 1,578 lb (716kg) of weapons aboard and a maximum speed of 426mph (685km/h) was to be achieved at 16,404ft (5,000m). The weapon loads could include the following – AO-9 cannon, two 256 lb (116kg) or two 736 lb (334kg) rocket projectiles, two 606 lb (275kg) bombs or ZB-360 napalm tanks. The drawing shows four underwing hardpoints plus another beneath the fuselage, but the project was not taken up. Span was 30ft 6½in (9.31m) and length 33ft 3in (10.13m).

Vertical Take-Off

The following vertical take-off aeroplanes are often classed as fighter types because both the Yak-36 and Yak-38 received Western 'fighter' codenames, when in truth these were respectively an experimental research aircraft and an attack aircraft. In the same way that the British Hawker Siddeley Harrier broke new ground for vertical take-off in the West, these designs led the way in the USSR.

Yakovlev Yak-36

In September 1960 Aleksandr Yakovlev visited the Farnborough Air Show where he witnessed the performance of the Short SC.1 VTOL research aircraft. By the end of year, Yakovlev OKB designers L M Shekhtyer and V K Tsvelev had produced a preliminary project for the Yak-30V VTOL aircraft, based on the Yak-30 trainer, with two lift engines installed in the rear cabin area. This project was soon terminated when development of the Yak-36 began, but the bureau's V/STOL research programme was under way.

A Council of Ministers directive issued on 30th October 1961 requested the OKB to design and build a single-seat fighter-bomber. It was intended that this should be powered by two 11,025 lb (49.0kN) R21M-300 engines and achieve a maximum speed approaching 746mph (1,200km/h), but in fact these power units were not good enough and the problem

The first prototype Yakovlev Yak-36.

was to find a better engine; fortunately Tumansky had developed the unreheated R27V-300 power unit, a version of the conventional R27-300, which gave 11,685 lb (51.9kN) of thrust. The resulting Yak-36 actually served as a research aircraft and was a little reminiscent of the 1950s Mikoyan MiG-15/MiG-17 series but fitted with vectored thrust nozzles. Four examples were built, the first making an initial untethered hover on 9th January 1963 followed by a full transition to horizontal flight on 16th September. These machines broke much new ground but they displayed a poor performance and did not inspire great optimism. One was displayed to the public over Domodedovo in July 1967 and the type subsequently received the Western codename *Freehand*.

Yakovlev Yak-38

Nevertheless Yakovlev continued its studies and by the mid-1960s was looking at a much improved design called the Yak-36M. The plan was to introduce a mixed powerplant of a lift/cruise engine with two rotating nozzles plus new separate vertically-mounted lift jets designed by the Kolesov OKB; an arrangement selected after other lift-jet alternatives had been considered. There were many doubts in official circles towards the idea of a VTOL aircraft but it was the Minister for Aircraft Industry, Pyotr Dementiev, who kept things going with a suggestion that the Yak-36 should be used as the start point for a two-stage development programme. The first part would comprise the Yak-36M light attack aircraft fitted with basic avionics and this would be followed by a much more sophisticated supersonic fighter called the Yak-36MF or Yak-36P (which was never built). The Air Force was not impressed but the Soviet Navy could see the benefits of VTOL for service on a new design of aircraft-carrying cruiser called Project 1143 (which was to become the *Kiev* class).

At the end of 1967 the Council of Ministers decided to go-ahead with the manufacture of five Yak-36Ms and in January 1968 the Air Force's Commander-in-Chief also approved a requirement for a Yak-36M light attack VTOL aircraft. The initial layout looked like a slimmed down Yak-36 and was fitted with side intakes, but the process of development

The first landing by a Yakovlev Yak-36M prototype onto a *Kiev* Class carrier was made on 18th May 1975.

meant that the design as built looked rather different. A mock-up was officially examined in April 1970 and the first of four prototypes made its first free hover on 22nd September 1970 and its first conventional flight on 2nd December. On 18th November 1972 one of the prototypes made the type's first landing on a warship when it touched down on the deck of the helicopter carrier *Moskva*. However, it was not until October 1976 that the aircraft was introduced into service under the designation Yak-38 while a later upgrade, called the Yak-38M, was given more power and could carry extra weapons. The large variety of stores available to the Yak-38 included a gun pack, air-to-air missiles, rocket projectiles, bombs of various sizes up to 1,102 lb (500kg), incendiary bombs and a 551 lb (250kg) free-fall nuclear weapon. A total of 193 single-seat Yak-38s were built, plus another 38 of a two-seat variant, and the type's service career ended in the 1990s.

Yakovlev Yak-39

In service the Yak-38 experienced plenty of problems and so it can only be classed as a moderate success. An attempt to improve the basic aeroplane was made in 1983 with the Yak-39 V/STOL fighter/bomber proposal. This looked similar to the Yak-38 but had a new and larger wing built in light composite materials and fitted with flaps and leading edge slats that could be set at three angles, a modified tail and it was to carry more weapons. One 14,770 lb (65.6kN) R28V-300 lift/cruise engine was fitted, with swiveling nozzles, together with two 9,040 lb (40.2kN) RD-48 lift jets and there was a large radome to cover an S-41D radar scanner. A 30mm cannon would be installed in the aircraft, which was expected to achieve a maximum of 559mph (900km/h) at low level, a cruise weight of 27,668 lb (12,550kg) and a range of 280 miles (450km). The engineering proposal was discussed in July 1983 by a special commission but rejected because the aircraft's air-to-air combat potential was limited and it still exhibited many of the Yak-38M's weaknesses; development was, therefore, discontinued. By this time Yakovlev was working on a supersonic VTOL fighter called the Yak-41, of which two prototypes were built and tested for naval operations. This also failed to reach service while a follow-on multi-role VTOL project called the Yak-43 stayed on the drawing board.

Ground Attack Aircraft – Data / Estimated Data

Project	*Span* ft in (m)	*Length* ft in (m)	*Gross Wing Area* ft² (m²)	*Max Weight* lb (kg)	*Powerplant* Thrust hp (kW) or lb (kN)	*Max Speed / Height* mph (km/h) / ft (m)	*Armament*
Alekseyev I-218	53 11 (16.43)	45 6 (13.88)	484 (45.0)	23,148 (10,500)	1 x VM-251 piston 2,000 (1,491)	289 (465) at S/L, 329 (530) at 6,562 (2,000)	6 x cannon, up to 3,307lb (1,500kg) of bombs or rocket projectiles
Alekseyev I-221 (I-218-II)	57 1 (17.4)	54 1.5 (16.5)	538 (50.0)	28,660 (13,000)	1 x TR-3 jet 10,140 (45.1)	457 (735)	3 x 23mm cannon, 8 rocket projectiles
Alekseyev I-220 (I-218-III)	60 4 (18.4)	55 9 (17.0)	591 (55.0)	26,455 (12,000)	1 x VD-4 piston 4,000 (5,364)	?	4 x 23mm cannon, 8 rocket projectiles
Ilyushin Il-20 (flown)	50 7.5 (15.43)	44 6.5 (13.58)	473 (44.0)	20,944 (9,500)	1 x M-47 piston 3,000 (2,237)	280 (450) at S/L, 320 (515) at 9,186 (2,800)	5 x 23mm cannon, up to 2,623lb (1,190kg) bombs + 8 rocket projectiles
Tupolev Tu-91 (flown)	53 9.5 (16.4)	52 4 (15.955)	510 (47.5)	29,211 (13,250) (normal)	1 x TV-2F turboprop 6,250 (4,661)	497 (800) at 26,247 (8,000) clean	3 x AM.23 23mm cannon, up to 3,307lb (1,500kg) bombs or torpedoes
Sukhoi Type 'N'	50 0 (15.23)	37 0 (11.58)	430 (40.0)	21,164 (9,600)	1 x VK-1 jet 5,950 (26.4)	?	7 x 23mm cannon, up to 882lb (400kg) of bombs
Ilyushin Il-40 (flown)	52 6 (16.0)	56 6 (17.215)	562 (52.3)	38,084 (17,275)	2 x AM-5F jets 4,740 (21.1) dry, 5,950 (26.4) reheat	595 (958) at 9,843 (3,000)	6 x 23mm + 1 x 37mm cannon, up to 3,038lb (1,378kg) of rockets/bombs
Ilyushin Il-40P (flown)	55 9 (17.0)	56 6 (17.215)	582 (54.1)	38,801 (17,600)	2 x RD-9V 5,730 (25.5) dry, 7,165 (31.8) reheat	617 (993) at 9,843 (3,000)	5 x 23mm cannon, up to 3,086lb (1,400kg) of rockets/bombs
Sukhoi T-8 (at March 1969)	32 0 (9.75)	39 4 (12.0) (excl pitot tube)	204 (19.0)	23,302 (10,570) 5,950 (26.4)	2 x R9-300 jet sea level	622 (1,000) at 6,614lb (3,000kg) of ordnance	Nose cannon,
Sukhoi Su-25 (flown)	47 1 (14.36)	49 5 (15.06)	324 (30.1)	38,801 (17,600)	2 x R-95Sh jet 9,040 (40.2)	603 (970) at sea level	1 x 30mm cannon, up to 9,700lb (4,400kg) of bombs
Ilyushin Il-102 (flown)	55 8.5 (16.98)	58 3 (17.754)	683 (63.5)	48,501 (22,000)	2 x RD-33I jet 11,465 (51.0)	Over 590 (950) at 3,281 (1,000)	2 x 30mm + 2 x 23mm cannon, max load 15,873lb (7,200kg) stores
Yakovlev Yak-36 (flown)	24 3 (7.4)	54 11.5 (16.75)	c170 (15.8)	20,723 (9,400)	2 x R27-300 jet 14,020 (62.3)	627 (1,009) at sea level	None fitted.
Yakovlev Yak-38 (flown)	23 0 (7.022)	53 8.5 (16.37)	201 (18.7)	24,912 (11,300) (rolling take-off)	1 x R27V-300 jet 14,550 (64.7) + 2 x RD36-35 lift jet 6,400 (28.4)	715 (1,150) at 656 (200)	Max 4,409lb (2,000kg) stores

Chapter Eight

Tactical Strike Aircraft

The Sukhoi T-6 prototype in its original form.

From the mid-1950s the Soviet Union put a considerable amount of effort into the development of tactical strike aircraft, known more often in the USSR as front-line bombers and operated by the regiments belonging to the Air Force's Frontal Aviation. Some of the aircraft actually chosen to fill these roles were fighter/bombers and their full story really belongs in a companion volume covering fighters, but they are included here briefly for completeness. (Note: The Soviet Air Force's tactical arm was known as the FA for *Frontovaya aviahtsiya* and the strategic bomber arm DA for *Dahl'nyaya aviahtsiya.*)

Early Studies

Tupolev 'Aircraft 112'

At the end of 1954 the Tupolev design bureau's Technical Projects Department began to examine a supersonic twin-engine front-line bomber which was intended to reach a maximum speed of Mach 1.7 to 1.9. Given the designation 'Aircraft 112', its primary task would be to deliver a tactical nuclear weapon. The design had much in common with Tupolev's 'Aircraft 105'/Tu-22 (Chapter 4) but was rather smaller and lighter and the initial sketches showed a high-wing aircraft with a 55° swept wing of thickness/chord ratio 7.2%. Two nacelles were placed over the rear fuselage around the swept all-moving fin to house Lyulka AL-7F turbojets and an all-moving swept tailplane was positioned beneath them.

The aircraft had a bicycle undercarriage, which meant that the trademark Tupolev main gear wing nacelles were not in evidence, and the wing was fitted with outriggers plus ailerons and two-section flaps (for comparison an alternative tricycle undercarriage with wing fairings was examined and showed a possible weight saving of 661 lb [300kg]). 'Aircraft 112' was intended to operate from concrete runways and the main wheels were designed to carry 77% of the aeroplane's weight and the nose gear the remaining 23%. Two alternative fuselages were considered, the principal differences being the positions of the fuel tanks. One contained part of the fuel in the fuselage while the rest went into the wing centre section torsion box and the outer wing, to give a total of 11,464 lb (5,200kg), while the second housed all of its fuel inside the fuselage for a maximum 11,243 lb (5,100kg). A large bomb bay was placed in the lower fuselage centre and rear.

Both the pilot and his navigator had their own pressurised cabins, placed in tandem in

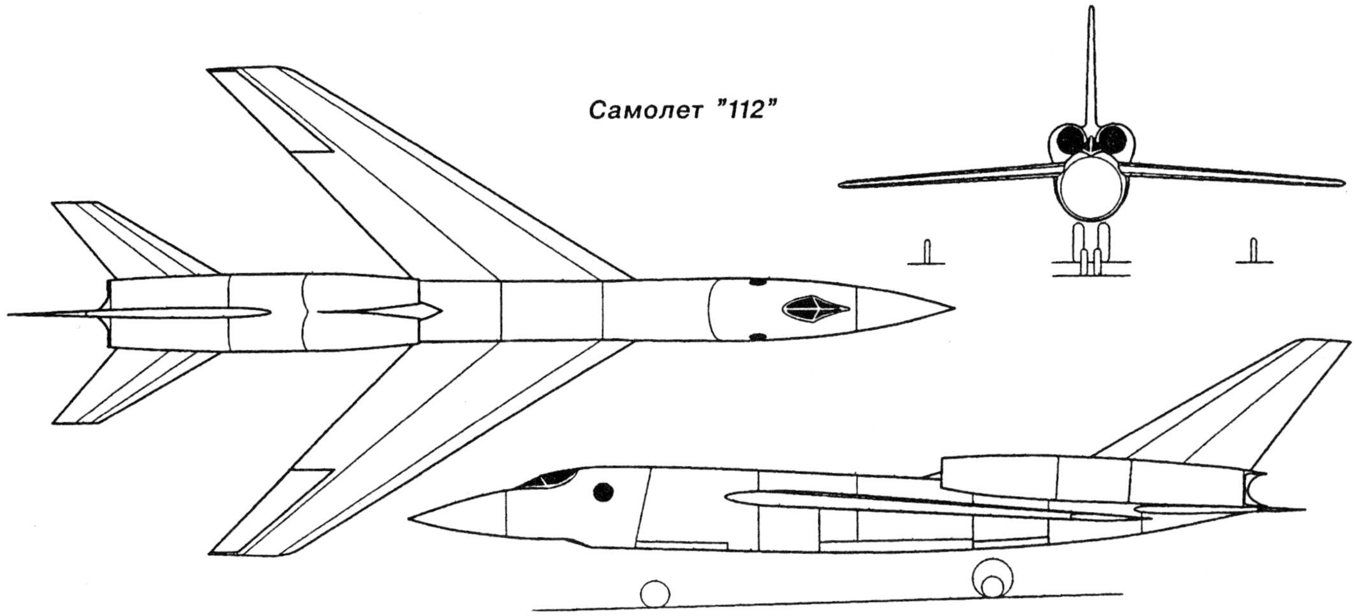

the forward fuselage, but the latter was prevented from seeing out because he was not given the luxury of windows. As a result he had to operate purely on his instruments and equipment, one part of which was the special Initsiativa navigation and targeting radar connected to an optical bombing sight. By May 1955 some of the material required to complete a preliminary project had been assembled but, despite discussions between Tupolev, the Air Force and other officials, this was as far as 'Aircraft 112' was to go. To fill its front-line supersonic requirements the Air Force plumped for the Yakovlev Yak-25 and its follow-on Yak-26 and Yak-28 developments (below), which spelt the end for 'Aircraft 112' and the lightweight 'Aircraft 98A' (Tu-24) version of the *Backfin* bomber proposed at around the same time and described in Chapter 4.

Yakovlev Yak-125B

During the 1950s and 1960s a long series of related interceptor fighter and tactical bomber types was built by Yakovlev, which began with the Yak-120 interceptor that first flew in June 1952 and entered service as the Yak-25 *Flashlight*. In due course a prototype tactical photographic reconnaissance derivative called the Yak-125 was built and then an official directive requested that a 'special-mission' version of this aeroplane, called the Yak-125B, should also be developed. The 'special mission' was to penetrate enemy defences and deliver a small tactical nuclear weapon behind their lines and, to achieve this objective, the aircraft had to be a two-seater. The bomb was housed in a bay near the aircraft's CofG and a ground-mapping and bomb-aiming radar was introduced. This aeroplane made its maiden flight in June 1955 but, by then, the Yakovlev design bureau was thinking in terms of supersonic performance and a more capable alternative was being considered. Consequently, the subsonic Yak-125B (which would have been designated Yak-25B had it entered production) remained purely a prototype.

Yak-123 (Yak-26)

Encouraged by the success of the Yak-25 fighter, Yakovlev continued to try to turn the type into a tactical bomber. By now Ilyushin's Il-28 (Chapter 2) was beginning to show signs of age and neither Tupolev nor Ilyushin had a ready replacement in the pipeline. For Yakovlev the solution was to increase the Yak-125B's speed by installing afterburning engines and the resulting project began life as a private venture. However, official support for it, and the associated Yak-122 reconnais-

Tupolev 'Aircraft 112' (1954/55).

This view shows the new nose and nuclear warload of the Yakovlev Yak-125B prototype.

A production Yak-28I *Brewer-C* tactical bomber.

sance project, was forthcoming on 30th March 1955 when the Council of Ministers issued a directive covering prototypes of each version. (Note: the Yak-121 interceptor project became the Yak-27.)

The main weapon was to be a single RDS-4 'special store', which in fact was a 30-kiloton nuclear warhead developed by the Khariton design bureau for fitting to both ballistic missiles and free-fall bombs. This was the Soviet Union's first production nuclear warhead and an improved version, the RDS-4T, was installed in the Tatiyana bomb which entered the Soviet Air Force's inventory in the mid-1950s; both looked like normal high-explosive bombs and weighed about 2,646 lb (1,200kg). The first Yak-123 was to be provided with two Tumansky RD-9AK turbojets and, in service, the aircraft would be known as the Yak-26.

The development of the Yak-123 light tactical bomber was completed in 1955 and the aeroplane looked very similar to the Yak-125B but with more streamlined engine nacelles and a new pointed nose. Rolled out in 1955, the prototype (also known as the Yak-123-1) completed a brief manufacturer's flight test programme and then moved on to its state acceptance trials. Even with interim engines the Yak-123-1 was much superior in speed and service ceiling to the subsonic Il-28, but the new type was also found to be suffering from several major aerodynamic weaknesses. Nevertheless, Yakovlev persuaded the Defence Industry Commission to order a pre-production batch of Yak-26s, for which a directive was issued on 5th January 1956. The aircraft's wings, tail unit and forward fuselage were then redesigned and a second prototype was built called the Yak-26-3. As modified the aircraft was much improved but some of the flaws remained and the Air Force would not accept the now completed pre-series aircraft, although some were eventually used for trials. Modification work continued, the upgraded aeroplanes reaching an all-up-weight of 25,353 lb (11,500kg), but by now Yakovlev was working on the far superior Yak-129.

Yakovlev Yak-129 (Yak-28)

The rapidly increasing power made available by the latest developments in jet engines eventually allowed Soviet officials to issue a directive on 28th March 1956 for a new high-altitude supersonic tactical bomber based on the Yak-26. This would employ two Tumansky R-11-300 engines and so, coupled with the design team's recent experience on the Yak-26, the project was able to move forwards reasonably quickly. Aleksandr Yakovlev himself was not keen to retain the basic layout but, by undertaking a thorough redesign, the bureau's engineers were able to re-use the twin-engine swept wing arrangement and rebuild one of the Yak-26s as the prototype Yak-129. This aircraft made its maiden flight on 5th March 1958 and the service designation Yak-28 was allocated to the type. Flight trials proved successful and, as a consequence, for the first time Aleksandr Yakovlev began to show real interest in the aircraft.

There were still problems to be overcome, not least with the engines, and a second prototype was constructed from another pre-series Yak-26. Factory No 39 at Irkutsk was tasked with the Yak-28's quite complex production and the aircraft received the Western codename *Brewer* after it had made a public debut at the 1961 Tushino Air Display. The bomber entered service soon after Tushino and was built in several versions; from the mid-1970s these were gradually replaced by the Sukhoi Su-24 described shortly. Like the Yak-26, this aircraft had a single defensive 23mm cannon mounted in a tail barbette. Initially the Yak-28 was seen, at least in some official circles, to be something of a stopgap. In service it proved disappointing because it suffered not only from a short range and a small war load but also from severe restrictions in the firing of its guns and in the accurate delivery of its bombs. In addition, bomb loading was made difficult and inconvenient because of the aircraft's extremely low ground clearance.

Yakovlev Yak-2VK-11

Another 'Yak-28-type' project, this was a proposed supersonic two-seat high-altitude front-line bomber with an estimated performance and capability superior to the Yak-28's specification. The powerplant was to comprise a pair of Klimov VK-11 engines, a unique power unit in that it used a centrifugal compressor followed by an axial compressor. The VK-11 gave 9,920 lb (44.1kN) of thrust dry and 19,840 lb (88.2kN) in reheat and the bomber's maximum speeds with dry and wet settings were expected to be 808mph (1,300km/h) and 1,554mph (2,500km/h) respectively. A normal bomb load would be 2,646 lb (1,200kg) and the maximum 6,614 lb (3,000kg), while the aircraft's estimated ceiling was 65,617ft to 68,898ft (20,000m to 21,000m) and range (with a 5% reserve of fuel) 1,554 miles (2,500km); a planned reconnaissance version would have the same range but was to be capable of reaching 72,178ft (22,000m).

A SovMin resolution was passed on 15th August 1956 which requested that five pre-production aeroplanes were to be built, with the first example ready to make its maiden flight in the first quarter of 1958. This would then be ready for its state acceptance trials in the final quarter of 1958, while the first reconnaissance variant would go for its Air Force evaluation during the third quarter of that year. However, work on the project did not progress much further because the engines were never delivered. Examples were bench tested in 1956 but, despite also being allocated to power the Mikoyan Ye-150 and Sukhoi T-37 heavy interceptors, no units were ever flight tested.

Yakovlev Yak-30/Yak-32/Yak-34

In 1957 the Yakovlev OKB began work on a two-seat supersonic reconnaissance aircraft which, in due course, was examined as a possible bomber. The new design had a configuration that was, overall, similar to the Yak-28, but the aerodynamics had been improved and the wing itself was bigger (up to 753ft^2 [70m^2]) and swept back 45°. The first wind tunnel testing of this project, the Yak-30, began in September 1957 while the design calculations suggested a maximum take-off weight of 76,940 lb (34,900kg), fuel load 24,250 lb (11,000kg), top speed 1,523mph (2,450km/h), service ceiling 58,727ft (17,900m) and range with internal fuel 2,300 miles (3,700km), or 3,294 miles (5,300km) when carrying external tanks. A further Yak-30V development was tunnel tested between August and October 1959 and was expected to carry 35,273 lb (16,000kg) of fuel.

The Yak-32 variant, the design of which was the responsibility of a team led by N G Kolpakov, actually reached the preliminary project stage and was approved by Aleksandr Yakovlev on 25th May 1959. Again intended to serve as an all-weather reconnaissance type for day and night operation from unpaved runways, the Yak-32's development was authorised by a SovMin resolution of 31st July 1958. Since it was to fly at very high supersonic speeds, the airframe was to make use of some titanium alloy and steel in those parts of the structure that were liable to be affected by kinetic heating. There were two possible alternative powerplants, a pair of Klimov VK-13 or Lyulka AL-7F engines, while the wing area was now quoted as either 645ft^2 (60m^2) or 753ft^2 (70m^2).

Thickness/chord ratio was 4.2% at the root and 4.0% at the tip and leading edge root extensions were fitted, together with forward extensions to the outer wing leading edges and a variable incidence tailplane, to improve the Yak-32's aerodynamics when flying at high angles of attack, improve its manoeuvrability and also its stability in the stall and at low speeds. Maximum internal fuel capacity was 16,534 lb (7,500kg), rising to 23,148 lb (10,500kg) with external tanks, and these gave respective normal take-off weights of 51,808 lb (23,500kg) and 59,524 lb (27,000kg) and ranges (with a 7% fuel reserve) of 1,616 miles (2,600km) and 2,237 miles (3,600km). Service ceiling was 68,898ft (21,000m) and the first machine was expected to undertake its state acceptance trials during the first three months of 1960.

This front-line reconnaissance project appears to have been the subject of a lot discussion at high level, primarily to see if it could be adapted as a bomber. Such a move was a definite possibility but Konstantin Vershinin, VVS Commander-in-Chief, wished to see the Yak-32 developed from the start as a bomber and then turned into a reconnaissance aircraft later on. He felt that this would be a much easier way of doing the job and, to back up his opinions, cited Yakovlev's own wartime BB-22, Yak-2 and Yak-4 twin-piston-engine programmes which all began as reconnaissance types before being turned into bombers – they were not a great success. The Yak-32's potential bomb load would have been about 6,614 lb (3,000kg) maximum.

In the event the Yak-32 was abandoned as part of the budget cuts described shortly, the gap being filled by the Sukhoi Su-7B. However, Yakovlev persevered with the more advanced Yak-34 which had an estimated top speed of 1,865mph (3,000km/h), a maximum range of 2,113 miles (3,400km), or 1,367 miles (2,200km) when flying at 1,554mph (2,500km/h), and a ceiling approaching 72,178ft (22,000m). A further design, the Yak-34R with two Metskhvarishvili R21-300 engines, was proposed in March 1962 and it was hoped that this aircraft would begin its flight testing during the last three months of 1965, but the programme was closed in 1963.

Sukhoi Su-7

After Pavel Sukhoi had re-established his own design bureau in May 1953, one of his first tasks was to design and produce two new supersonic fighters for front-line and all-weather duties. The resulting aircraft were the swept wing Su-7 and delta wing Su-9, both of which entered production. Development of Sukhoi's S-1 design began in August 1953 and the first prototype flew on 7th September

The Sukhoi S-22-2 was the second Su-7B prototype.

1955. The second prototype, the S-2, had a longer fuselage with more fuel and it was this version that was put into production in 1957 as the Su-7 front-line fighter. However, during 1956 new VVS requirements were raised for a fighter-bomber type and the Air Force's experts declared that the ideal solution would be a specialised version of the Su-7. As a result, in March a SovMin resolution was passed authorising the go-ahead for this variant with a warload composed of various bombs and missiles up to a maximum of 4,409 lb (2,000kg).

Design work began in 1957 and the first prototype, a converted early production Su-7 given the in-house designation S-22-1 by Sukhoi, was finished in the autumn of 1958. However, delays and some extra work prevented the aircraft's maiden flight from taking place until 24th April 1959 but this machine and a second prototype successfully completed their state testing in April 1960. In due course the project was cleared for production and in early 1961 the first example was delivered for service evaluation. The type was built in large numbers and several more versions of the basic aircraft, with additional fuel, and the like, and for export, were developed. However, the Su-7's configuration could only be improved up to a certain point and it was not until the introduction of a variable sweep wing (on the S-22I prototype described later) that a new generation of fighter-bombers could follow.

The Su-7B arrived during an important period in Soviet aviation 'politics', which has been discussed previously but proved critical to the USSR's tactical aircraft forces. During 1956 and 1957 an assessment was made covering the re-arming and upgrading of the various arms of the Soviet Air Force. For the heavy strategic and long-range bomber force the objective was to complete the development of the Myasishchev M-56 and Tupolev Tu-105 (Chapter 4). In addition the introduction of medium-range cruise missiles aboard the Tupolev Tu-16 (Chapter 3) plus new types of land-based long-range cruise missiles (the Lavochkin Burya and Myasishchev Buran) would cover any possible offensive target. However, the VVS's front-line strike forces were not so healthy.

In May 1957 Marshal V D Sokolovsky reported that the 'new generation' Yak-26, Ilyushin Il-54 and Tupolev Tu-98 (Chapter 4) all fell short of their requirements, the Yakovlev because it was too slow and could not deliver enough ordnance and the others because their airframes were too heavy which restricted the bases from where they could operate. During 1956 the Air Force had put together some new requirements but neither of the preliminary proposals produced by Ilyushin (which project this was is currently unknown) and Tupolev (possibly 'Aircraft 122' described in Chapter 4) offered speeds and ceilings that were considered suitable. As noted above, Yakovlev carried on with its Yak-28 studies and also produced a new project, the Yak-32.

Because of this relatively bleak picture, the modified Su-7 appeared to be a good move. Potentially it offered a slightly smaller warload than the Yak-26 but it could fly at high speeds at low level and also make use of unpaved runways. Yakovlev answered with a proposed fighter-bomber development of its Yak-35MV low-level fighter interceptor project, which itself was still a paper design, but this was seen to be 'little different' from the Su-7B when Sukhoi's aircraft was much further ahead in its development and was, in fact, 'almost ready'.

The successful testing of the first ballistic missile systems further complicated the issue and, as described in previous chapters, several strategic bomber programmes were duly terminated while Tupolev Tu-16 production was cut back in favour of the Tu-105 missile launcher. However, prior to this K A Vershinin, from January 1957 the VVS Commander-in-Chief, insisted that all of his fighter-bomber regiments, thirty in total, should convert to the Su-7B while the VVS's Headquarters itself estimated that a full re-equipping of the Air Force's fighter regiments over five years could require 14,000 new aeroplanes. Incredibly, on 22nd February 1958, the Soviet Defence Council approved in principle all of the VVS's 'wish list', together with other major programmes for land-based surface-to-air missiles, nuclear submarines, and so on, but by the middle of the year Khrushchev and his staff had realised that these acquisitions would be phenomenally expensive.

It was here that Khrushchev decided to introduce his massive cut-backs by reducing expenditure for aircraft development and production, closing several OKBs and factories and disbanding some Air Force regiments. The VVS's front-line units suffered accordingly – for example an Il-28 bomber division was calculated to cost 1.5 times the money of three FKR-1 cruise missile regiments (a weapon which began testing in 1959) but the argument was lost that aircraft were far more flexible than missiles. (At this time these same arguments were on going in several other Western countries, not least the United Kingdom). Those ambitious re-equipment plans were indeed heavily curtailed and the Yak-32, Yak-35MV and the proposed strike version of the Tupolev Tu-28 heavy interceptor were all cancelled. Only the Yak-28 and Su-7B survived, although many in the VVS did not like the Yakovlev type; R Ya Malinovsky for example, when chairing a VVS Military Council meeting in February 1959, declared that 'in the near future we are about to receive a new front-line bomber, the Yak-28, which, however, is not what we really want'.

Nevertheless, although both Yak-28 and Su-7B were initially considered as stopgaps, they both achieved long careers. Exported examples of Sukhoi's bomber saw combat during the Arab-Israeli and India-Pakistan conflicts and many of the Soviet Union's own front-line Su-7Bs were not withdrawn until the mid-1980s.

Sukhoi Su-17

Thanks to the need for multi-mission aircraft like the Su-7B to be capable of high supersonic speeds, such types brought with them a long take-off run and, as a consequence, needed rocket-assisted take-off gear to help them get airborne in a suitably short distance. At this time high speed and good airfield performance were a difficult combination to put together into a single type of aeroplane, but there were alternative ways of dealing with the problem and one was to fit variable geometry wings. In the forward position these gave more lift and allowed the aeroplane to become airborne within a reasonably short distance; when swept fully back the aircraft could fly at high speed.

Sukhoi's solution was to redesign the Su-7 with a variable geometry wing and a development programme, with Sukhoi's deputy N Zyrin as the Chief Designer, got moving in 1965. A good deal of wind tunnel research and preliminary design was required before the team settled on the most suitable wing arrangement. Then a production Su-7BM, which had already been used in a variety of experimental roles, was rebuilt as the S-221 experimental fighter-bomber prototype and, as the Soviet Union's first swing-wing aeroplane, this made its maiden flight on 2nd August 1966. The resulting flight test reports showed that the aircraft presented a considerable improvement over the Su-7B and, despite being something of an interim aircraft, an official decree was duly issued clearing the way for the type to enter production. Known in-house as the S-32, the first production aeroplanes were manufactured in 1969 and the type received the service designation Su-17; modified versions went for export as Su-20s and Su-22s. In the meantime the basic design was improved and upgraded until the S-52 first

Early production example of the Sukhoi Su-17 fighter bomber.

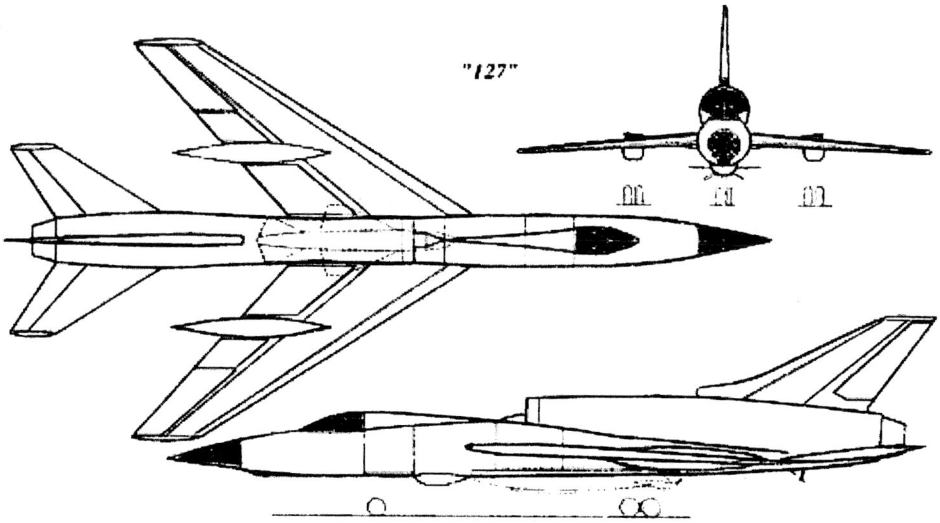

Tupolev 'Aircraft 127'.

flew in 1976 with a modified forward fuselage. Series production of this variant also began in 1976 and the type entered Soviet service as the Su-17M3; it was exported as the Su-22M.

Tupolev 'Aircraft 127'

At the beginning of 1958 Tupolev began some preliminary studies into a new front-line supersonic light bomber and missile-carrier project called 'Aircraft 127'. The concept stipulated an intermediate class strike aircraft that would fall between the Sukhoi Su-7B (S-22) fighter-bomber and Tupolev's own 'Aircraft 98' (Chapter 4). In February S M Yeger's department defined the '127's dimensions, layout and estimated flight performance and operational characteristics. The preliminary calculations suggested a two-seat single-engined multi-purpose aircraft with a mid-position 55° swept wing and a low swept tail. 'Area rule' was applied to the airframe and a single VD-7M turbojet, giving 35,270 lb (156.8kN) of thrust in reheat, occupied the entire rear fuselage with its dorsal air intake positioned behind the cockpit cabin.

There was a mid-fuselage cargo 'compartment' situated beneath the wing centre plane which, depending on the required tactical mission, housed either a P-15 air-to-ground missile and an auxiliary fuel tank, a total of 4,409 lb (2,000kg) of bombs, or a set of reconnaissance equipment. The front part of the fuselage contained the pressurised pilot and (fully glazed) navigator's cabins, the latter equipped with an optical bomb sight, and in addition a fairing was mounted under the navigator's cabin to house the missile targeting system. All of the fuel, 12,125 lb to 13,228 lb (5,500kg to 6,000kg) in total, would be contained inside the fuselage and a Tupolev-style tricycle landing gear retracted into the forward fuselage and wing-mounted fairings. The undercarriage had high-pressure tyres (two-wheel nose and four-wheel main gears) which allowed the designers to keep the size of the retracted wheel compartments to a minimum. An Initsiativa targeting and navigation radar was included, a standard feature of Tupolev's previous front-line bomber projects, and this was supplemented by the required missile guidance and targeting equipment.

The design bureau also prepared a further development of the '127' called 'Aircraft 129',

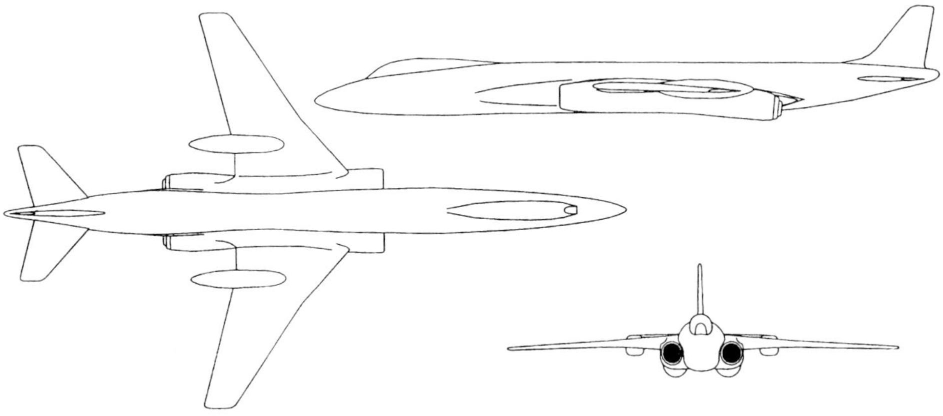

Tupolev 'Aircraft 132' (1958).

Artist's impression of Tupolev's 'Aircraft 132'.

the layout for which resembled the American Republic F-105 Thunderchief tactical fighter-bomber; however, neither 'Aircraft 127' or '129' progressed beyond the preliminary design stage. 'Aircraft 127's estimated service ceiling was 59,055ft (18,000m) while its range would be 1,492 miles (2,400km) when flying at subsonic speed and 497 to 684 miles (800km to 1,100km) at supersonic speed.

Next Generation

From the late 1950s and well into the 1960s, Soviet proposals for new tactical strike aeroplanes brought forth some very capable designs and eventually led to Sukhoi's Su-24, which proved a great success.

Tupolev 'Aircraft 132'

By the late 1950s the ever improving capability of enemy ground-based anti-aircraft missile defence systems, and the air-to-air missile-equipped supersonic high-altitude interceptors that would go alongside them, had greatly reduced the chances that bombers and strike aircraft flying at middle and high altitudes could safely break through to their targets. In consequence, the world's leading aviation powers began to develop bombers and cruise missile carrier aircraft that were capable of reaching their targets when flying at transonic speeds and at low or even extremely low altitudes. The capability to do this would significantly reduce the chances of a successful interception by the enemy's defences and also improve the survivability of the strike aircraft themselves.

As a result, the United Kingdom began its TSR-2 strike aircraft programme while, in the USSR, Pavel Sukhoi's design bureau began its first studies into what was to become the T-6 (below). In 1958 the Tupolev design bureau began work on the Tu-132, or 'Aircraft 132', low-altitude transonic bomber. On 31st July of that year the USSR Council of Ministers issued a Decree which ordered the Tupolev OKB, as the leading organisation, to develop in co-operation with TsAGI, TsIAM and the NII-17 and NII-2 Scientific Research Institutes, a preliminary project for a long-range bombing system capable of undertaking missions at low altitudes between 656ft and 1,640ft (200m to 500m) and, when flying at these altitudes, to have a maximum range of 2,797 and 3,108 miles (4,500km to 5,000km).

This work lasted for two years and was carried out under the leadership of S M Yeger. The '132's preliminary project was ready by the beginning of 1960 and the resulting aerodynamic layout was close to that used in the initial projects prepared for Tupolev's 'Aircraft 105' powered by two VD-5F engines (Chapter 4). However, because 'Aircraft 132's main cruise condition was to be transonic flight at low altitudes, its structure was designed to withstand the long-term effects of high dynamic air pressures, intensive short-term vertical loads and long-term vibration loads, all of these being features that came with prolonged flight at low altitude and which had to be taken into account.

'Aircraft 132' had a clean area-ruled airframe with a cranked low aspect ratio swept wing, the inner portion being swept back 45° and the outer 35°. The front fuselage housed a radar and a pressurised cabin for the two crew and the main landing gear nacelles were attached to the wing. The aircraft's powerplant was to be two reheated by-pass jet engines (it appears that a specific engine was never actually chosen) while its flight-control and navigation systems were intended to allow low-level flight when the machine was in either an automated or manual semi-automated mode.

The following versions of 'Aircraft 132' were considered – a bomber to deliver general purpose and nuclear bombs, a cruise missile carrier, tactical strike aircraft, reconnaissance aircraft and an anti-submarine aircraft. All of the offensive stores were to be carried inside a weapons bay which, depending on the version concerned, could house a cruise missile, general purpose or nuclear bombs, rocket projectiles in retractable launchers, a container with photo and electronic reconnaissance equipment, anti-submarine buoys, torpedoes and depth charges or chaff/decoy flare dispensers. 'Aircraft 132' was not provided with any defensive cannon because it was thought that flight at low-level and high speed, combined with a low degree of radar reflection from its relatively small size plus the use of radiation absorption materials and a radar jamming system, would give the aircraft a sufficient degree of protection.

The new technology required for this type of aircraft included the preliminary development of special radars by the Soviet Union's

Two versions of Sukhoi's S-6 strike aircraft project (early 1960s). Note the different air intakes.

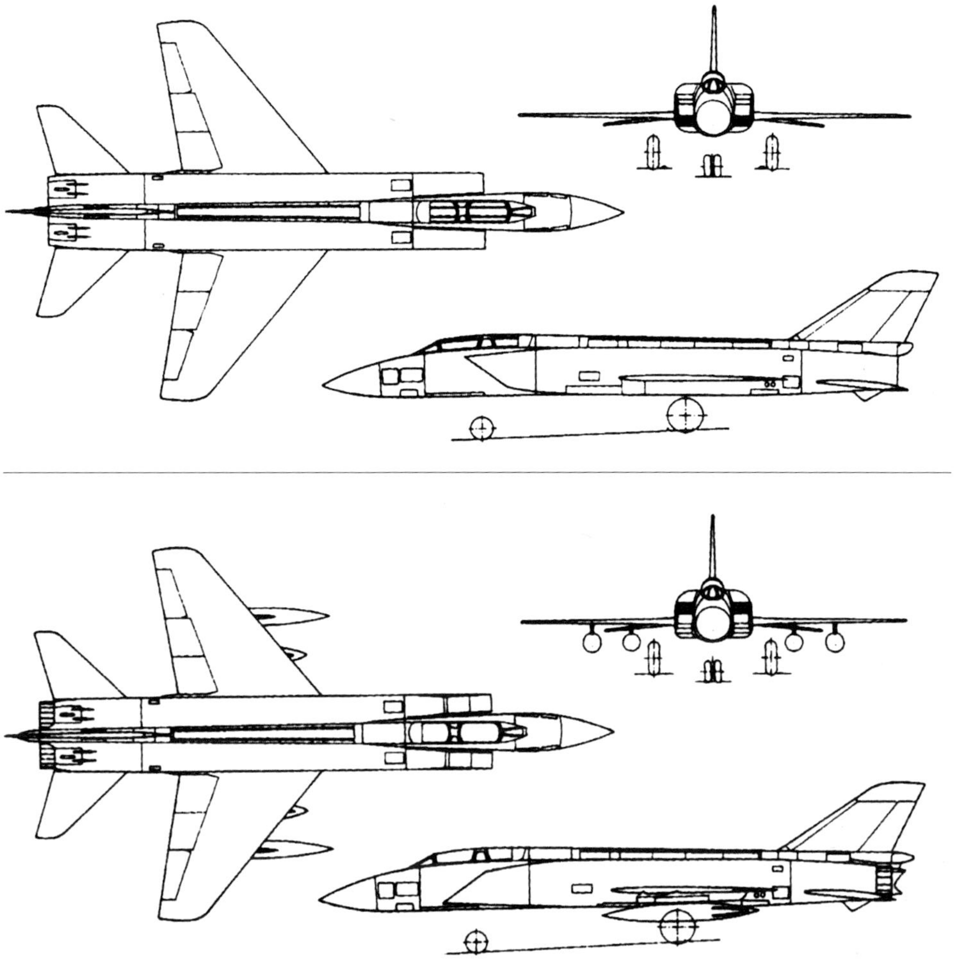

radio-electronic industry for low-level and extra-low-level flight. In addition, in co-operation with the Air Force, research was undertaken into the various possible applications of the aeroplane which, since it was such a new concept, would break a lot of new ground. However, the work required to bring the 'Aircraft 132' system to fruition, particularly in the area of low-level flight, considerably surpassed the then current capabilities of the USSR's electronics industry and so the project did not move beyond the stage of preliminary development. However, many important lessons were learnt from this research and these were used to good effect in the design of the combat aircraft that followed. The '132's normal take-off weight was estimated to be between 88,183 lb and 99,206 lb (40,000kg and 45,000kg) and the aircraft was indeed expected to cruise at an altitude of around 656ft and 1,640ft (200m to 500m); dependent on the weapon load, its range when flying at these heights would be between 1,865 and 3,108 miles (3,000km to 5,000km).

Sukhoi S-6

By the mid-1960s, two important factors were becoming clear. First was the undoubted superiority over Soviet equipment of equivalent USAF types like the General Dynamics F-111 first flown in 1964. Such aircraft could outperform any current Soviet type, carry a wider range of weapons and, perhaps most important of all, they possessed much superior avionics. The second point, already touched on, was rapid development in defensive surface-to-air missile capability. Active radar-homing missiles were now capable of destroying any aircraft at the altitudes at which current Soviet types had to fly and, therefore, new methods had to be found for flying safely at supersonic speeds underneath the enemy's radar screen and also to locate and destroy a target from such a low altitude.

The solution was to produce an aircraft with a significantly stronger airframe capable of withstanding the buffeting imposed by low-level tactical manoeuvres, although a reduction in range and weapon load could not be tolerated and so the extra structure weight would have to be offset by carrying less fuel and using more fuel-efficient engines. In addition, low-level flight at supersonic speeds, even for short periods, imposed a massive strain on the pilot and to persevere under these conditions long enough to reach the target would not be possible without automatic systems to navigate and control the aeroplane. At this time such equipment did not exist in the USSR and it was evident that new 'weapons systems', combining the aircraft, avionics and weapons together, were needed urgently.

Two design bureaux, Mikoyan and Sukhoi, were invited to meet the challenge. The former focused on short-range fighter-bomber derivatives of existing interceptors and ultimately produced the MiG-27 *Flogger-D*. In addition a new specialised air defence suppression (Wild Weasel) variant of the MiG-25 *Foxbat* long-range heavy interceptor was developed called the MiG-25BM, and both of these types are described shortly. Sukhoi was already working on a replacement for the best current Soviet fighter-bomber, the Su-7BKL, and the result was the Su-17 above, but the bureau also decided to design a longer-range interdiction bomber.

The objective was to find a match for the F-111 and the initial research included variable-geometry wings and auxiliary lift engines, but the OKB's limited experience in both of these fields at that time led them to favour a configuration similar to that of the successful Su-7B with a fixed wing of moderate aspect ratio but two engines. The Soviet Government wanted the new type to be capable of day, night and all-weather operations, which meant that a radar would have to be housed in the nose, so an Su-7B-style nose intake was out. In 1963 Oleg Samoilovich, one of the bureau's designers, completed four draft designs for what the OKB called project S-6 (S for *strelovidnoye* or swept wings) and one of these was then produced as a full-scale mock-up. This showed a mid-fuselage fixed 40° swept wing, an all up weight of 44,092 lb (20,000kg) and an estimated top speed at sea level of 870mph (1,400km/h), rising to 1,554mph (2,500km/h) at height.

There were two crew seated in tandem and two lateral air intakes fed the engines placed side-by-side at the rear. Fuselage cross section was rectangular with rounded corners, the integral cockpit canopy had two hinged doors and the nose radome housed the antennae for the specially-developed

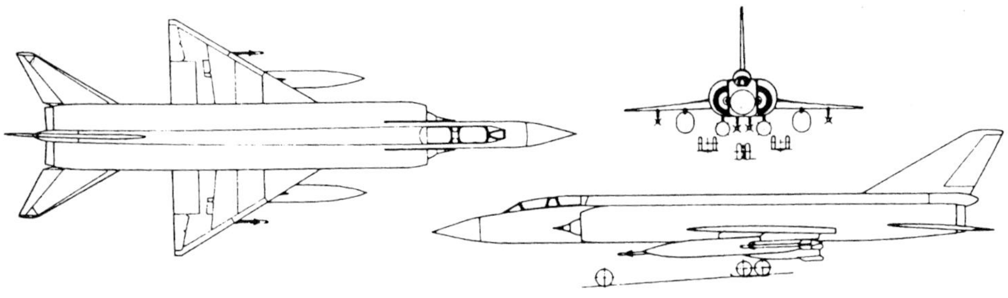

Sukhoi T-58M (mid-1960s). Russian Aviation Research Trust

Puma computerised navigation and attack system. The S-6 was fitted with a tricycle landing gear which had single mainwheels and twin nosewheels and power was supplied by two R-21F-300 two-spool turbojets rated at 15,870 lb (70.5kN) thrust in full reheat; two large additional rocket boosters were also provided for short take-off assistance.

The R-21F-300 was a new engine developed by N Metskhvarishvily as a derivative of the Tumansky R-11F and had also been used by the experimental Mikoyan Ye-8 fighter. However, the engine's reliability on the Ye-8 had been poor (an engine explosion caused the loss of the first prototype) but Metskhvarishvily assured Sukhoi that the engine had now been modified to eliminate this problem. The S-6 could carry three drop tanks, two under the wings and one under the fuselage, and four underwing and one underfuselage hardpoint were provided for bombs and guided or unguided missiles up to a maximum weight of 6,614 lb (3,000kg). After a protracted but unsuccessful struggle the designers decided that the S-6 configuration could not be made to satisfy all of the customer's requirements and they began to look at some other alternatives.

It was now clear that a conventional layout was inadequate for this project and attention moved towards variable geometry wings. The first exploratory steps were now being taken with the S-22I (Su-17) prototype (above) with the refitting of a modified Su-7B with swing wings placed at approximately half span, but the principle was still unproven. The other promising area was still the lift engine and trials were also now under way using Sukhoi's T-58VD technology demonstrator, a much modified Su-15 *Flagon-A* interceptor (T-58 was the in-house designation for the Su-15, T stood for *treoogol'noye* or delta wing and VD for *vertikahl'nye dveegateli* or vertical engines or lift-jets). This aircraft used three 5,180 lb (23.0kN) Kolesov RD36-35 lift units housed in its centre fuselage with air supplied through two intakes closed by aft-hinged doors positioned on the fuselage upper surface; vanes for vectoring the engine thrust were located under the belly.

Although the use of fixed vertical lift engines reduced the fuel capacity, and thus the aircraft's range, this was felt to be a reasonable trade-off against the calculated reduction in take-off and landing distances of 57% and 40% respectively. A key factor here was the Air Force's complex requirements which included a take-off run of no longer than 1,312ft (400m) and Air Force command was recommending VSTOL capability to meet it. (Mikoyan also evaluated vertical lift engines with prototype versions of the MiG-21 and MiG-23 fighters called the MiG-21PD/23-31 and MiG-23PD/23-01 respectively. Both machines used two RD36-35 lift jets installed in the centre fuselage and first flew on 16th June 1966 and 3rd April 1967 respectively.)

Sukhoi T-58M

Inspired by this Sukhoi's designers looked at several new layouts, including the T-58M, an enlarged T-58VD with semi-circular air intakes plus rectangular-shaped ducts further aft. The aircraft's two main propulsive units were housed side-by-side and the Su-15's delta wing was preferred together with a conventional low-position swept tailplane. Again the crew were seated in tandem and the tricycle undercarriage main gears had four wheels, the nose gear two. There were positions for four drop tanks and the T-58M was intended to carry two Kh-56 air-to-surface missiles on the inner wing pylons plus two Kh-24s, for which the exact location was never finalised. These would either have been loaded under the fuselage on an 8U46 ejector rack, or on the outer wing pylons but, in the event, no final decision was required because not only were the Kh-24s themselves never built, but the entire T-58M project was discarded on cost grounds.

Sukhoi T-58T/T-6

The result was that work now began on another completely new design, the T-6 research aircraft fitted with lift engines. This project was first proposed in 1963 as the T-58T and included a Zazorin Puma radar, two AL-21 drive engines and four RD36-35 lift units. However, very soon it was upgraded to Top Secret status and accordingly redesignated T-6. For the first time the design bureau used a new technique to co-ordinate the various assemblies where a drawing board nearly 50ft (15.2m) in length held a reduced-scale side view of the aircraft drawn onto a reference grid. On this were the positions for all of the components and control circuits, together with their attendant ducts, piping and wiring, and even the undercarriage was included in the retracted position; full-size detail cross sections were also drawn.

A mock-up was completed in 1966, the first prototype (designated T-6-1) was completed in the late spring of 1967 and the aircraft flew for the first time on 2nd July. It was originally intended that T-6-1 should participate in the Domodedovo Air Display held on the 9th June but the aircraft was not ready; however, the show had to go on so a substitute Su-15 fighter was painted all-black, displayed at very low heights and described as the 'new Soviet Attack Aircraft'. Two Tumansky R-27F2-300 turbojets formed the T-6-1's primary powerplant, each fed by lateral variable air intakes, while, behind the cockpit, came the four RD36-35 lift engines inclined 15° forwards and mounted in two compartments, one to either side of the aircraft's CofG. Air entered the lift engines through two extended scoops on the upper fuselage and exited via nozzles closed by rotatable doors; instead of radar, the nose cone contained test instruments. T-6-1 was intended to carry, on four underwing and two underfuselage hardpoints, guided and unguided air-to-surface missiles, air-to-air missiles and all types of bombs and other stores. The performance figures for this aircraft have never been released.

One area of the T-6-1's design that gave a lot of trouble was the undercarriage. However, Oleg Samoilovich visited the 1967 Paris Air Show where he was able to see some of the latest Western technology, including the Anglo-French SEPECAT Jaguar light strike aircraft. Samoilovich was impressed by the Jaguar's undercarriage arrangement and persuaded Pavel Sukhoi to fit a similar format to the second T-6 (T-6-2/Su-24) prototype. This was done and proved a success, and some elements of the design also found their way onto the Su-25 attack aircraft (Chapter 7).

The original engine selected to power the T-6-1 had been the Lyulka AL-21F three-spool turbojet but delays forced the interim fitting of

The Sukhoi T-6 prototype, showing the downturned wingtips.

This production Sukhoi Su-24 strike aircraft is 'armed' with a selection of dummy Kh-29 series air-to-surface missiles.

two R-27F2-300s. The Air Force had wanted the aircraft to be capable of five minutes low-level flight at supersonic speed and Lyulka was instructed to design and produce a new engine to match the specification. The resulting AL-21F was expected to give 19,620 lb (87.2kN) of thrust and was based on the American General Electric J-79 which powered the McDonnell-Douglas F-4 Phantom fighter, and for which some examples were available for examination. When it appeared that the T-6-1's take-off weight would be more than originally estimated, Lyulka was asked to increase the AL-21F's full afterburner power to 24,690 lb (109.7kN). Then the low-level supersonic requirement was dropped altogether, which meant that Lyulka had wasted a considerable amount of time creating engine types that were not required.

By 1969 the AL-21F's were ready, each of them producing another 2,200 lb (9.8kN) of thrust over the earlier Tumansky units. In addition, while fitting the new engines, the designers were able to reduce the aircraft's drag and fit downward-pointing tips to the wings (the latter were required to solve problems with the aircraft's transverse stability). Various other modifications were introduced but the extra thrust ensured that there was no loss of performance. However, the operation of the vertical lift engines did make landing the aircraft very difficult.

During the T-6-1's trials programme the Soviet Air Force once again changed its requirements, the most important item being a substantial increase in combat load which meant that lift engines were no longer viable. Two further requirements, seemingly contradictory, were the need to attack at ground level at transonic speeds and the large wing area demanded for STOL, which would mitigate against a smooth ride for the attack. Fortunately, for the sanity of the Sukhoi design team, TsAGI had just completed a new and rigorous study into the aerodynamics of twelve different airframe shapes and the variable-geometry wings compared so favourably against the others that Sukhoi now designed a version of the T-6-1 with swing-wings (Mikoyan also switched to variable geometry for its new interceptor project, the future MiG-23).

The decision to concentrate on variable sweep wings was taken before the end of 1967, less than six months after the T-6-1's first flight. In December work began on building a new prototype and most of the subsequent test flying carried out between 1967 and 1970 on the T-6-1 was centred on testing and developing the new AL-21F engines. After this the sole T-6-1 served as a testbed for new avionics before being retired to the Air Force Museum in Monino in 1974.

Sukhoi Su-24

The second T-6 prototype, T-6-2, introduced the variable sweep wings. During his visit to the Paris Show Samoilovich had also been able to take close-up photographs of the variable geometry wings fitted to the American F-111 and these helped him to complete the design used on the T-6-2. He also wanted the crew to be seated in tandem, which left room to fit avionics along the fuselage sides and would also help with maintenance, but Pavel Sukhoi ordered a side-by-side arrangement to be adopted which held other advantages such as removing the need for a training version of the aircraft, reducing the duplication of instruments, and the like.

Such was the Air Force's need for this new aircraft that in August 1968, around the time design work really got moving and well before first flight, the swing-wing T-6 was ordered into production as the Su-24. The Sukhoi OKB worked very hard to get it ready, trying as much as possible to retain some of the structure and as many parts as possible from the fixed-wing T-6-1, which proved to be difficult. As a result of this pressure a lot mistakes were made and numerous development problems surfaced, not least a steady increase in weight (in part because the customer wanted the aircraft to carry so many different types of weapon and electronics) so that the original wings and elevators had to be redesigned. In fact some extraordinary innovation was required to get the first two prototypes to fly and in due course ten Su-24s were lost during the development programme, but in 1975 the type was cleared for operational service.

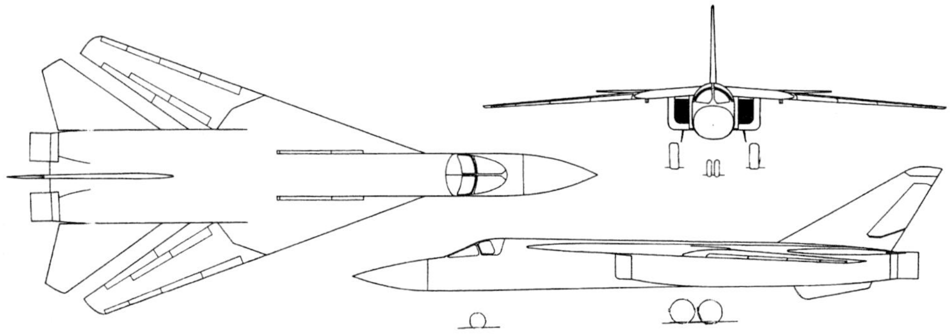

The first Sukhoi Su-24BM design, fitted with variable geometry wings (1979). Russian Aviation Research Trust

Model of the Sukhoi Su-24BM.

Initially the aircraft had been categorised as a ground-attack aircraft but the Su-24 entered service as a frontal bomber and some examples still serve today. The offensive weapon load could include any of the following – 1,102 lb (500kg), 551 lb (250kg) or 220 lb (100kg) bombs, Kh-25 or Kh-58 air-to-surface guided missiles, 370mm, 85mm or 57mm unguided rockets, R-60 air-to-air missiles, and one internally-mounted 23mm six-barrel cannon or three six-barrel 23mm mounted in external pods (all of the missiles and bombs were carried externally). When the Su-24's existence had been established in the West it was codenamed *Fencer* and in 1977 the first prototype of an upgraded version, the Su-24M with improved avionics and a slightly longer fuselage, made its maiden flight.

Sukhoi Su-24BM

In 1979 V Sarov started work on a much larger development of the Su-24M which had an internal weapons bay that could hold most of the weapons intended to be used by the new aircraft. Known within the OKB as the T-6BM (*Bolshaya Modifikatsiya* or large modification), and to the VVS as the Su-24BM, this was a much bigger aeroplane than the Su-24. The weapons bay was located between the engines and the shortened engine air intake channels allowed the intakes themselves to be located below the wings under the wing roots in a similar way to the American General Dynamics F-111; tandem main wheels were also introduced. As the design progressed and the overall weight increased, concern grew as to whether a strong enough wing swivel mechanism could be designed within the specified weight limit. Theoretically it was possible to produce such a mechanism that would work effectively but, with the types of engine currently available, the resulting aircraft may have been too heavy to take-off. The Su-24BM was intended to replace Tupolev's Tu-22M3 (Chapter 11).

General Designer Pavel Sukhoi died in 1975 and his deputy Yevgeny Ivanov was not confirmed as his successor as until 1977. In 1979 the bureau's Mikhail Simonov was seconded to the Ministry of Aircraft Industry as Deputy to Minister Ivan Silaev and, while there, Simonov was impressed by TsAGI's suggestion that the Sukhoi T-4 could be developed into a long-range missile carrier (with the designation T-4MS – Chapter 11). Following Ivanov's decision to retire at the age of 72, Simonov returned to Sukhoi as General Designer in 1983 and he instructed a team to start a new project called the T-60S (Chapter 11), which was to be a supersonic medium-range bomber based on the T-4MS. Thus there were now two project teams within the OKB competing to create the Tu-22M3's replacement – the Su-24BM based on the Su-24M and the T-60S based on the T-4. Powered by two new turbofans mounted above the rear fuselage as on the original Tupolev Tu-22, the T-60S had a delta wing instead of a swing-wing. Canards were fitted just ahead of the wing and the new design was described as 'stealthy'.

Supporters of the Su-24BM were aware of the TsAGI proposal and came to the conclusion that a radical redesign was needed if they were to win the internal OKB competition. A new layout was sketched out that showed fixed-geometry wings with large leading edge extensions, which allowed the troublesome variable geometry wing to be dropped, plus a twin tail unit and also the state-of-the-arts avionics suite destined for the T-60S. In 1983, representatives of the VVS and Ministry of Aircraft Industry accepted the Su-24BM full-scale mock-up and authorised the OKB to proceed with the design but, by this time, the rival T-60S team were slowly but surely gaining ground; in due course, more resources were moved onto this more advanced project. However the Su-24BM was not terminated until the President of the new Russian Federation, Boris Yeltsin, stopped all work on it in 1992.

The dissolution of the Soviet Union in 1991 and the subsequent economic problems that followed resulted in a hiatus in military aircraft production; this situation was, undoubtedly, also a factor in the decision to abandon the Su-24BM. Since funds were scarce and likely to remain so for some time, plans for the Tu-22M3 replacement were delayed. In addition, nothing further has been heard of the T-60S, so it can only be assumed that either this project suffered the same fate as the Su-24BM or that work continues in great secrecy. Approximate dimensions for the VG Su-24BM were span with wings in the minimum sweep position 37ft 5in (11.40m), at maximum sweep 61ft 10in (18.84m) and length 78ft 11½in (24.07m). The model shows a single air-to-air missile and two air-to-surface missiles beneath each wing and two rear-facing defensive cannon in the end fuselage.

Sukhoi Su-24MM

In 1985 Sukhoi proposed a heavier and more powerful version of the Su-24M fitted with the latest navigation and attack systems. An

Model of the Mikoyan Ye-155MF strike aircraft with four Kh-58 anti-radar missiles aboard.

increase of 4,409 lb (2,000kg) to the maximum take-off weight was to be accommodated by fitting the new Lyulka AI-31F engine, which in 1984 had entered series production for Sukhoi's Su-27 fighter. This four stage turbofan had a maximum thrust of 16,770 lb (74.53kN) dry and 27,585 lb (122.60kN) with afterburner and required a large increase in the quantity of air to be supplied by the intakes, which was solved by installing a third intake on top of the Su-24's fuselage. Although rumoured to have received the service designation of Su-24MM (*Malaya Modifikatsiya* or small modification) this is unlikely because no equivalent OKB designation has ever been quoted and the design never progressed beyond the project stage. The most likely reason for its abandonment was that the Su-27 was considered a more logical design on which to base a replacement for the Su-24M.

Mikoyan Ye-155MF

This tactical bomber project, first suggested in about 1969/70, was a development of the Ye-155MP interceptor and was intended to be capable of puncturing enemy air defences at high supersonic speed, neutralising enemy radars and attacking high priority targets with bombs and air-to-ground missiles. The aircraft, designated Ye-155MF (F for *Frontovoy* or frontline), was quite similar to the Ye-155MP fighter prototype (which was also known as the MiG-25MP and eventually became the MiG-31 *Foxhound*) but it had a wider forward fuselage with the two crew seated side-by-side in a manner similar to the Su-24; this give the navigator/weapon system operator a better view. The armament was carried on four underwing hardpoints (typically four Kh58 [AS-11] *Kilter* anti-radar missiles) and in fuselage bays (twelve 551 lb [250kg] bombs). One of the Ye-155MF's main tasks was to have been air-defence suppression but this bomber was never built because Sukhoi's Su-24 was considered to be the more attractive solution. Mikoyan's design was not produced as specific competition for Sukhoi's T-6 series; in fact there were no direct competitors to the T-6 and most of Mikoyan's attack aircraft were more lightweight designs.

Mikoyan MiG-25BM

There was also a specialised defence suppression variant of the very fast MiG-25 *Foxbat*

Mikoyan MiG-25BM defence suppression aircraft.

Mikoyan MiG-27 armed with three bombs plus two rocket-launching pods.

interceptor, the MiG-25BM, which carried four Kh-58 missiles and a powerful set of ECM equipment. This machine was intended to pierce enemy defences to allow MiG-25RB reconnaissance aircraft to pass through and also to search for and destroy enemy radars within a selected area of a battle region. However, particularly through problems with adapting the missiles for anti-radar work, this programme took thirteen years to reach fruition. The prototype was completed in 1976 and all of the production aeroplanes, fewer than one hundred in total, were manufactured by 1985, but problems with training crews delayed the type's service entry until 1988. The MiG-25BM's top speed showed virtually no deterioration over other variants of the MiG-25, that is, around 1,865mph (3,000km/h) at 42,651ft (13,000m).

Mikoyan's first attempt at a defence suppression MiG-25 variant was in fact the Ye-155K project from the early 1970s (K for *Kompleks* to describe air-to-ground capability) which had ECM gear and two Kh-58U air-to-air missiles, but no official interest was forthcoming. However, the design bureau kept trying and the resulting MiG-25BM was finally accepted.

Tupolev 'Aircraft 148'

Tupolev also produced a strike version of one of its fighter designs. Studies for a new heavy interceptor to replace the bureau's own Tu-28 *Fiddler* really belong in an equivalent volume covering fighters, but work began in 1965 and the resulting design had a variable geometry wing. The type's primary task was to be long-range interception but 'Aircraft 148' was also expected to be able to perform attacks on enemy transport operations to the rear of his front lines, tactical bombing operations with both nuclear and conventional stores and to serve as an attack aircraft for hitting areas of the battlefield with poor quality air defences. Reconnaissance was another possibility and for these operations the alternative weapon loads included air-to-surface guided missiles, two tactical nuclear bombs or unguided rockets. The '148' had a length of 106ft 7in (32.5m), maximum span 87ft 3in (26.6m), maximum weight up to 132,275 lb (60,000kg) and an estimate top speed of 1,554mph (2,500km/h); power was to be supplied by two 31,305 lb (139.1kN) thrust Rybinsk RD-36-41 engines which were also used to power the Sukhoi T-4 bomber prototype (Chapter 10).

In fact 'Aircraft 148' was essentially a multi-role aircraft and, at this time, such a type did not fit in with the normal pattern of Soviet Air Force design. The service usually looked for a specialist type to fill a single primary function and, as a result, the VVS inventory always had several different types of aircraft on strength. The value of a multi-function aeroplane was questioned by the Air Force and this weakness was one reason why, later in the year, work was discontinued on 'Aircraft 148' in this form. Instead Frontal Aviation plumped for Sukhoi's Su-24 while PVOS Command declared that, had the '148' been allowed to continue as a multi-role aircraft, its interceptor performance would have been degraded in order to take on board the additional strike roles. Another key reason was that the Soviet electronics industry would be unable to produce the required standard of avionics for some time. Soon afterwards, however, the project was resurrected as a single-role type for long-range interception but this time the Air Force opted for developments of Mikoyan's MiG-25 fighter which eventually led to the MiG-31.

Into the 1970s

Mikoyan MiG-27

During the second half of the 1960s the Soviet Union's leaders finally realised that, besides needing heavy strategic bombers for a possible confrontation with America, a smaller tactical strike aircraft with conventional weapons was a must for minor localised conflicts – in fact the Vietnam War had made this point very clear indeed. With the rejection of the Il-40 (Chapter 7) it had been left to types like the Su-7B and fighter-bomber versions of the Mikoyan MiG-19 fighter to fill the gap, but these carried a limited weapon load.

The Mikoyan design bureau had proposed some new lightweight designs to compete for the attack aircraft requirement won by the Sukhoi T-8 (Chapter 7), but TsAGI did not support the idea of a light combat aircraft while, at this point, Mikoyan had a strong preference for the variable-geometry swing wing. In due course it was realised that an attack version of the OKB's MiG-23 *Flogger* fighter would be cheaper than an all-new aeroplane, and capable of supersonic speeds, and so work began in 1969. The result was an aircraft called 'Project 32-24', and then MiG-23B, which differed primarily from the standard fighter in having an all-new nose that replaced the radar with the new PrNK-23S *Sokol* nav/attack system and also gave the pilot a much better forward view. A new engine was fitted, the Lyulka AL-21F-3 used by the Su-24, and a large amount of offensive stores could be carried.

The first prototype first flew on 20th August 1970 and the type quickly passed through its state testing programme, but only 24 were built before the upgraded MiG-23BN ('32-23') followed. This had a modified wing with a dog-tooth and more area, plus other changes, and was built in large numbers, many going abroad. It also used a Khachaturov R-29B-300 engine, a version of the MiG-23's power unit simplified for ground-attack operations.

Next came the MiG-23BM (the '32-25') which got rid of any surviving elements of the fighter role to give a full-time attack aircraft that was relatively simple and inexpensive. Much of the equipment was upgraded, including fitting the new six-barrel GSh-6-30 30mm cannon, and there were seven underwing hardpoints to take a whole variety of stores including tactical nuclear bombs, heavy rockets, Kh-23M missiles, napalm tanks, bomblet dispensers or self-defence air-to-air missiles.

In addition, limiting the top speed of these fighter-bombers to Mach 1.7 allowed the removal of complex air-intakes. A MiG-23B was converted into a prototype and flew during 1974, and the first service deliveries were made in early 1975. The West allocated the codename *Flogger-D* and all production fighter-bombers to serve with the VVS were redesignated MiG-27. The type was used extensively and proved to be very effective.

Model of the Sukhoi S-37. George Cox

Up to Date

During the 1990s some new Sukhoi have designs appeared, one of which was stillborn but another is intended to replace Su-24s in the Russian Air Force inventory.

Sukhoi S-37

At the November 1991 Dubai exhibition Sukhoi released details and displayed models of this proposed multi-role fighter. The S-37 (also initially known as the Su-37), like many combat aircraft designs from the 1980s, was a canard delta and really belongs in an equivalent fighter volume. However, the design had 17 underwing and underfuselage hardpoints and one of the models was shown carrying four air-to-air missiles, two air-to-ground missiles and nine bombs. The powerplant was to be a single Tumansky/Soyuz AL-41F engine giving 40,780 lb (181.2kN) of thrust, although at the time there was still the possibility that two engines might be fitted, and the S-37 was to be capable of supersonic flight at low level and Mach 2 at height. Gross weight was 55,115 lb (25,000kg) and maximum weapon load 17,637 lb (8,000kg). In fact the Russian Air Force favoured two engines which did not help the S-37's cause and neither did the chronic shortage of hard currency that has affected so much of Russia's programme funding since the early 1990s. Despite attempts by Sukhoi, the S-37 also failed to attract any foreign investment to help with its development and work on the project eventually ground to a halt.

The prototype Sukhoi Su-27IB.

Sukhoi Su-30 and Su-32

The single-seat Sukhoi Su-27 *Flanker* has proved to be one of the most successful fighters of the last twenty years and versions are still winning export orders. Indeed each new version of the Su-30 multi-role variant has grown steadily more capable in terms of the amount of ground-attack weaponry it can carry. The two-seat Su-30 was initially intended to be a 'control' fighter commanding a group of single-seat Su-27s during a combat. In 1993, however, the prototype was modified into the Su-30MK multi-role demonstrator aircraft capable of carrying, besides air-to-air missiles, a powerful selection of guided air-to-ground missiles and bombs plus anti-ship and anti-radar missiles. Versions have now been supplied to India and China.

On 13th April 1990 Sukhoi flew the prototype of a long-range theatre bomber development called the Su-27IB which was fitted with a new two-seat side-by-side cockpit, canards, a nose containing nav/attack and terrain-following/avoidance radar and, on later examples, twin main wheels. The first aircraft had been rebuilt from an Su-27UB airframe and was later redesignated Su-34-1, and more recently Su-32; thanks to its very distinctive nose the bomber was nicknamed *Platypus*. The development programme for this variant began in the early 1980s and the new type was intended to replace the Su-24, with perhaps 200 aircraft in service by 2000. However, the end of the Cold War and Russia's recent monetary difficulties have put paid to the original programme and, by 2003, only two prototypes and four pre-production aeroplanes had flown and the type's state acceptance trials had been postponed until 2004/5. At the time of writing the programme's future is in doubt – that the Soviet Air Force wants and needs the bomber is in no doubt but, so far, the Government has been unable to find sufficient finance to get things moving. The next few years will be critical.

Tactical Strike Aircraft – Data / Estimated Data

Project	*Span ft in (m)*	*Length ft in (m)*	*Gross Wing Area ft² (m²)*	*Max Weight lb (kg)*	*Powerplant Thrust lb (kN)*	*Max Speed / Height mph (km/h) / ft (m)*	*Armament*
Tupolev 'Aircraft 112'	40 0 (12.2)	74 2 (22.6)	516 (48.0)	?	2 x AL-7F	Mach 1.7-1.9	One large nuclear weapon
Yakovlev Yak-123 (Yak-26) (flown)	36 0 (10.964)	56 3.5 (17.16)	311 (28.94)	24,691 (11,200)	2 x RD-9AK 12,630 (56.1) dry, 15,795 (70.2) reheat	764 (1,230) at height	1 x 23mm cannon, 1 x 2,646lb (1,200kg) nuclear weapon or up to 6,614lb (3,000kg) bombs
Yakovlev Yak-28 (flown)	38 2 (11.64)	?	379 (35.25)	30,049 (13,630) normal	2 x R-11AF-300 12,675 (56.3)	1,181 (1,900) at 39,370 (12,000)	1 x 23mm cannon, 1 x 2,646lb (1,200kg) nuclear weapon or up to 6,614lb (3,000kg) bombs
Yakovlev Yak-32 (recce aircraft)	47 7.5 (14.52)	80 8.5 (24.6)	645 (60.0) or 753 (70.0)	59,524 (27,000)	2 x VK-13 or AL-7F	1,554 (2,500)	None carried (see text)
Sukhoi Su-7B (flown)	30 6 (9.309)	54 6 (16.607)	366 (34.0)	23,940 (10,859)	1 x AL-7F-1 14,990 (66.6) dry, 20,280 (90.1) rh	777 (1,250) at S/L, 1,318 (2,120) at 42,651 (13,000)	2 x 30mm cannon, 4 x 1,102lb (500kg) bombs or 64 S-5, 28 S-3 or 4 S-24 rockets
Sukhoi Su-17 (flown)	44 9.5 (13.656) 30° sweep 31 7 (9.64) 63° sweep	53 10 (16.415) without pitot	414 (38.52) 30° sweep 385 (35.85) 63° sweep	37,366 (16,949)	1 x AL-7F-1-250 14,990 (66.6) dry, 21,165 (94.1) rh	746 (1,200) at S/L, 1,336 (2,150) at height	2 x 30mm cannon, up to 6,614lb (3,000kg) bombs and stores
Tupolev 'Aircraft 127'	43 11.5 (13.4)	78 4 (23.875)	645 (60.0)	50,705 (23,000) (normal)	1 x VD-7M 35,270 (156.8) rh	1,168 (1,880) at 36,089 (11,000)	4,409lb (2,000kg) of bombs
Tupolev 'Aircraft 132'	56 9 (17.3)	90 3 (27.5)	?	Up to 99,206 (45,000) normal	2 x jet engines	559 (900) at S/L	Cruise missile or nuclear or conventional stores (see text)
Sukhoi T-58M	34 2 (10.41)	73 8 (22.45)	487 (45.33)	?	?	Supersonic	2 x Kh-24 + 2 x Kh-56 ASMs
Sukhoi T-6-1 (flown)	34 2 (10.41)	77 10 (23.72)	487 (45.33)	57,540 (26,100)	2 x R-27F2-300 22,490 (100.0) rh + 4 x RD36-35 lift units 5,180 (23.0)	?	See text
Sukhoi Su-24 (flown)	57 10.5 (17.638) 16° sweep 34 0 (10.366) 69° sweep	74 3.5 (22.67)	593 (55.16) 16° sweep 548 (51.0) 69° sweep	87,522 (39,700)	2 x AL-21F-3 24,690 (109.7)	870 (1,400) at S/L	23mm cannon, large mix of unguided bombs, guided missiles, rockets, AAMs, up to 17,637lb (8,000kg) stores
Mikoyan MiG-27 (flown)	45 10 (13.965)	56 0 (17.08)	402 (37.35)	45,326 (20,560)	1 x R-29B-300 17,635 (78.4) dry, 25,355 (112.7) reheat	839 (1,350) Mach 1.1 at S/L 1,119 (1,800) Mach 1.7 at 26,247 (8,000)	1 x 6-barrel 23mm cannon, up to 6,614lb (3,000kg) or stores
Sukhoi Su-32 (flown)	48 3 (14.70)	76 7 (23.34)	667 (62.04)	99,427 (45,100)	2 x AL-31F 16,760 (74.5) dry, 27,585 (122.6) rh	808 (1,300) at S/L 1,181 (1,900) at 36,089 (11,000)	1 x GSh-301 30mm cannon, wide variety of self-homing and guided missiles incl anti-ship missiles and laser-guided bombs, air-to-air missiles

Chapter Nine

Maritime Patrol

Production Beriev Be-10 flying boat.

The biggest part of the Soviet Union's maritime aircraft story centres on the design bureau led by Georgy M Beriev, which was first opened in October 1934 as the Central design bureau of Seaplanes Manufacturing (TsKBMS). This was responsible for several wartime designs including the MDR-5 long-range maritime reconnaissance aircraft first flown in 1938 and the short-range MDR-7 of 1939 (neither entered production, the MDR-5 losing out to its Chetverikov MDR-6 rival). On several occasions the war forced the OKB to move to another home but a short production run of Be-2 ship-based reconnaissance aircraft was completed. Wartime project designs included a twin piston-engined bomber (the BB-288) but the bureau's priority was still seaplanes, in particular the MDR-10 maritime reconnaissance flying boat.

Beriev LL-143 and Be-6

The MDR-10 project was eventually turned into the LL-143 prototype powered by two Shvetsov ASh-72 piston engines (LL = *Letayushchaya Lodka* or flying boat). The construction of two prototypes began at Factory No 477 in Krasnoyarsk, the OKB's base at that time, during 1944 and the first completed aeroplane was moved to Taganrog where it made its maiden flight on 6th September 1945. In February 1946 the Beriev design bureau also moved to Taganrog and on 21st June the organisation became the State Union Experimental Plant No 49, the only establishment in the whole of the Soviet Union devoted to the design and manufacture of flying boats; Beriev himself became the plant's Director and General Designer.

The great advances in the technology of aircraft and their equipment and weapons that followed the end of the Second World War forced the bureau to modify the second LL-143 considerably, and the experience gained in doing this allowed the bureau to produce a new service flying boat, fitted with ASh-73 engines and a new radar, called the Be-6. The first all-new Be-6 flew on 2nd June 1948 and the later Be-6M variant carried five cannon, plus bombs, torpedoes or mines beneath the wings. Between 1952 and 1957 a total of 123 Be-6s were built, the first entering service in June 1952, and the West allocated the recognition codename *Madge* to the type.

Beriev Be-10

In 1948 the OKB proposed the Be-10 amphibian which, compared to the Be-6, had a tricycle undercarriage for land operations that retracted into the hull. In every other respect it was identical to the earlier aircraft but the project only progressed as far as the preliminary design stage. The powerplant was two ASh-73 radials, estimated maximum weight 58,862 lb (26,700kg) and cruise speed 250mph (403km/h).

Beriev R-1

The new jet engine powerplants used in certain fighter and bomber aircraft during the latter war years gradually spread to other types once the conflict was over, including flying boats. While the development of the Be-6 was ongoing, Beriev made his first tentative assessments of a reconnaissance flying boat powered by jet engines and in 1947, without waiting for an official directive, he put together a private venture design. The R-1 would use two of the British-supplied Nene jet engines, mounted on the upper wings to keep them clear of spray, and, after backing had been received from naval command, a SovMin resolution was passed on 12th June 1948 giving Beriev permission to go ahead with the project. The boat's state acceptance trials were earmarked for December 1949 but the Beriev OKB was relatively small and, having to devote so much effort into completing the Be-6 programme as well, it proved impossible to achieve this target.

In June 1950 the programme was revised, the powerplant switched to home-built VK-1s and new 23mm cannon replaced the 20mm in the tail barbette. The mock-up was approved on 21st March 1951 but controversy continued in regard to the range and bomb load combination. The first taxi runs were made on 22nd November 1951 but the aircraft suffered severe porpoising at 80% of the take-off speed caused, it was discovered, by a new phenomenon called the 'hydro-dynamic instability barrier' (which almost resulted in the loss of the aircraft). This problem was essentially unstable aquaplaning at high take-off speeds and the elevator compensation and tailplane incidence angle were both corrected before the River Azov froze over for the winter. When taxying resumed in April 1952 the trouble persisted until modifications had been made to the planing bottom; finally the problem was brought to manageable level (but not cured) and the R-1 made its first flight on 30th May.

Further modifications followed but the R-1 porpoised very badly during an attempted take-off on 3rd October, water was ingested by the engines and, although repaired, the aircraft did not fly again before the next winter freeze. In the meantime there was considerable discussion as to whether the project should continue, MAP Minister M V Khrunichev for example proposing that it should be abandoned (a move which also threatened the design bureau's future). Others, however, gave support, not least Admiral of the Fleet and Commander-in-Chief of the Navy N G Kuznetsov, and on 18th July 1953 flying restarted. After further changes, including Fowler-type flaps rather than slotted flaps plus a revised bottom step venting system, the problem was finally cured and the R-1 rode steadily over the water during both take-off and landing. The effort needed to solve the R-1's aquaplaning had been painstaking and was undertaken with the help of scientists from TsAGI. Flight testing for the manufacturer, and then for research, continued until February 1956 when the R-1 was damaged during a landing. Although it had given a lot of trouble, and fell well short of the requirements laid down by the Soviet Defence Ministry, this aircraft proved the potential and concept of the jet engine in hydroaviation.

Beriev R-2

This modified version of the R-1 had a revised nose section, two (3,100kg) VK-5 engines and the internal fuel reduced from 1,863gal to 1,803gal (8,470lit to 8,200lit), but the rest of the hull plus the wings and tail were unaltered. It remained a preliminary project because of the desire to have more range (1,492 miles [2,400km] was declared insufficient) and due to the R-1's teething troubles. A mock-up was inspected on 30th July 1952 but the project did not proceed further. Length was 66ft 6½in (20.285m), span and wing area unchanged from the R-1, take-off weight 44,092 lb (20,000kg) and estimated maximum speed 503mph (810km/h).

Beriev Be-10

As a result of the bureau's newly won experience with jet-powered flying boats, the next step was to create an aircraft that would be suitable for service use and the result was the second aircraft to receive the Be-10 designation (the initial designation was 'Article M'). This began as a proposed reconnaissance/strike torpedo-carrying flying boat capable of dealing with enemy warships and transports and came into existence after the OKB had received an official request for such a type from E N Preobrazhensky, the Commander of the AVMF (naval aviation) – Admiral of the Fleet N G Kuznetsov also offered his support. This move was followed by an official resolution dated 8th October 1953, the receipt of which was a big step forwards for the Beriev design bureau.

Unlike previous flying boats, the offensive weaponry was housed in an internal bomb bay with doors in the bottom hull behind the planing step. The R-1's straight wing was replaced by an alternative with moderate sweepback and the Be-10's full design had been completed by 15th May 1954. It showed an aircraft that was much bigger than the R-1 and the mock-up was officially examined between 7th June and 15th July; the first prototype had been completed by October 1955. However, the OKB's home base presented a problem in that every winter Taganrog Bay froze over, which automatically stopped the flight testing of flying boats and seaplanes for a good proportion of the year. An alternative was required that would remain open all year round and the final choice was Gelendzhik.

As a result the Be-10 was towed to Gelendzhik in a floating dock and made its maiden flight from there on 20th June 1956; in due course it successfully completed its state acceptance testing, although not without some teething troubles. However, the 27 production aeroplanes that followed, manufactured between 1958 and 1961, were all assembled at Taganrog. The type entered service in mid-1959 and for many years was the world's only all-jet flying boat to become fully operational; in the West the Be-10 was codenamed *Mallow*. (A few production examples of the American Martin XP6M-1 Seamaster flying boat, powered by four jet engines, were taken on charge but were not a success). In August 1963 the Be-10 was grounded for two reasons – three boats had recently crashed, with some fatalities, while, in addition, cracks had been found in the Al-8 alloy structural parts of several other airframes; after sitting idle for five years, all of the survivors were eventually scrapped.

Beriev Be-10N

The Be-10's offensive stores comprised three RAT-52 rocket-assisted torpedoes, up to twelve 551 lb (250kg) or one 6,614 lb (3,000kg) bomb, or three mines but, overall, this was considered to be a relatively limited warload. Consequently a follow-on Be-10N version was designed to carry two K-12BS anti-ship cruise missiles on underwing pylons, thus giving the type the ability to

The Beriev LL-143 prototype.

The Beriev R-1 jet-powered flying boat.
Russian Aviation Research Trust

A production Beriev Be-6.

Artist's impressions of the Beriev Be-10N anti-ship missile carrier.

Beriev Be-12.

attack larger warships, other surface vessels and coastal targets. The K-12BS could receive conventional or nuclear-tipped warheads and on 31st July 1958 an official Ministry resolution was passed to cover the design. The structural changes made to the basic Be-10 came in the form of a much larger nose cone (for the massive K-12U Shpil targeting radar scanner) and in the weapon bay, but the avionics were also upgraded and improved. Maximum take-off weight was 106,922 lb (48,500kg) and combat radius about 901 miles (1,450km), while the missile would be launched at 32,808ft (10,000m) about 62 miles (100km) from its target. The draft design was submitted to officials and gained SovMin approval on 10th June 1959, but no instructions to proceed were ever issued. A later Be-10S preliminary proposal, which was intended to carry the SK-1 Skalp nuclear depth charge, was abandoned in August 1960.

Beriev Be-12

Beriev's next flying boat to reach the hardware stage was turboprop-powered. The Be-12 was initially developed as a search and attack aircraft for which it carried a radar, a detection and sighting system, a magnetometer, sonobuoys, anti-submarine torpe-

Beriev Anti-Submarine Flying Boat (1962).
Russian Aviation Research Trust

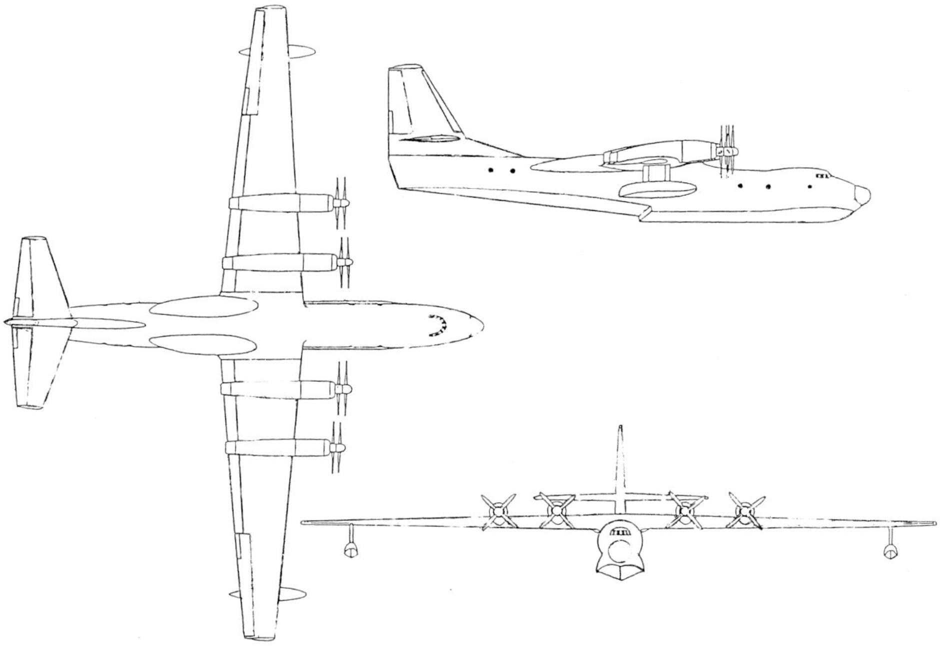

Artist's impression of the Beriev Anti-Submarine Flying Boat.

does and depth charges. Research and development work began in 1956 for a design to replace the Be-6 and the first flight, from a hard runway, was made on 18th October 1960. Aerodynamically this aircraft was quite similar to the Be-6 but it was bigger and heavier. The Initsiativa radar was initially placed in the bottom fuselage ahead of the first step but, if it malfunctioned, the Be-12 would then have to alight on dry land to get it repaired; consequently the design team moved the equipment to a housing on the nose. Fitting a land undercarriage also allowed the aircraft to operate at all times, and not just during the periods when its sea bases were free of ice. A total of 143 aircraft, including search and rescue examples, was built between 1963 and 1973, the first examples entering service in the spring of 1964; during the 1970s the Be-12 was upgraded. NATO called the boat *Mail,* the Soviets named it Tchaika and this flying boat became the workhorse of their naval aviation operations; in 1989 alone 29 foreign submarines were detected by Be-12s.

Beriev Anti-Submarine Flying Boat

In 1962 the bureau began to design a new heavy flying boat intended for anti-submarine warfare and co-operation with specially built submarine tankers. In cross section the aircraft's hull shape was a figure-of-eight and power was supplied by four Kuznetsov NK-12M turboprop engines, plus two supplementary Lyulka AL-7PB jets mounted in wing root fairings to give extra thrust for take-off. This machine was to have three roles, military transport (with a hinged nose for loading), search and rescue or anti-submarine and, for the latter task, its tanker submarines were to be the specially-built diesel-powered Class 648 or nuclear-powered Class 664 designs. However, as the early development work moved ahead it became clear that producing such a large flying boat with a performance that matched land-based types would be very difficult and so the entire project was soon abandoned. Estimated service ceiling was 42,651ft (13,000m) and maximum range 7,520 miles (12,100km); up to seven crew would have been carried.

This concept was replaced by a huge project called the LL-600 which was to serve either as a bomber or an airliner (with 2,000 seats!). This aircraft's maximum weight was estimated to be 2,204,590lb (1,000,000kg) and top speed 559mph (900km/h), but in the mid-1960s this too was cancelled.

Beriev Be-26

This long-range oceanic anti-submarine amphibian was intended to be able to take-off from land or water. Studies began in 1963 and it was hoped to make the aircraft capable of operating from poor quality unpaved runways since, it was predicted, the better quality paved strips were likely have been destroyed by a pre-emptive nuclear strike. To give the type STOL capability a total of sixteen RD35-36 lift jets were fitted, eight per side in clusters around the wing root leading and trailing edges, and the ability to refuel in the open sea from submarine or surface tankers was another key facility. The cabin for the Be-26's seven crew was to be pressurised, the weapon bay was positioned mid-fuselage with all weapons carried internally (there were no defensive guns) and the aircraft was to have a tricycle undercarriage with four-wheel main gears retracting into wing nacelles and a two-wheel nose gear. Service ceiling would have been 42,651ft (13,000m), maximum range 7,272 miles (11,700km) and range with 6,614lb (3,000kg) of stores aboard 6,607 miles (10,630km). The Be-26 stayed on the drawing board.

Two artist's impressions of the Beriev Be-26 (1963).

Beriev A-150

This spectacular delta wing project for a large ocean-going jet-powered amphibian began life in 1965. It was intended to be a true multi-role flying boat capable of undertaking long-range anti-submarine work, reconnaissance, search and rescue, anti-shipping and in-flight refuelling and, to make the transition from one role to another as quick and simple as possible, special detachable containers housing equipment specific for each task were to be carried in two sections of the centreplane. The aircraft had to be able to operate from land or sea and, in the Arctic, even off ice runways. Just like the Be-26, the type was to have STOL capability but this time twelve RD36-35P lift jets were installed in two lines in the wing leading edge root extensions ahead of the CofG; in addition the nozzles for the main engines, four NK-8 turbojets mounted in paired nacelles above the wing trailing edge, were to be capable of thrust vectoring at angles between 0° and 65°.

The multi-spar wing had both ailerons and flaps, with the outboard sections also carrying retractable floats, and a tricycle undercarriage was fitted with four-wheel main gears retracting into the centrewing and two wheels on a nose leg underneath the flight-deck. There were five crew in a pressurised cabin (two pilots, navigator plus operators for sonar and radar), all of the fuel (maximum 220,459 lb [100,000kg]) was housed in the

Beriev A-150 (1965). Jens Baganz

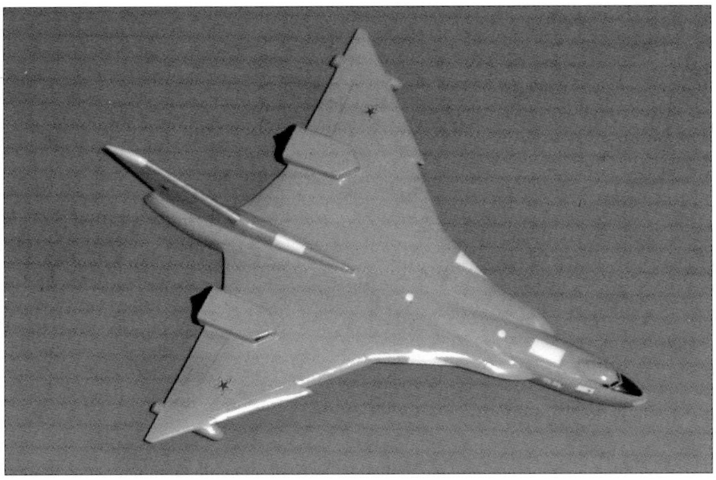

Model of the Beriev A-150 flying boat. George Cox

wing and the avionics included the Polyot long-range navigation aid, Zubr anti-submarine weapon control system and the Uspyekh target indication system. Defensive cannon would be housed in both nose and tail barbettes, the aircraft's service ceiling was estimated to be 49,213ft (15,000m), range 7,303 miles (11,750km) and flight endurance 10.3 hours. This was another highly advanced Beriev project to stay on the drawing board and there was also a proposed A-150TD assault transport derivative.

Beriev A-40

For a period from the mid-1960s the Beriev design bureau's work on military flying boats fell away because official interest moved towards alternative types. However, in 1976 some preliminary studies were initiated for a new generation anti-submarine flying boat. The project was eventually proposed to the Government by Chief Designer A K Konstantinov and in 1983, following his prompting, the Soviet Union's Defence Industry Committee issued an order to go ahead with the design, called Be-42 or A-40, to replace the Be-12 and the Ilyushin Il-38 land-based maritime aircraft (below). The A-40 maritime patrol and surveillance aircraft was designed to deal with submarines and surface ships at close and medium distances from home bases and also provide access to remote areas in the eastern sectors of the Soviet Union; it was to be powered by two Soloviev D-30KPV main engines and two Klimov RD-60Ks.

Two prototypes were built and the first made its maiden flight from a land runway in December 1986, following this with its first water take-off in November 1987. Publicly revealed at the Tushino airshow in August 1989, the type was eventually named Albatros (and designated *Mermaid* by NATO) and was proposed in numerous military and civil versions. By 2002 a prototype of the military A-42 variant was planned with more advanced avionics and powered by D-27A propfan engines that offered improved fuel efficiency.

Although, in general, flying boats remained the domain of the Beriev design bureau, several other OKBs had a go which resulted in some extraordinary designs. In addition two important land-based anti-submarine aeroplanes were produced and these are summarised below.

Beriev B-40 Albatros.

Tupolev Design Bureau

Tupolev 'Project 504'

This project used the bureau's Tu-85 design (Chapter 1) as a basis for a long-range flying boat bomber and work on it lasted from 1950 until 1953. Once again it was intended that the aircraft should touch down during its mission to be refuelled by a submarine tanker, which would enable the bomber to hit targets in North America and still be able to return to its base. Initially there were two versions, both with four 4,500hp (3,356kW) Shvetsov ASh-2K piston engines, one having fixed outboard floats and the other having floats that folded outwards towards the wingtips; however, different hull forms were also examined. There were to be twelve crew, the maximum range was approaching 6,215 miles (10,000km), service ceiling 36,089ft (11,000m) and the defensive armament comprised three twin 23mm cannon turrets. A Rubidy-MM radar was fitted and during the later stages of these studies alternative TVD-1 or TV-2 turboprop engines were also considered.

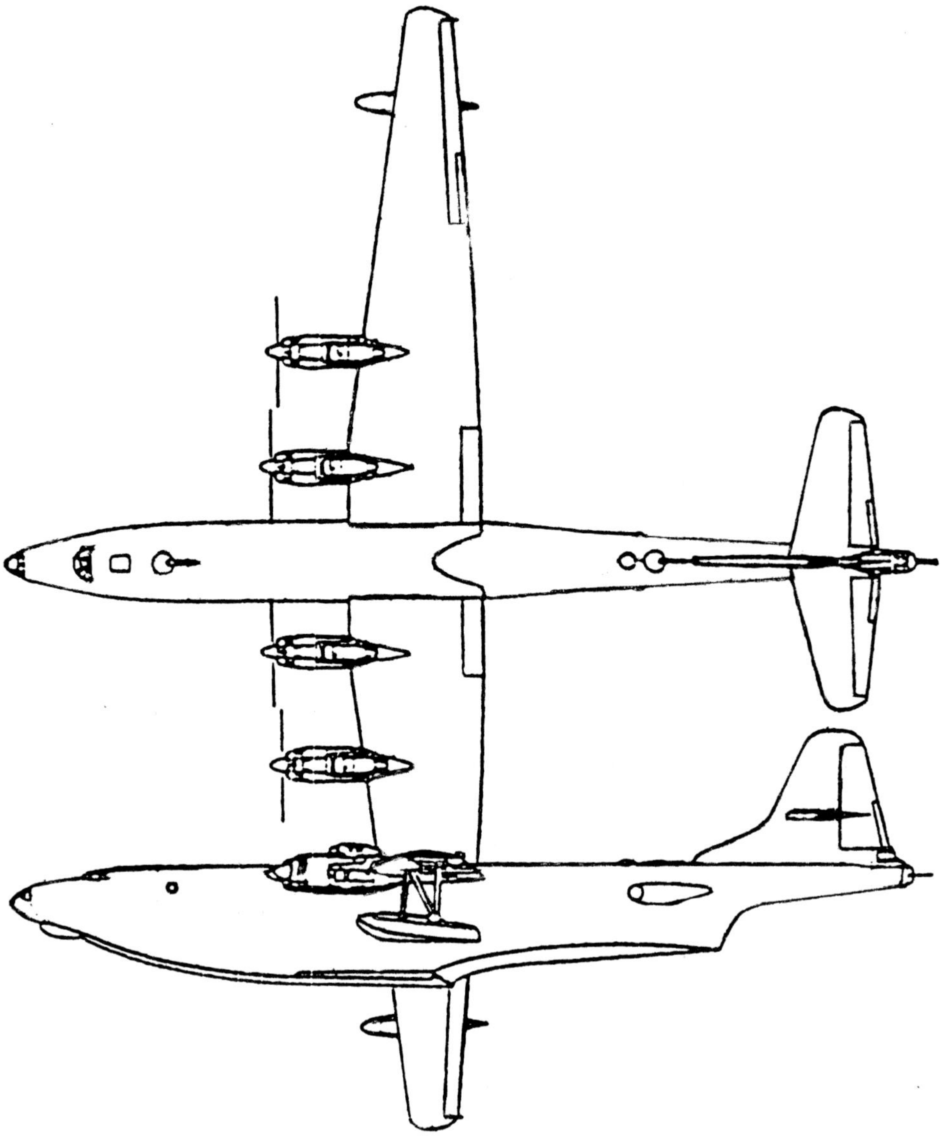

Tupolev 'Project 504' (1950). Russian Aviation Research Trust

Prototype Tupolev Tu-142M.

Tupolev Tu-142

The appearance of Western nuclear ballistic missile submarines armed with the Polaris system meant that the Soviet Union had to find a way of detecting and destroying these vessels at a distance which exceeded the range of the submarine's weapons. In the early 1960s several design bureau began research into long-range anti-submarine aircraft and Tupolev's solution was 'Aircraft 142', a development of the Tu-95M *Bear* bomber described in Chapter 3, which was capable of carrying sonobuoys and various anti-submarine bombs and mines plus self-guiding torpedoes with either nuclear or conventional tips. The design was initially designated Tu-95PLO (*protivolodochnaya oborona* or anti-submarine defence) and the maximum patrol time was expected to be ten and a half hours.

Eventually the proposal was accepted and a USSR Council of Ministers Decree was issued on 28th February 1963 ordering Tupolev to produce the Tu-142 long-range anti-submarine warfare aircraft equipped with the 'Berkut-95' search and targeting system. The wing area was increased and the fuselage was stretched slightly. The first prototype made its maiden flight on 18th June 1968 and the first service deliveries were made in May 1970. Later aircraft modified with a wider cabin and more modern equipment were designated Tu-142M; in due course over 100 Tu-142s were built, the last being rolled out in 1994. In 1972 Beriev issued production drawings for the avionics sections of the Tu-142M's development as part of that OKB's support to the Tupolev design bureau.

Tupolev 'Aircraft 146'

To follow and replace the Tu-142M, during the second half of the 1980s Tupolev proposed a new anti-submarine aircraft for Soviet naval aviation called 'Aircraft 146'. The intention was to eliminate the older aeroplane's weaknesses, particularly by increasing the fuselage diameter and length to accommodate more and better avionics and weapons, although the overall aerodynamic layout was retained. A new wing would be fitted and four NK-12MP turboprops formed the powerplant. The project was proposed to Naval aviation command but was abandoned when no support was forthcoming.

Ilyushin Il-38.

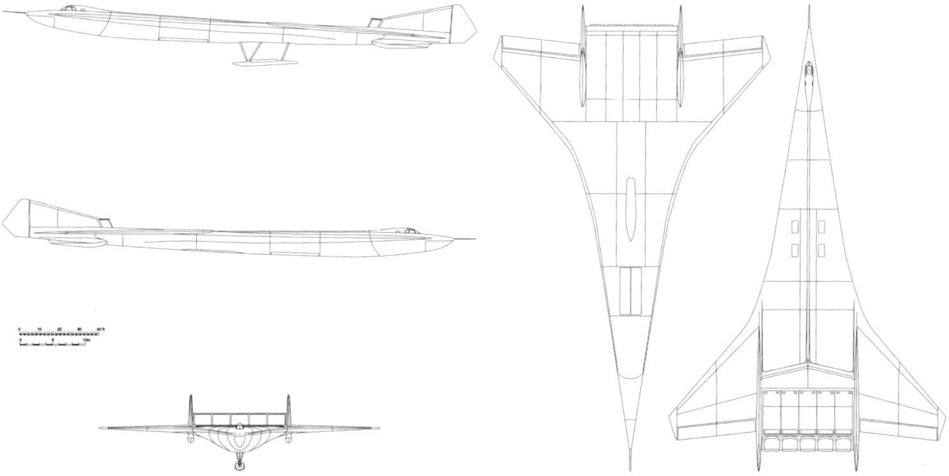

Bartini A-57 (1957). Jens Baganz

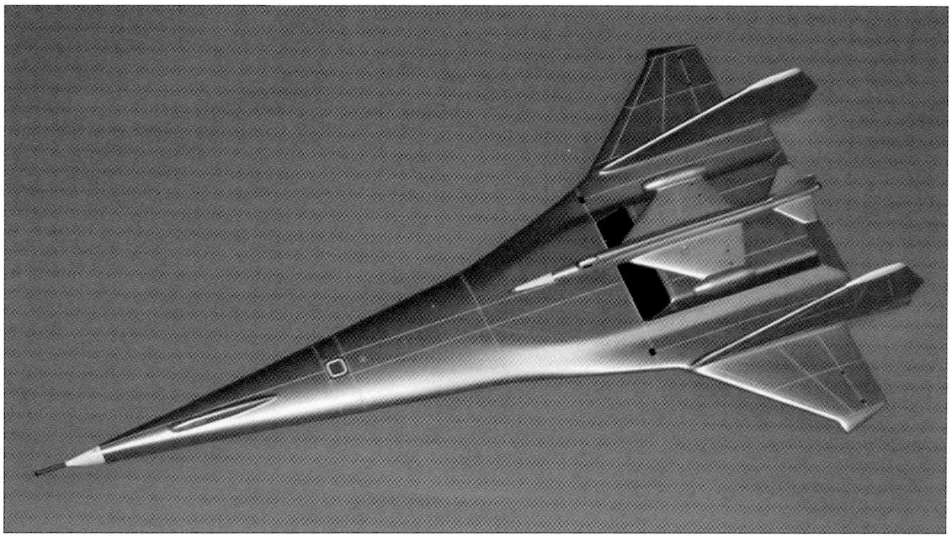

This model of the Bartini A-57 is carrying a Tsybin RSS weapon on its back. George Cox

Ilyushin Il-38

In August 1958 the American Lockheed company flew the first prototype of its new P3V-1 Orion long-range maritime reconnaissance and anti-submarine aircraft as a much modified development of the Electra airliner. In the Soviet Union the Ilyushin design bureau covered very similar ground when it took its Il-18 airliner and converted it into the Il-38 maritime aircraft which, in appearance, has much in common with the Orion. The Il-38 made its first flight on 28th September 1961 and was fitted with two bomb bays and powered by four Ivchenko AI-20 single-shaft turboprops. NATO codenamed the type *May* and the aircraft is still in service today.

Supersonic Strategic Flying Boats

Bartini A-55 and A-57

The ability to land on water could allow an aircraft to be refuelled on the surface by submarine tankers operating reasonably close to enemy coasts and, indeed, from the middle of any ocean or sea. Supersonic performance would aid the aircraft's ability to evade enemy defences while the fact that the aircraft was a flying boat removed the need to

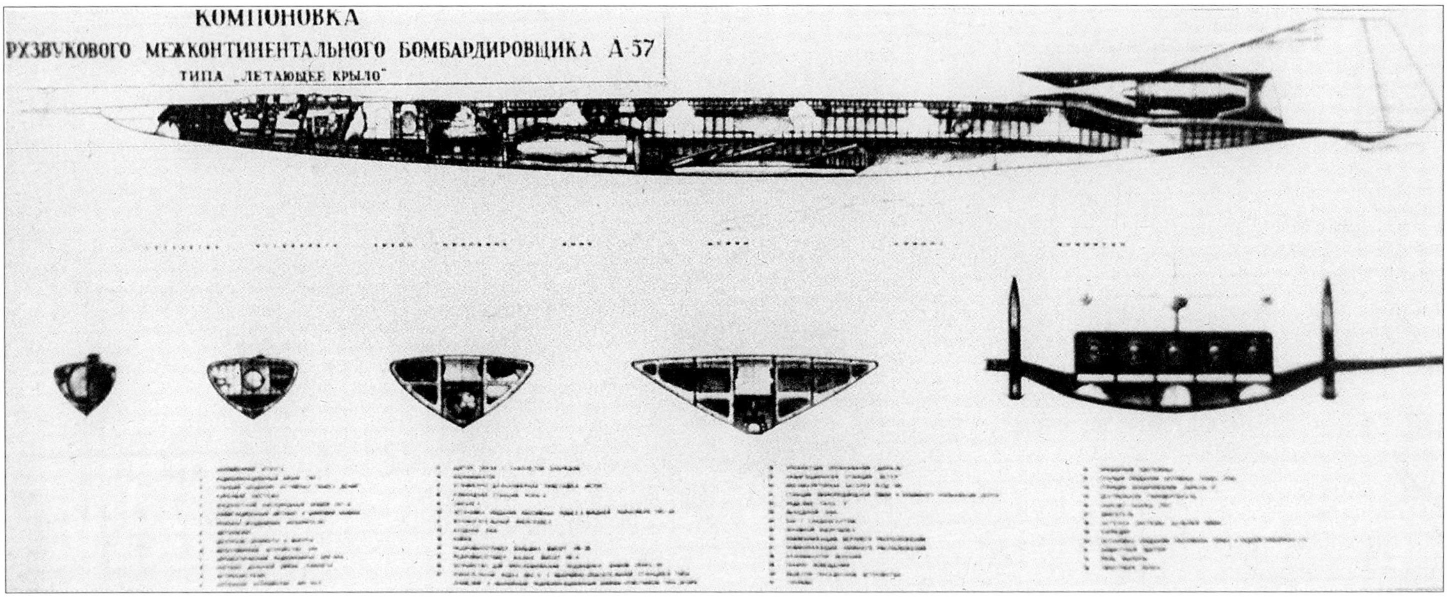

Side view cutaway of the A-57 showing the bomb bay.

build expensive runways that were susceptible to enemy attack. It also gave a practically unlimited take-off run which allowed heavier loads to be taken aloft.

The concept of the supersonic flying boat bomber was the brainchild of Robert L Bartini who from 1952 lived in Novosibirsk where he worked as the head of Future Projects at the Siberian Scientific Institute of Aerodynamics (SibNIA). Bartini's department conceived a new type of supersonic wing, of compound sweepback with reduced sweep further out and a very low aspect ratio, that would automatically maintain stability when flying at transonic speeds. To ensure that this wing possessed good take-off and landing characteristics, coupled with aerodynamic efficiency at high speeds, it was given variable geometric camber. Starting at the root the wing was set at a positive angle of incidence but working along the surface this was gradually reduced until a negative angle was achieved at the tips. The resulting shape had a self-stabilising effect when the aircraft passed through Mach 1 which eliminated the need to move any fuel around inside the fuselage. The 'Bartini Wing' was tested in SibNIA's wind tunnels on a large number of scale models. In 1952 Bartini designed the first of his tailless flying wing aeroplanes fitted with this wing, the unbuilt T-203, and then followed it with a supersonic amphibian designated A-55 which had all of its engines housed in one module in the rear fuselage. This was to be followed by the A-57 series.

The A-57 was a proposed long-range strategic bomber capable of carrying conventional or nuclear bombs or stand-off missiles. These including the '244N' nuclear store carried in a lower fuselage weapon bay placed just over a third of the way back from the nose or, for longer-range operations approaching 6,215 miles (10,000km) range, the unmanned RSS version of the Tsybin OKB's reconnaissance aircraft (described in Chapter 5) on the A-57's back. The combination of these two aeroplanes produced a potentially impressive weapon system which could threaten cities on America's Pacific coast; in fact any point on the earth's surface could be reached. The flying boat was also expected to be able to operate on snow and ice and was seen as an opportunity to use advanced air stations set up in the Arctic on drifting ice floes.

Bartini and Pavel Tsybin worked together on this combined project along with Tsybin's colleague Vadim Shavrov, but the Tsybin OKB gave the RSS relatively low priority compared to its manned aircraft projects. The Tsybin RSS was little different from the 2RS/RSR aircraft except for its nose which dispensed with the pilot cabin and replaced it with a '244N' nuclear warhead. It was to be powered by two Bondaryuk RD-013 ram jets, giving a top speed of 1,740mph (2,800km/h), and its range was 3,108 miles (5,000km); however, a pilot's cockpit was to be installed for the flight test programme.

The A-57 was a massive hydroski flying boat. It was to have a triangular cross-section hull, which was integrated with the lower wing surface, and five Kuznetsov NK-10 engines were placed side-by-side in a line across the top of the rear wing. Twin vertical fins were mounted to each side of the powerplant and the RSS was intended to sit on top of the engines before being released to proceed on to its target (five 49,600 lb [220.4kN] Kuznetsov NK-6 units were to be installed in the A-57 initially as a temporary measure). When carrying an RSS, the A-57 was limited to subsonic speeds, but after release at about 497mph to 528mph (800km/h to 850km/h) the amphibian could then accelerate to 1,554mph (2,500km/h) in its clean condition; its estimated ceiling was around 59,055ft (18,000m) and range 8,701 miles (14,000km). The large array of avionics included an SVR-1 radar, Veter navigation system and Roza and Venik jamming systems and, in case an emergency landing needed to be made on dry ground, the A-57 had a 32ft 9¾in (10m) long skid fitted beneath the hull bottom. Six crew would be carried.

Having formulated the idea in Novosibirsk, in April 1957 Bartini was moved to Moscow to lead a Special Design Group that would undertake the A-57's detail design at No 938 OKB and GAZ; TsAGI and TsIAM were also involved. However, the whole project was cancelled in 1960 and a variant called the R-57, proposed as a reconnaissance aeroplane, was never started.

Bartini E-57 and A-58

Further developments of the A-57, the first of this pair of designs was intended to be used for operations closer to home in Europe. The scaled-down E-57 (or Ye-57) was a medium-range flying wing bomber and cruise missile launcher and could carry one '244N' store internally and a K-10S cruise missile externally (the latter could use either a nuclear or conventional warhead). It was to be powered by two Kuznetsov NK-10s housed in a single nacelle on top of the rear hull and these had flat air intakes controlled by horizontal diffusion wedges.

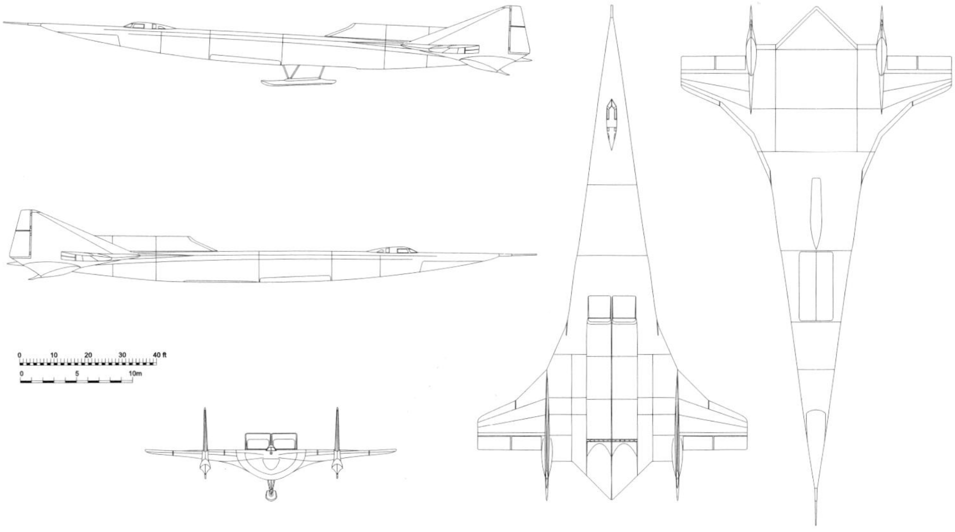

Bartini E-57 (c1957). Jens Baganz

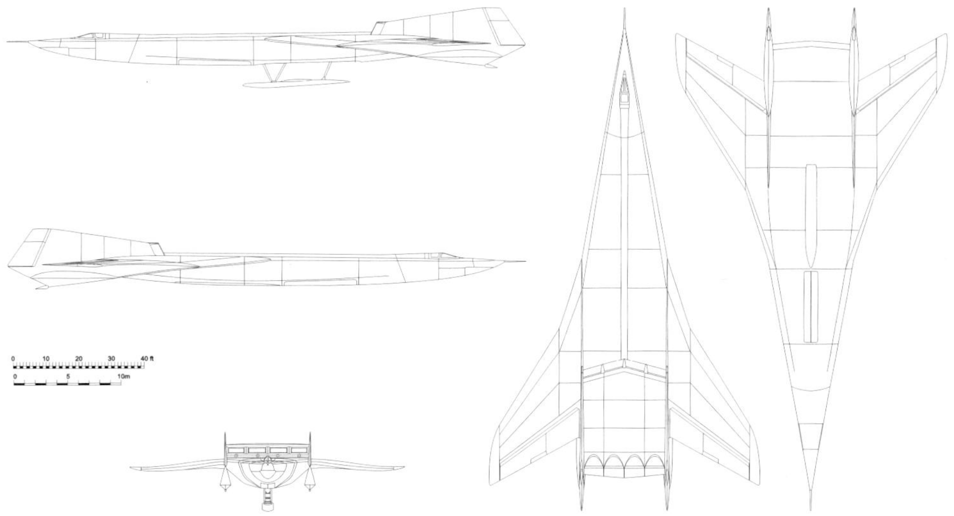

Bartini A-58 (1958). Jens Baganz

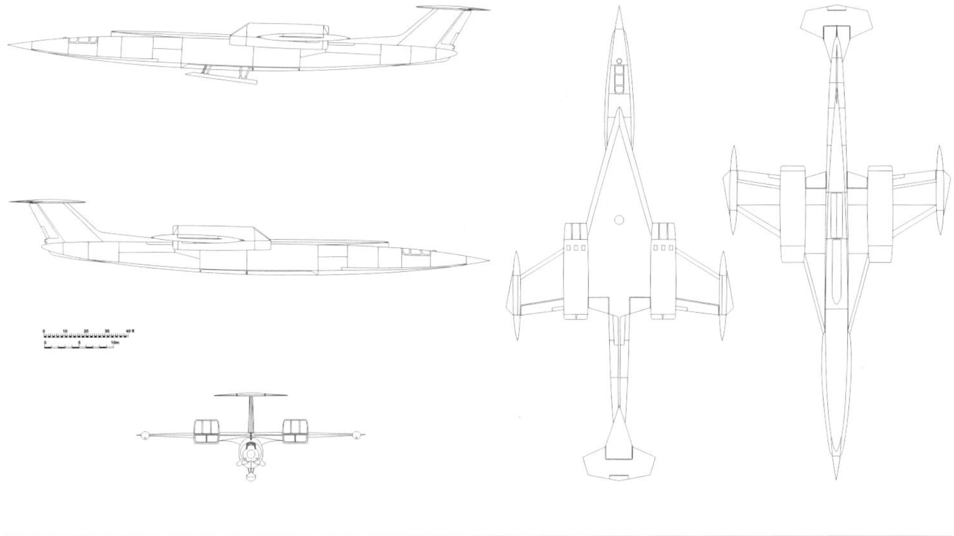

Moskalyov (LLVVIA) GSB-1 (GS-1) (1957). Jens Baganz

E-57 had the same compound sweepback on the wing leading edge, with the same geometric form, as Bartini's earlier designs and its twin triangular tailfins (with tapered rudders) extended below the wing into the balancing floats. The wing trailing edge was formed by two-part differential ailerons while flaps were placed between the engines and fins; the latter would also control the airflow when the E-57 was close to or on the water surface by helping to provide an air cushion. Like the A-57, a special extending skid was available to make an emergency landing on land. E-57's estimated ceiling was 59,055ft (18,000m) and range 2,797 miles (4,500km). The type would be deployed from Soviet waters or from the Atlantic or Arctic oceans.

The A-58 was a long-range flying boat bomber based on the A-57, again using a similar configuration and the same aerodynamics, but of a size, and carrying a warload, which corresponded to the Tupolev Tu-16 or Tu-22 land-based bombers (Chapters 3 and 4); it could take one Tsybin RSS with a nuclear warhead. Maximum range was 3,729 miles (6,000km) and ceiling 59,055ft (18,000m) and the aircraft was to be fitted with four Kuznetsov NK-6 engines but the cancellation of the A-57 also brought work on the A-58 to an end. The A-58 had a longer range than the E-57, which is why it retained the 'A' designation (for 'America') rather than the latter's 'E' (for 'Europe').

Moskalyov GSB-1 (GS-1)

A project designed and assessed during 1957 to 1960, this was the work of a special research laboratory based at the *Leningradskaya Krasno'znamyonnaya Voyenno-Vozdushnaya In'zhenernaya Akademiya imeni A F Mozhaiskovo* (the LKVVIA or Leningrad Red Banner Engineering Academy of the Air Force named after A F Mozhaiskovo) led by Aleksandr Sergeyevitch Moskalyov. This group was responsible for several very advanced bomber projects including the DSB-LK long-range strategic bomber described in Chapter 6. For the GSB-1 flying boat design several different layouts were examined, including delta wing, flying wing or tailless, before the team settled on a high moderately-swept wing with large leading edge root extensions, a slim fuselage and a high tail. A top speed of Mach 2.8 was expected to generate considerable kinetic

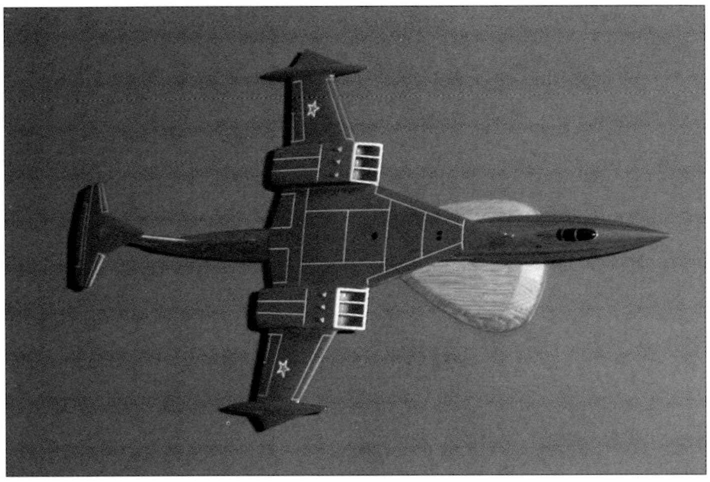

Model of the Moskalyov GSB-1 supersonic flying boat. George Cox

Model of the Beriev SD-MBR supersonic flying boat. George Cox

heat and so some of the structure was to be built in titanium. Up to five 6.2 mile (10km) range air-to-air missiles, plus dorsal and belly position gun turrets, would be carried for defence while a winged thermo-nuclear missile would form the offensive load.

This design used a hydroski near the centre fuselage together with wingtip floats and employed eight Klimov VK-15F or VK-15M jets for power, with solid-fuel rocket motors also available to give more thrust on take-off. The engines were mounted in an unusual design of 'boxed' wing nacelle placed outside the sharply swept inner leading edge. Estimated maximum range was 9,323 miles (15,000km) and service ceiling 68,241ft (20,800m), but the GSB-1 was essentially just a research exercise.

Tsybin RGSR

Besides designing the unmanned RSS, Tsybin also produced a 'seaplane' variant of the RSR called the RGSR. This was to serve as a Mach 2 reconnaissance aircraft and had its fuselage heavily redesigned and transformed into a hull with a planing bottom and keel. There was no step and the nose now included a wave cutter. Span was reduced to 26.25ft (8.0m) and the Solovyev D-21 (D-20F) engines were now mounted above the wing, close to the trailing edge at the position of maximum chord, to keep them clear of water spray. The tailplane too was moved upwards to reduce the degree of contact with water, which required the fin to have greater chord to accommodate it, and the external underwing fuel tanks were intended to act as bal-

Beriev SD-MBR (1957). Jens Baganz

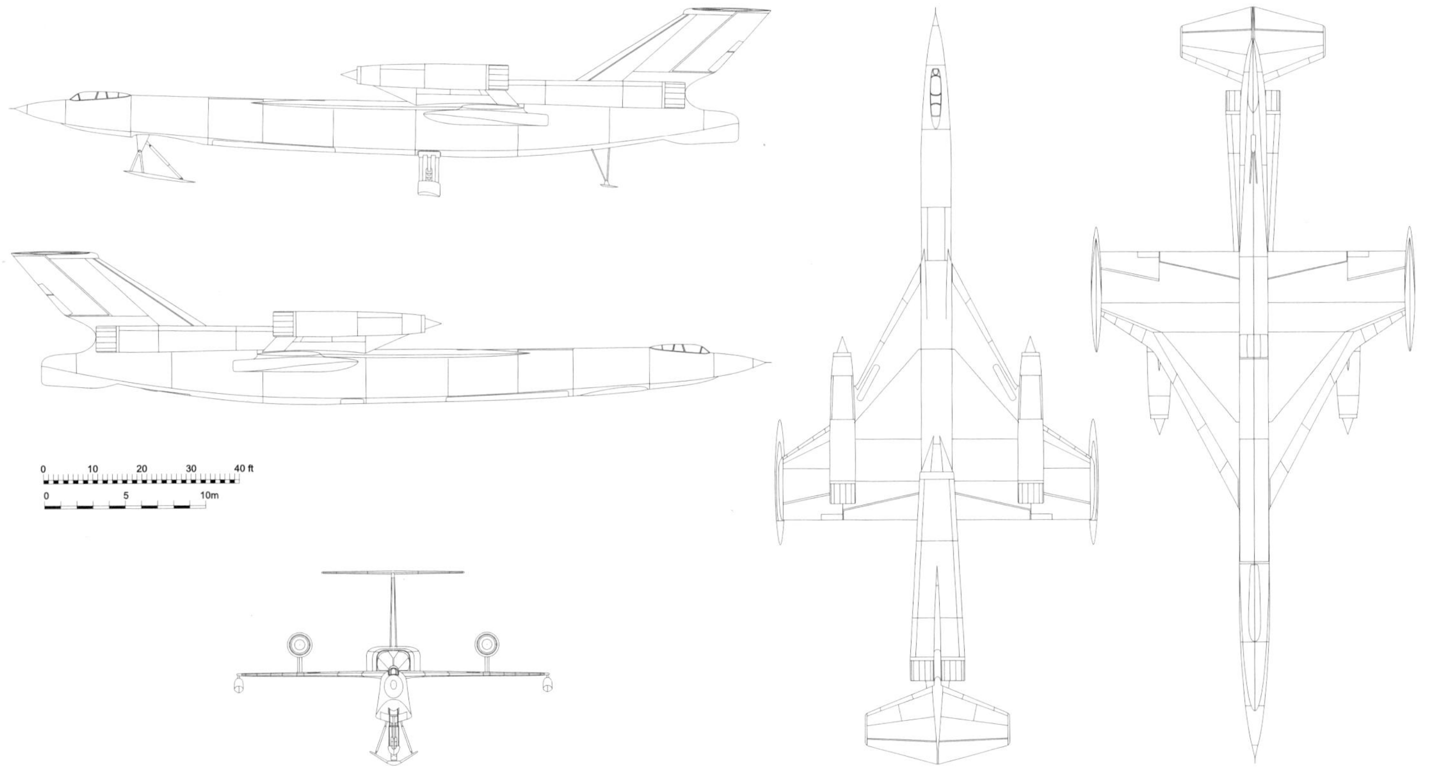

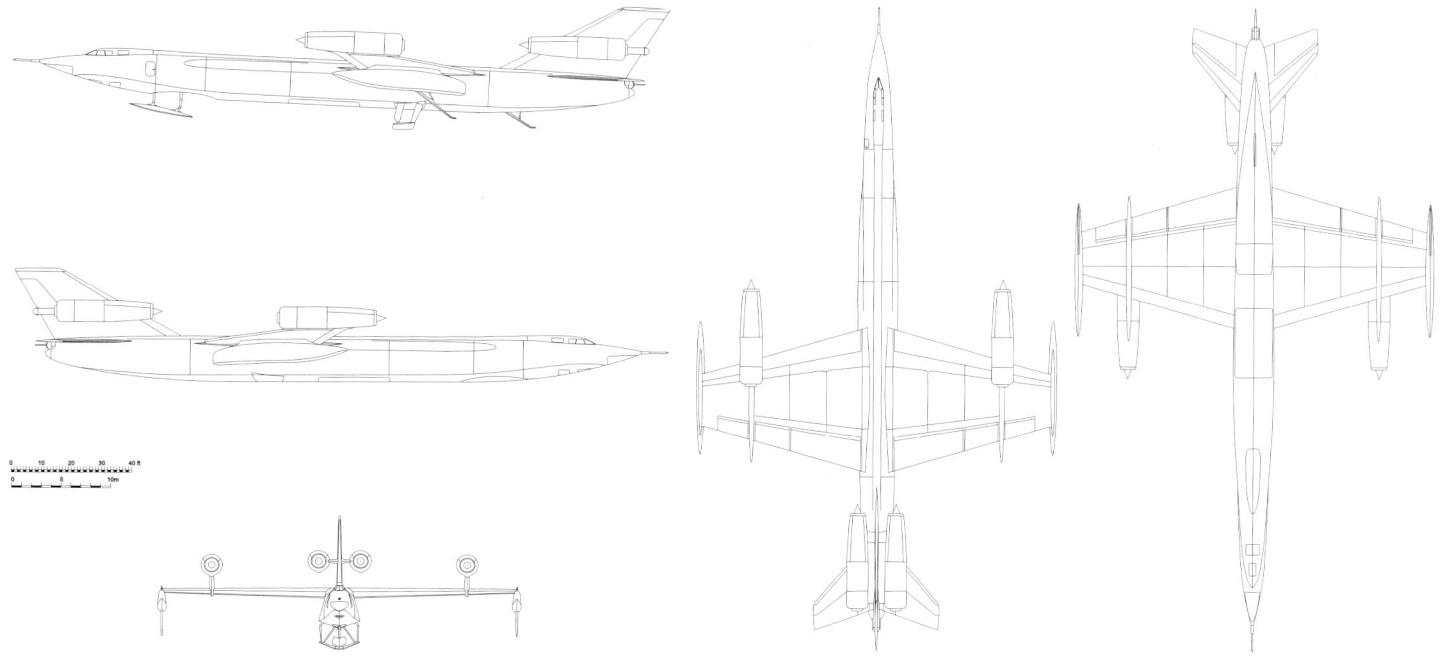

Myasishchev M-70. Jens Baganz

ancing floats. Hydrofoils would lift the RGSR's hull clear of the water for take-off; the balancing floats had winglets which would act as stabilisers during take-off and landing and these winglets were swept back 45°, which meant that they should produce relatively little drag during supersonic flight. The RGSR was not built.

–

Preliminary research into the design and development of a supersonic strategic flying boat bomber began in November 1955 with the backing of an order made by a MAP department chief, and this was given greater importance by a SovMin resolution dated 15th August 1956. The specification called for a normal warload of 11,023 lb (5,000kg), a top speed approaching 1,119mph (1,800km/h) and a minimum range of 4,661 miles (7,500km). Operations were to be independent or with submarines and three OKBs are believed to have responded, although no material has yet come to light as to any of the work that Tupolev might have carried out; the known contenders are Beriev and Myasishchev.

Beriev SD-MBR

The designation for this aircraft stands for *Sverkh'zvukovoi Dalnii Morskoi Bombardirovshchik-Razvedchik* or Supersonic Long-Range Sea-Based Bomber/Reconnaissance Aircraft and it was originally intended to be powered by four Kuznetsov NK-6 turbofans; however, these were soon replaced by the same OKB's more powerful NK-10 which improved the estimated performance figures by a considerable margin. For example top speed at 36,090ft (11,000m) rose from 1,119mph (1,800km/h) to 1,492mph (2,400km/h) and cruise altitude to 65,617ft (20,000m) from 52,493ft (16,000m), although the range was about the same or slightly less.

The SD-MBR had a long slim fuselage with a cranked-delta wing and had two of its engines housed in nacelles above the wing, with the intakes forward of the leading edge, plus the other pair placed either side of the fin. Its wing had slotted flaps and conventional ailerons and was to be built using a honeycomb structure. To take off the boat used a nose hydroski, a hydrofoil and a tail damper (which meant that the usual hull step was no longer required), all retracting flush with the hull, and there was a central weapon bay. The hydroski would only take about 10% of the take-off load, so the hydrofoil just rearwards of the CofG had to carry the biggest portion. There were two crew, a substantial set of avionics would be fitted, a total of 317,460 lb (144,000kg) of fuel filled most of the available space in the airframe and wings and the estimated range with a normal 11,023 lb (5,000kg) load was about 4,786 miles (7,700km).

Myasishchev M-70

After the MAP order of November 1955, the M-70's preliminary project was actually ready in 1956. TsAGI had begun studies into hull shapes and their performance both aerodynamically and in the water while a team of VVS engineers were attached to the OKB to help with the project's definition and design. The aircraft was to perform cruise missile or conventional bombing attacks on enemy shipping, reconnaissance for Soviet submarines (with which it could also rendezvous when operating at large distances from shore), refuel from a submarine, find its target at heights approaching its ceiling when in very hostile airspace and launch a cruise missile at supersonic speeds in any conditions, day or night, at any height up to its ceiling.

Myasishchev lacked experience in designing flying boats and several decisions had to be made concerning the optimal configuration and the type and strength of the structure to be used, although experience here with the M-50 supersonic bomber (Chapter 4) was a help. The resulting design, with two engines in nacelles above the wing and another adjacent to either side of the fin, was quite elegant and shared some features with Beriev's competitor (once again placing the engines high up kept them out of water). The possible engine choices centred on the 49,600 lb (220.4kN) Kuznetsov NK-6 or the 48,500 lb (215.6kN) Zubets M16-17F, both of which were expected to ensure that the M-70 could meet its requirements. Another study examined wings with a lower thickness/chord ratio of 3%, a powerplant of four of the projected 58,420 lb (259.6kN) thrust Kuznetsov P-10B (NK-10) engines and had the tail barbette deleted. This version was expected to achieve a maximum speed of 1,554mph

Model of the Myasishchev M-70 flying boat.

(2,500km/h), a range of 4,661 miles (7,500km) and a 'missile release' ceiling of 68,898ft (21,000m) while the equivalent figures for the NK-6 variant were 1,119mph (1,800km/h), 4,351 miles (7,000km) and 55,774ft (17,000m).

After a great deal of research it was concluded that Myasishchev's flying boat design was feasible; however, some of the OKB's staff felt that a supersonic seaplane would have an inferior flight performance overall to an equivalent land-based bomber while the M-70's main targets (for instance, at 5,594 miles [9,000km] from Soviet territory) could also be tackled by Myasishchev's own M-52K and M-56K strategic bomber systems (Chapter 4). In addition targets much closer to home, at say 1,865 miles (3,000km) distance, could be engaged by conventional bombers or cruise missiles. As a result of these criticisms the OKB had to produce a detailed memorandum expounding the M-70's future operations. This included a statement that the flying boat's range could be extended to around 13,052 miles (21,000km) by two in-flight refuelling operations while, using the M-70 as an offshore tanker, the M-52 bomber could receive fuel from the flying boat to allow that aircraft also to fly deeper over enemy territory; in addition, after taking a 'detour', the attacking aircraft would be able to approach their targets from less well defended directions.

The principal method for refuelling the M-70, however, was to land on the sea so that it could take its fuel onboard from a submarine. Part of the overall plan was to build some special submarine tankers and such a vessel was expected to carry as much as 3,968,250lb (1,800,000kg) of fuel and refuel up to four or five flying boats at least twice, dispensing fuel at a rate of 1,320gal (6,000lit) per minute. To back this up, from November 1956 through to August 1957 three series of submarine refuel trials were carried out (one each with the Black Sea, Northern and Pacific Fleets) using a Beriev Be-6 test aircraft and these laid down a set of procedures that were needed to make the concept work.

With a low aspect ratio thin wing and a streamlined hull form of semi-monocoque structure, the M-70 presented a striking impression. The nose contained the radar and IFR probe while the three crew were housed behind in a pressurised cabin – portions of the structure, for example the cockpit canopy frames, employed welded titanium construction. Fuel was housed in the space between the crew and the bomb bay and in the rear fuselage, and the main hydrofoil well sides were placed alongside the bomb bay. The three spar wing, with a box spar made in V-95 panels, had a drooped leading edge at 15% chord and housed more fuel and there were inner and outer ailerons with flaps between the hull and ailerons. Outer stabilising floats, again of semi-monocoque construction, were fitted and the tailplane was of the 'all-flying' type.

The M-70 used the hydroski method for take-off and landing and the 'undercarriage' comprised a nose ski (in D16AT alloy), welded titanium main hydrofoil, stabilising skis and a tail bumper. The hydrofoil retracted flush into the hull planing bottom, as did the bottom surface of the nose ski, while the hinge-mounted stabilising skis retracted forwards into a special well. Plenty of avionics were carried, including a Veter radar, and the defensive equipment included a tail barbette with two 23mm cannon, chaff dispensers and ECM equipment; AFA-45 cameras could be carried for reconnaissance operations.

The M-70's preliminary project was completed, on time, in September 1957, but it was never reviewed by the VVS because later that month the project was cancelled. The principal reason for the closure was that the USSR's first nuclear warhead to be specifically designed for an inter-continental ballistic missile, the Korolyov R-7, had been successfully tested on 15th May and this suggested that spending so much money on large and expensive supersonic flying boats and bombers could be a huge waste. Another factor, however, had been those members of the Myasishchev OKB who had not given their full support to the project; instead they had concentrated on the reconnaissance and in-flight refuelling roles when the latter did not even form part of the VVS's requirement. This political environment also put paid to Beriev's contender.

Maritime Aircraft – Data / Estimated Data

Project	Span ft in (m)	Length ft in (m)	Gross Wing Area ft² (m²)	Max Weight lb (kg)	Powerplant Thrust hp (kW) or lb (kN)	Max Speed / Height mph (km/h) / ft (m)	Armament
Beriev LL-143 (flown)	108 3 (33.0)	75 5.5 (23.0)	1,290 (120.0)	46,958 (21,300) normal	2 x ASh-72 2,250hp (1,678kW)	231 (371) at S/L	2 x 2 + 1 x 1 defensive cannon, up to 8,818lb (4,000kg) bombs
Beriev Be-6 (flown)	108 3 (33.0)	75 5.5 (23.0)	1,290 (120.0)	47,035 (21,335)	2 x ASh-73 2,400hp (1,790kW)	241 (388) at S/L	5 x 23mm cannon, up to 8,818lb (4,000kg) bombs
Beriev R-1 (flown)	65 7 (20.0)	65 3.5 (19.9)	624 (58.0)	44,753 (20,300)	2 x VK-1 5,950lb (26.4kN)	497 (800)	2 x 2 x 23mm cannon (nose and tail), 2,205lb (1,000kg) bombs carried externally
Beriev Be-10 (flown)	93 10 (28.6)	103 4 (31.5)	1,398 (130.0)	106,922 (48,500)	2 x AL-7PB 16,205 (72.0kN)	566 (910)	1 x 2 + 2 x 1 23mm cannon, up to 7,407lb (3,360kg) of torpedoes, anti-submarine bombs or mines
Beriev Be-12 (flown)	99 1 (30.2)	98 9 (30.1)	1,065 (99.0)	79,365 (36,000)	2 x AI-20 5,180hp (3,863kW)	342 (550)	Up to 6,614lb (3,000kg) of torpedoes, bombs or depth charges
Beriev Anti-Submarine Flying Boat 1962	230 7.5 (70.3)	173 10.5 (53.0)	4,839 (450)	462,963 (210,000)	4 x NK-12M 15,000hp (11,186kW) + 2 x AL-7PB 16,205 (72.0kN)	Cruise 404 (650)	Mix of sonobuoys, torpedoes and depth charges
Beriev Be-26 1963	186 4 (56.8)	146 0 (44.5)	2,688 (250)	220,459 (100,000)	2 x NK-15MV 15,000hp (11,186kW) + 16 x RD36-35 5,510lb (24.5kN)	Cruise 404 (650)	Sonobuoys, torpedoes, mines and depth charges up to 14,991lb (6,800kg) max
Beriev A-150 1965	137 9.5 (42.0)	163 3.5 (49.8)	5,376 (500)	374,780 (170,000) off land, 330,688 (150,000) off water	2 x NK-8 20,945lb (93.1kN) 12 x RD36-35P 8,160lb (36.3kN)	Cruise 559 (900)	Sonobuoys, bombs and mines
Beriev A-40 (flown)	136 6.5 (41.62)	143 8 (43.8)	2,151 (200.0)	189,594 (86,000)	2 x D-30KPV 26,483lb (117.7kN) + 2 x RD-60K 5,510lb (24.5kN)	472 (760)	Up to 14,330lb (6,500kg) of bombs, torpedoes or mines
Tupolev 'Project 504'	183 9 (56.0)	144 4 (44.0)	2,903 (270)	229,277 (104,000)	4 x ASh-2K 4,500hp (3,356kW)	360 (580)	3 x 2 x 23mm cannon, normal bomb load 13,228lb (6,000kg)
Tupolev Tu-142M (flown)	164 2 (50.04)	174 1.5 (53.07)	3,117 (289.9)	407,848 (185,000)	4 x NK-12MP	497 (800)	2 x 23mm cannon, up to 19,500lb (8,845kg) of anti-submarine weapons
Ilyushin Il-38	122 8 (37.4)	130 7 (39.8)	1,507 (140.2)	143,298 (65,000)	4 x AI-20M 4,190hp (3,124kW)	450 (724)	Up to 19,841lb (9,000kg) of anti-submarine stores
Bartini A-57 1957	119 9 (36.5)	228 0 (69.5)	8,118 (755)	705,467 (320,000)	4 x NK-10 22,050 (98.0) dry, 55,115 (245.0) reheat	1,554 (2,500) at c36,089 (11,000)	1 x Tsybin RSS, 1 x '244N' nuclear weapon or various conventional stores
Bartini E-57	70 6.5 (21.5)	148 3.5 (45.2)	2,903 (270)	264,550 (120,000)	2 x NK-10 22,050 (98.0) dry, 55,115 (245.0) reheat	1,554 (2,500)	1 x '244N' atomic bomb, 1 x K-10S missile or up to 6,614lb (3,000kg) bombs
Bartini A-58	84 2 (25.65)	151 7 (46.2)	?	220,459 (100,000)	4 x NK-6 44,090 (196.0) reheat	1,243 (2,000)	1 x '244N' bomb, 1 x RSS missile or up to 11,023lb (5,000kg) bombs
Moskalyov GSB-1 (GS-1)	108 3 (33.0)	229 8 (70.0)	3,602 (335)	up to 661,376 (300,000)	8 x VK-15F 22,050 (98.0) dry, 34,870 (155.0) reheat	1,864 (3,000) Mach 2.8	Up to 5 AAM, 2 gun turrets, one nuclear missile.
Beriev SD-MBR 1957	86 7.5 (26.4)	201.5.5 (61.4)	3,495 (325)	529,101 (240,000)	4 x NK-10 22,050 (98.0) dry, 55,115 (245.0) reheat	1,516 (2,440) at 33,000 (10,058)	1 x Kh-21 cruise missile or, (normal load) up to 11,023lb of freefall nuclear bombs
Myasishchev M-70 1956	118 9 (36.2)	209 0 (63.7)	3,688 (343)	529,101 (240,000)	4 x NK-6 49,605lb (220.5kN) maximum	1,057-1,119 (1,700-1,800)	2 x defensive cannon, up to 11,023lb (5,000kg) of bombs or 1 x Chelomei P-6 ('Article 44') missile

Chapter Ten

Ultimate Performance

The Sukhoi T-4 prototype.

This chapter looks at the Soviet Union's answers to the North American XB-70 Valkyrie supersonic strategic bomber prototype, which did not enter service. That American aeroplane was rather a political animal and perhaps it is no coincidence that none of the following Soviet designs entered service either, and only one of them actually flew.

Tupolev 'Aircraft 125'

Besides the effort put into the design of the advanced developments of the Tu-22 supersonic bomber described in Chapter 4, the Tupolev OKB also looked at some all-new layouts as a possible replacement for this type. One of these was 'Aircraft 125' on which work started in 1958 and this was intended to have a top speed approaching Mach 3. The preliminary calculations put together by S M Yeger's Technical Projects department suggested that the lift/drag ratio when flying at supersonic speeds would have to be at least 6, and 12 in subsonic flight, which meant that at least 88,180 lb (391.9kN) of thrust would be needed to get airborne an aeroplane that weighed between 220,459 lb and 275,573 lb (100,000kg and 125,000kg). In addition the Tu-22's VD-7M engines used too much fuel – something offering a much better rate of consumption would be needed for supersonic cruise operations.

Mach 3 performance automatically means that kinetic heat will be a problem for an airframe in that the strength of traditional aluminium alloys will deteriorate when operated in the temperatures generated by these speeds. Consequently the '125' employed a large proportion of steel and titanium in the areas most affected by heat, although aluminium was retained in the less critical spots. After analysing several dozen different layouts and engine combinations, the designers settled on a canard arrangement and, as the work progressed, the concept of a Tu-22 replacement gradually faded away as the '125' was turned into a pure strategic bomber intended to penetrate enemy airspace and defences at high speed and high altitude. In this form it matched Tupolev's 'Aircraft 135' described shortly, albeit at a total weight around 30% less, and from the early 1960s both projects ran pretty well in parallel because their designs had a great deal in common.

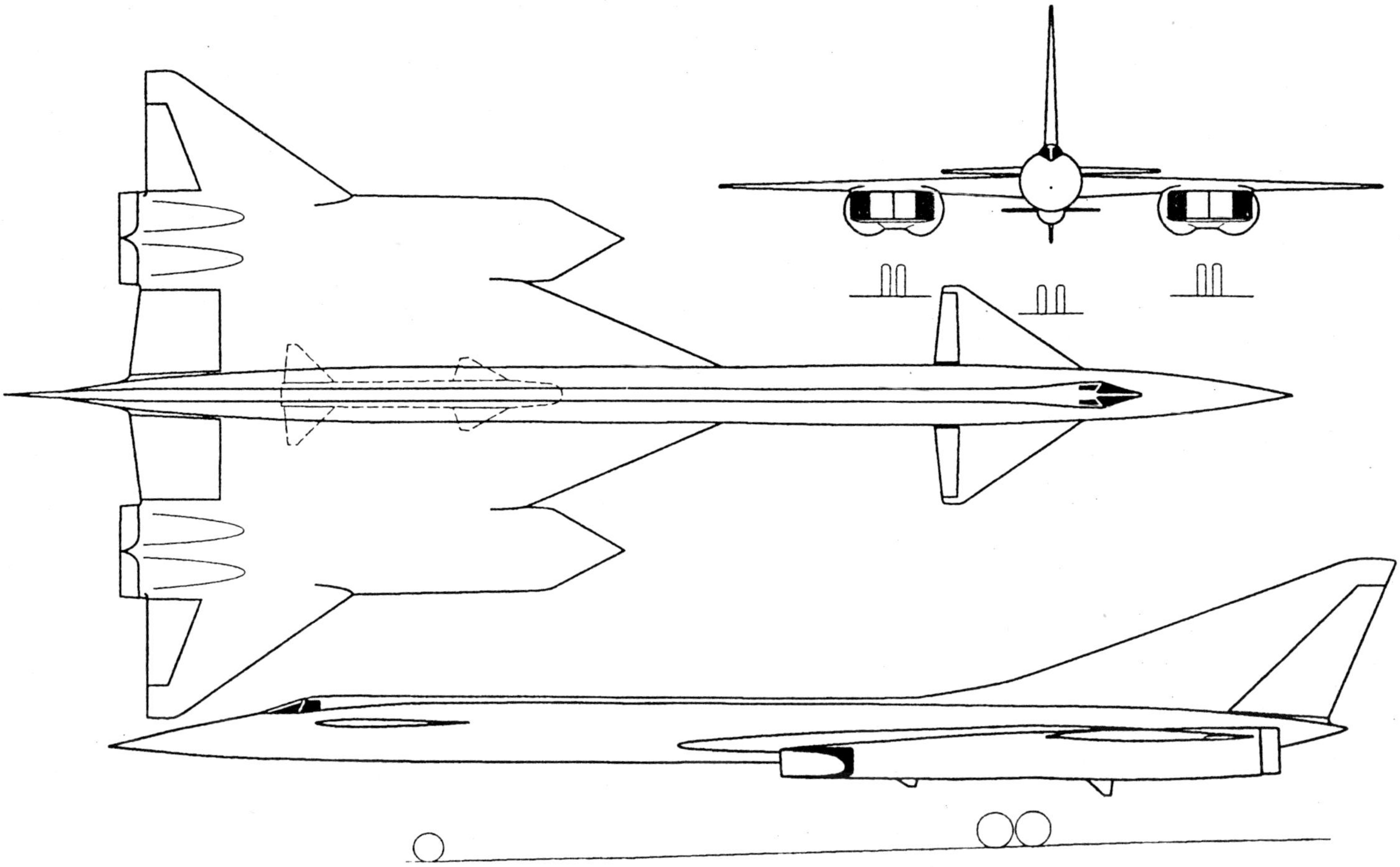

Tupolev 'Aircraft 125' in its 1960s form with four engines mounted in two nacelles. Note the cruise missile carried beneath the fuselage between the nacelles.

Another version of Tupolev's 'Aircraft 125'.
Russian Aviation Research Trust

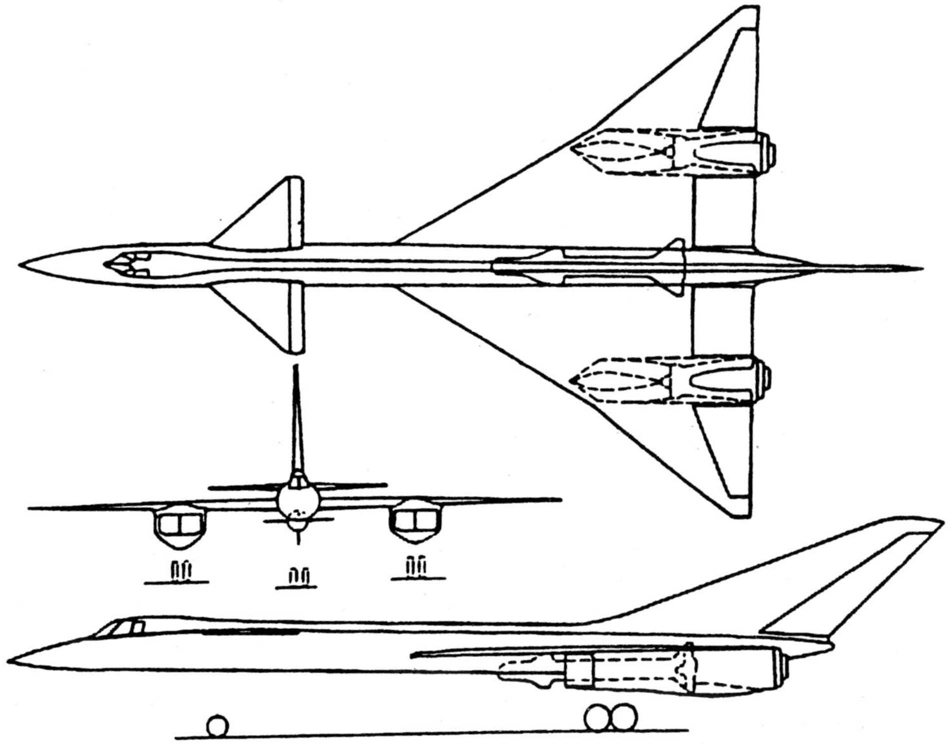

In its earliest '1958' form, the canard '125' had a long slim fuselage, a large cranked high-mounted delta wing and two Kuznetsov 52,910 lb (235.2kN) NK-6 or NK-10 engines mounted alongside the fuselage beneath the wing roots, the box intakes starting ahead of the wing. There were two fins placed about halfway out along each side of the wing, top speed was 1,678mph (2,700km/h), service ceiling 82,021ft (25,000m) and range 4,350 miles (7,000km). By the 1960s the aircraft had been altered and the layout, in general, was now quite similar to the '135' but with either two or four engines depending on the power of the chosen unit – the Tumansky R-15B-300 for example offered 33,070 lb (147.0kN) of thrust which meant four of these would be required instead of two 54,675 lb (243.0kN) NK-6Bs; the engines were housed in two underwing nacelles with adjustable super-

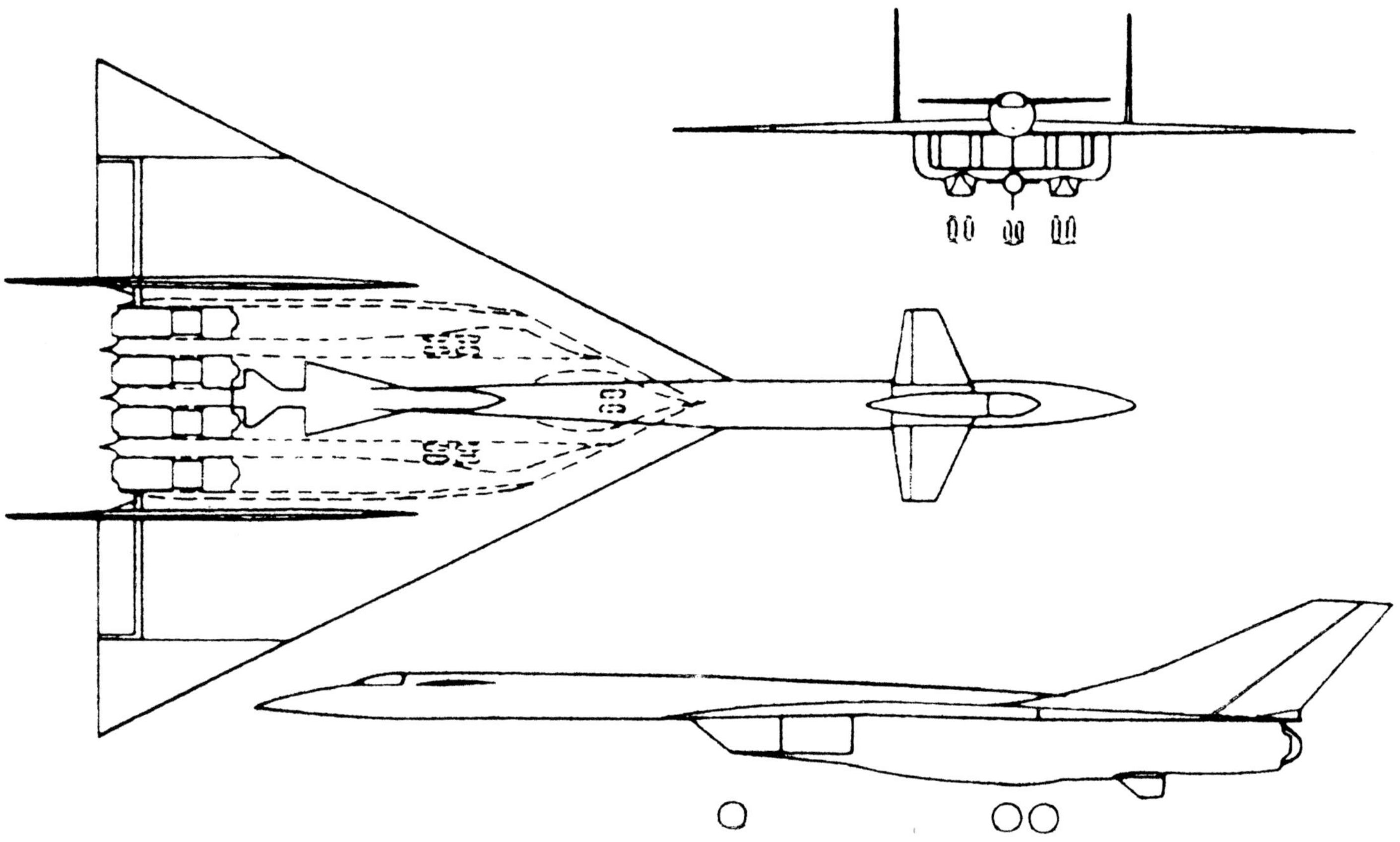

The Tupolev 'Aircraft 135' Version 'A' described in the text. This design had a span of 91ft 10½in (28.0m), length 147ft 0in (44.8m) and carried a cruise missile beneath the centre wing.

sonic intakes and vertical ramps. The Tumansky gave an estimated service ceiling of 82,021ft (25,000m) and a range at supersonic speed of 2,983 miles (4,800km), the Kuznetsov NK-6B about 65,617ft (20,000m) and up to 3,729 miles (6,000km) respectively, or as much as 5,594 miles (9,000km) when flying at subsonic speeds.

'Aircraft 125' was eventually offered in bomber and missile-carrier versions, as a reconnaissance aircraft, long-range interceptor and even as an anti-submarine aircraft, and was to be equipped with the latest state-of-the-art avionics. The primary weapon for the cruise missile version was a single store weighing between 7,716lb and 8,818lb (3,500kg and 4,000kg) carried beneath the fuselage and loaded with a 1,984lb (900kg) warhead; this had a range of 311 to 373 miles (500km to 600km). By the mid-1960s however, progress with the deployment of ground-based strategic missiles throughout the USSR was well advanced and officials turned their attention to a long-range multi-role bomber, which resulted in the Tupolev Tu-22M *Backfire* described in Chapter 11. In addition Sukhoi was building the T-4 (below) and so the '125' project was eventually abandoned without ever moving beyond the drawing board or wind tunnel testing stages. The '125' was actually a very radical design and its tail-first layout did reveal some inherent shortcomings such as poor directional stability.

Tupolev 'Aircraft 135'

The official merger of the Myasishchev design bureau with, and as a branch of, the Chelomey design bureau was confirmed by a USSR Council of Ministers Decree issued on 3rd October 1960, which also released Myasishchev from undertaking further work on the M-56 missile carrier (Chapter 4). However, despite the politics and Khrushchev's general lack of interest in large aeroplanes, the loss of the M-56 did leave a gap in the Soviet Union's studies into potential future military aircraft and, as a consequence, the same Decree requested that Tupolev should put together a new proposal for a supersonic long-range missile carrier and reconnaissance aircraft; in due course this project was developed into a full multi-purpose aeroplane.

The result of this was that the Tupolev OKB's Technical Projects department began studies under the designation 'Aircraft 135' and this work eventually lasted nearly five years and examined dozens of designs. Initially no less than nine different engines were considered (plus another version with a nuclear powerplant) and their different thrust ratings meant that either four or six power units would be required. Fourteen different models were wind tunnel tested, including six different wing shapes and ten engine positions. Alternative engine nacelles and engine intakes were tried and, in particular, the effects of interference between the nacelles, wing and fuselage were studied closely.

Several configurations were developed in parallel and three of these had four NK-6 turbofans located in a common nacelle under the wing centre-section, each engine having a separate air intake with a vertical intake ramp. Where the three versions differed was in the wing and canard planform – Version 'A' had trapezoidal canards and pure delta wings with 65° leading-edge sweep, 'B' had triangular canards and double-delta wings with 65° leading-edge sweep on the inboard panels and 45° on the outboard panels, and 'C' had triangular canards and delta wings with swept trailing edges and 65° leading-edge sweep. Their respective estimated ranges when flying at 1,554mph (2,500km/h) were

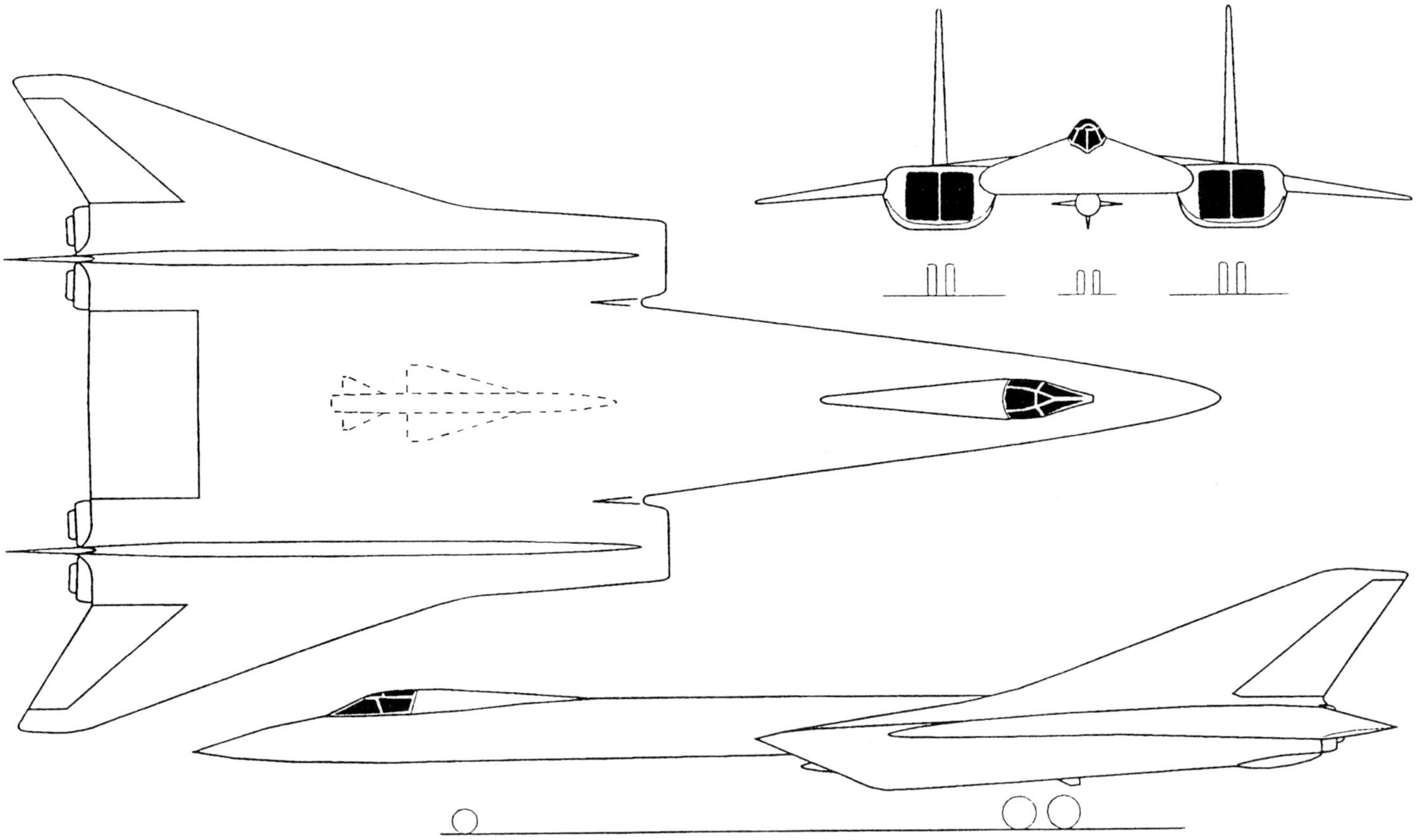

Tupolev Tu-135 in 'flying wing' form (early 1960s?).

4,941 miles (7,950km), 4,879 miles (7,850km) and 5,158 miles (8,300km), and service ceilings in full afterburner 73,819ft (22,500m), 74,475ft (22,700m) and 74,475ft (22,700m). A fourth version envisaged an aircraft with paired NK-6s under the outer wings. In the end, just like Tupolev's '125', the final choice had a canard foreplane, although here it was all-moving, and the favoured wing shape was a cranked delta but with just a single fin; in addition the paired nacelles were placed well apart beneath the wing.

The '135' had much in common with the XB-70 Valkyrie experimental bomber and was generally similar in appearance, so Tupolev also kept a close eye on the progress of his American rival. A recent arrival within the OKB was L L Selyakov, who had previously worked on the M-50 series at Myasishchev and who had been transferred after the latter's closure, and it was Andrei Tupolev's intention to let Selyakov lead the '135' programme. Concurrent with the airframe research, a lot of effort also went into designing the '135's navigation and flight control systems and its missile armament. The problem of kinetic heat meant that the resulting 'Aircraft 135' was actually speed limited to 1,864mph (3,000km/h) Mach 2.82, with long-term cruise limited to 1,554mph to 1,647mph (2,500km/h to 2,650km/h) Mach 2.35 to 2.5. This allowed a greater volume of aluminium light alloys to be used in the airframe with steel and titanium only employed for the most highly stressed components, a move which also ensured that the manufacture of the airframe would be an easier and less time-consuming task using well-known technology.

The selection of the Kuznetsov NK-6 bypass turbofan offered a better range over the alternative engines during both subsonic and supersonic flight, plus the possibility of long flights at low altitudes; in addition a multi-wheel undercarriage was required to ensure that the aircraft could operate from unpaved strips or airfields built with lower grade concrete runways. Over water 'Aircraft 135' was expected to attack and destroy enemy warships and aircraft carrier groups, ballistic missile submarines or merchant ships and convoys, for which (as the Tu-135K) it would carry a selection of cruise or guided missiles capable of a range of over 311 miles (500km); in addition depth charges or torpedoes could be carried and the '135' would be fitted with navigation and search radars. The aircraft's range had to be 3,108 miles (5,000km), or 3,729 miles (6,000km) with IFR, and the '135' was expected to undertake long patrols of up to eight hours when 1,243 miles (2,000km) from base, or 2.7 hours when 2,486 miles (4,000km) from base.

Other roles included the harassment of enemy air transport lines, for which at least four air-to-air missiles would be carried, electronic and photographic reconnaissance at heights in excess of 65,617ft (20,000m) (and possibly as much as 78,740ft [24,000m]), and the elimination of small heavily defended strategic targets with two cruise or ballistic missiles having a range of 93 to 218 miles (150km to 350km) or with nuclear stores; for the latter operation the '135' itself would require a range of 1,864 miles (3,000km). The '135' was not intended to be an intercontinental bomber. The reconnaissance version was based on the Tu-135K missile carrier and differed only in having cameras and electronic intelligence (ELINT) equipment housed in the missile bay. A supersonic passenger transport variant was also proposed, the '135P', which offered a capability that came very close to the Tu-144 supersonic transport begun in 1963 and flown in 1968; in fact some of the '135's engineering was later used in that aircraft. During its life 'Aircraft 135' was constantly modified and a later ver-

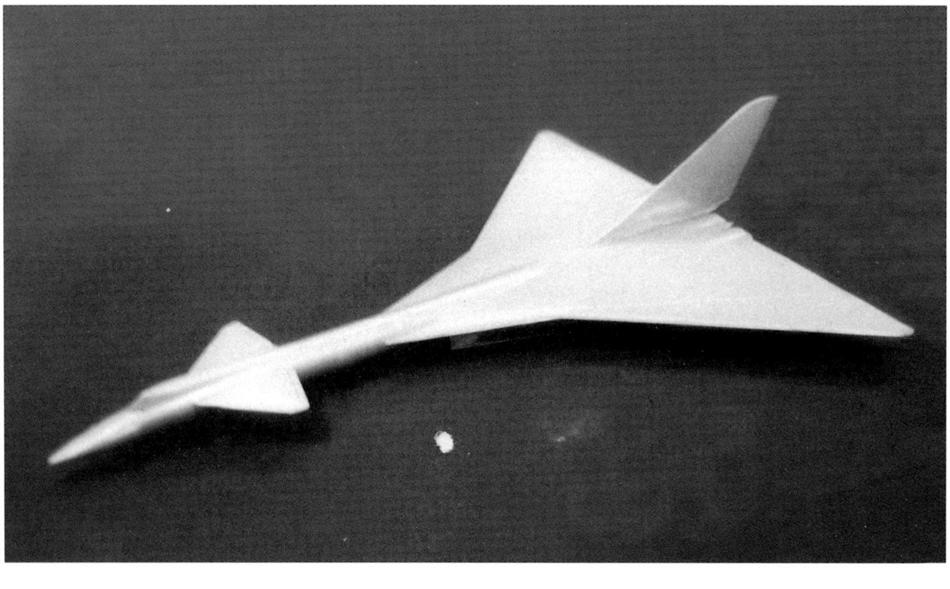

Three wind tunnel models showing two arrangements considered for the Tupolev 'Aircraft 135'. The first pair show a design with twin fins plus the four engines paired in nacelles; the third reveals an alternative with a single fin and the four engines grouped together under the centre wing.

sion showed a near flying-wing layout where the canard had disappeared to be replaced by long leading-edge root extensions which stretched as far as the nose.

By the mid-1960s work on 'Aircraft 135' had been abandoned, once again because the Air Force now wanted a heavy multi-role bomber and missile carrier type which was eventually fulfilled by the Tu-22M and the Tu-160 *Blackjack* (Chapter 11); a very similar step was taken in America when the B-70 was abandoned and replaced by the Rockwell B-1 Lancer. However, from a design point of view the '135' also presented plenty of problems, particularly in terms of creating sufficiently advanced avionics for targeting, navigation and flight control, and this eventually pushed up the cost quite considerably. That said, the progress and results achieved on the '135' proved of great benefit in the design of the follow-on Tu-22M and Tu-160. Both of Tupolev's 'Aircraft 125' and '135' also became involved in a design 'competition' with Sukhoi and Yakovlev, which is described shortly.

Tupolev 'Aircraft 137'

This long-range supersonic strategic missile carrier represented a larger and heavier development of 'Aircraft 135'. It was a tailless design and, to reduce the lift/drag ratio in cruise flight, the six NK-6 or NK-10 engines were housed in nacelles, and used air intakes, of the type employed for the first time on the Myasishchev M-56 (Chapter 4). 'Aircraft 137' represented only a brief study.

Sukhoi T-4

As noted several times already, Soviet leader Nikita Khrushchev had a soft spot for missiles and this biased defence policy set his nation's military aviation back a good deal with many promising designs being killed off. However, the rapid development of the USA's quick reaction forces in the early 1960s, especially for carrier-borne aviation, forced Khrushchev to change his point of view and agree to the development of a new strategic bomber/reconnaissance aircraft. Following a careful analysis of the tasks this aircraft would need to undertake, the basic requirements for the type were put together by the State Research Institute of Aviation Systems (GosNIIAS) in Moscow. Special importance was attached to attacking non-strategic enemy strike forces

Model of the Tupolev Tu-135. George Cox

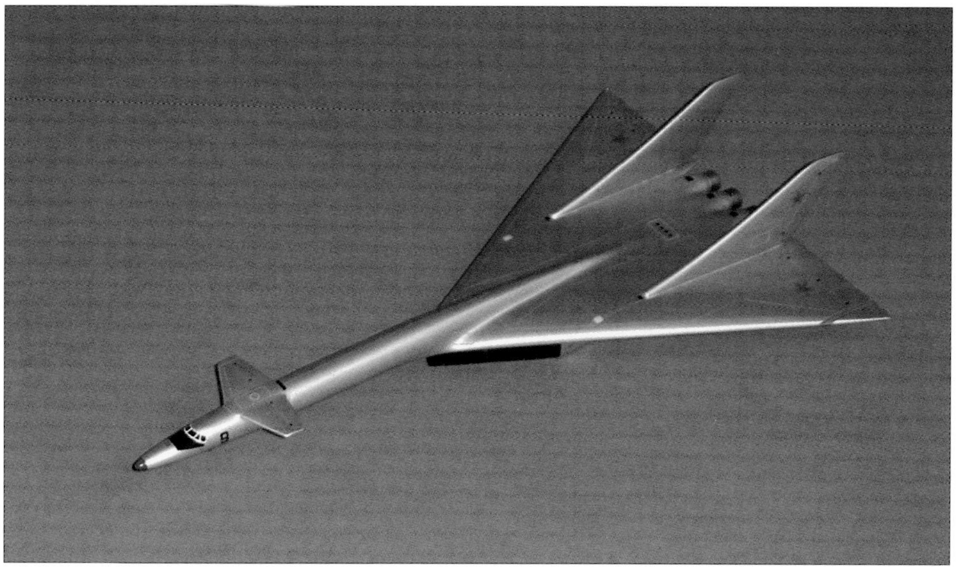

Sukhoi's I-2 (T-4) study as designed by Oleg Samoilovich (3.62). The drawing shows that the wingtips could be drooped. Russian Aviation Research Trust

(primarily large surface ships such as aircraft carriers and other naval and merchant vessels) as well as land-based strategic targets such as ICBM silos and command posts; strategic reconnaissance was regarded as a secondary role.

The bomber was to carry two air-to-surface missiles to be launched 311 miles (500km) from their target, and was to have a top speed of Mach 3.0, a cruise speed of Mach 2.8 and a range of 3,729 miles (6,000km). It also had to be capable of operating from dirt strips as well as paved surfaces and its structure was to be built of heat-resistant steel and titanium. The Sukhoi OKB responded with the T-4 or 'Project 100' the second designation being derived from the aircraft's projected gross weight of 100 tonnes (220,459 lb); hence it was often referred to as the *sotka* – 'the hundred' and the alternative name acted as a cover after the T-4 was raised to Top Secret status. As a result many Western publications have incorrectly called the aircraft Sukhoi '100'; in fact the T-4's weight eventually rose to over 130 tonnes (286,596 lb).

Work began at Sukhoi in September 1961, the new type being known as the 'aircraft carrier killer', and, without authorisation, the bureau's Oleg Samoilovich produced his own design to the requirement which was later accepted for further development. Sukhoi's overall proposals did include Samoilovich's I-2 design of March 1962 which had all of its engines grouped together beneath the wing to be fed by a large intake, plus a canard and a cranked delta wing. Initially two different turbojets appeared suited to power the aircraft, the Tumansky R15BF-300 (an uprated version of the engine fitted in the Mikoyan MiG-25 *Foxbat* fighter) rated at 22,045 lb (98.0kN) thrust dry and up to 31,965 lb (142.1kN) in full afterburner, or the Zubets RD17-15. In the meantime Aleksandr Bereznyak's design group was tasked with developing the anti-ship missile.

By now Andrei Tupolev was established as the leading Soviet designer of large bomber aircraft and, having had to cope recently with the rival Myasishchev OKB as described elsewhere, he was far from pleased to see yet more competition in the shape of Sukhoi. Tupolev knew Pavel Sukhoi well since the

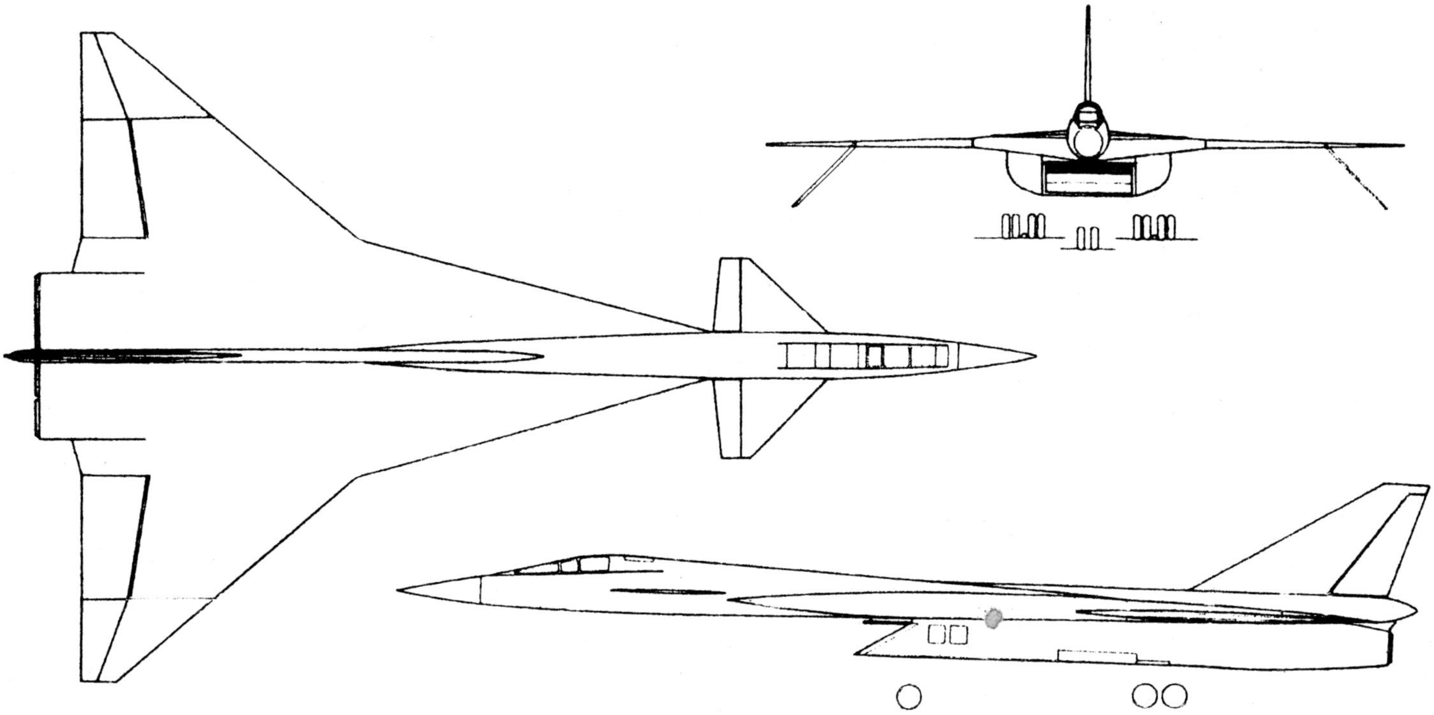

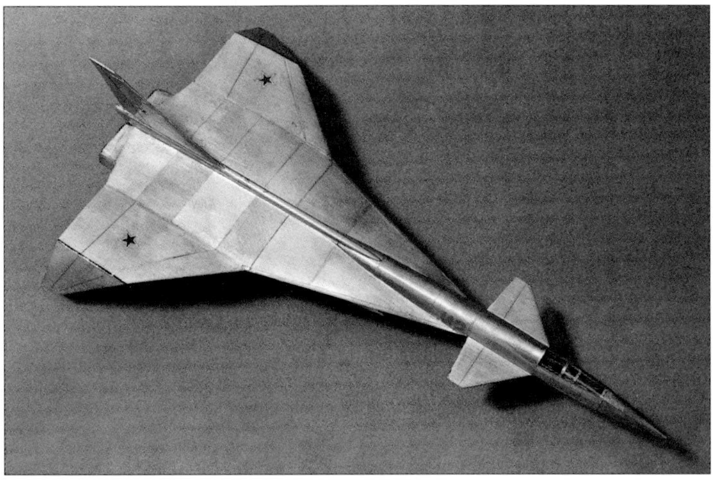

Model of the Sukhoi I-2. John Hall

latter had been his aide for many years. Annoyed by the young upstart challenging his authority, Tupolev even remarked at a session of the State Committee for Aircraft Equipment (GKAT) 'Sukhoi won't cope with such an aircraft. I know it because he was my student.' Sukhoi was quick to retort. 'I *will* cope because, as you have rightly pointed out, I was your student!'

During the spring of 1962 a design competition was organised for the new requirement and three Bureaux submitted proposals – Tupolev's group tendered the Tu-135, Yakovlev the Yak-33 (below) and Sukhoi the T-4. All three submitted their projects to GKAT for review but Yakovlev's was rejected pretty quickly because it was too small and did not meet the VVS requirements. However, the duel between the Tupolev and Sukhoi contenders continued for a long time with much of the debate centred on the choice of structural materials (aluminium versus titanium and steel), but Sukhoi's T-4 eventually won and a go-ahead was given for advanced design studies to proceed. (In truth this competition had not been simply just a decision to select the best design, it was much more involved and further details follow at the end of the chapter). However, to make the go-ahead binding the T-4 needed to be approved by a Government Decree and after close scrutiny by numerous research institutes, plus the efforts of Sukhoi's engineers campaigning on the bomber's behalf, the decree was finally signed. The next job would be to find the optimum aerodynamic layout that would permit prolonged Mach 3 flight at high altitude.

Although Pavel Sukhoi took the T-4 project very seriously, many of his colleagues did not. Yevgeni Ivanov, Sukhoi's first deputy, was not in favour of the OKB working on large bomber

The attractive Sukhoi P-4 represented the T-4's preliminary design and was completed in April 1963. Russian Aviation Research Trust

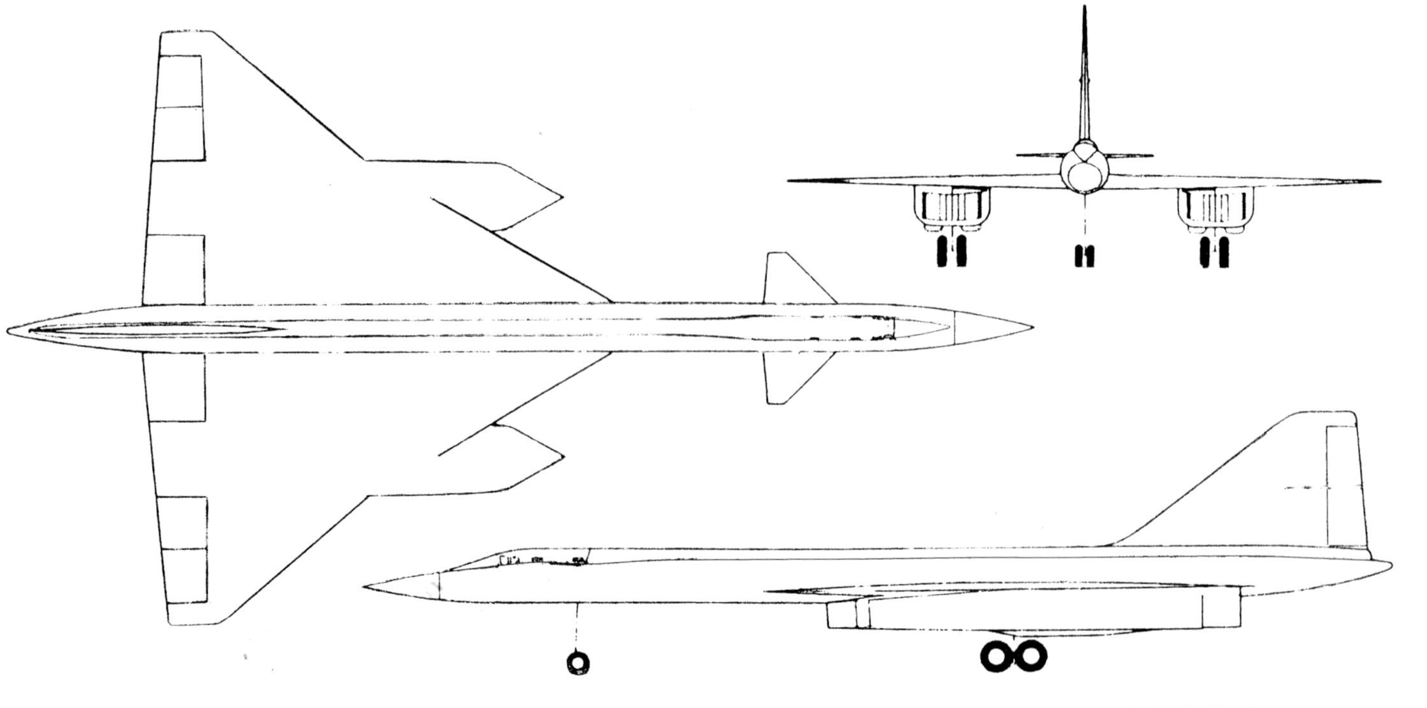

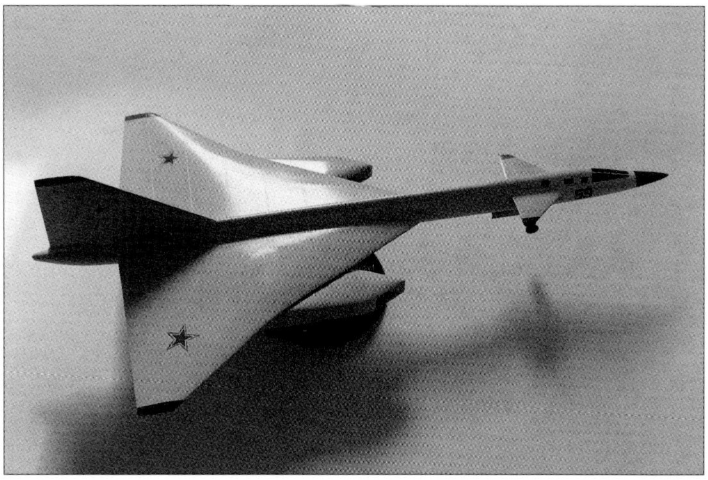

Two views of a model of the Sukhoi P-4. George Cox

projects and he was set against the T-4 right from the start, while the heads of some other sections within the design bureau also looked upon this departure from their traditional area of fighter design with some hostility. Nevertheless, the T-4 and its development programme stayed and proposals were also suggested for reconnaissance, long-range interceptor and airliner versions.

During its early studies the Sukhoi Bureau considered about 130 different designs, including nuclear and hydrogen-powered bombers, and 34 of these were semi-officially approved with Pavel Sukhoi's signature. The canard P-4 layout represented the preliminary design completed in April 1963 and had a long slim fuselage, a small blended canopy and two underwing nacelles with two engines in each fed by divided air intakes placed ahead of the wing leading edge. The main wheels also retracted into these nacelles and the weapons were housed in three parallel bomb bays with the middle one placed along the longitudinal axis. Sukhoi's designers had been interested in a canard format right from the start, although there were regular attempts to revert to a tailless design, but a conventional tailed arrangement was rejected very early in the research.

The T-4's advanced preliminary project was submitted to the Soviet Air Force and GKAT for review in April 1963 and, between 21st May and 3rd June, the design was thoroughly examined. Eventually the commission ruled that the T-4 project met the VVS requirements in terms of task, performance, armament and avionics, and indeed it was considered to be far superior to all contemporary aircraft in its class. A T-4 powered by either four Tumansky R15BF-300 or Zubets RD17-15 engines was now declared accept-

The Sukhoi K-2 was another design to be examined under the T-4 umbrella (1964). Russian Aviation Research Trust

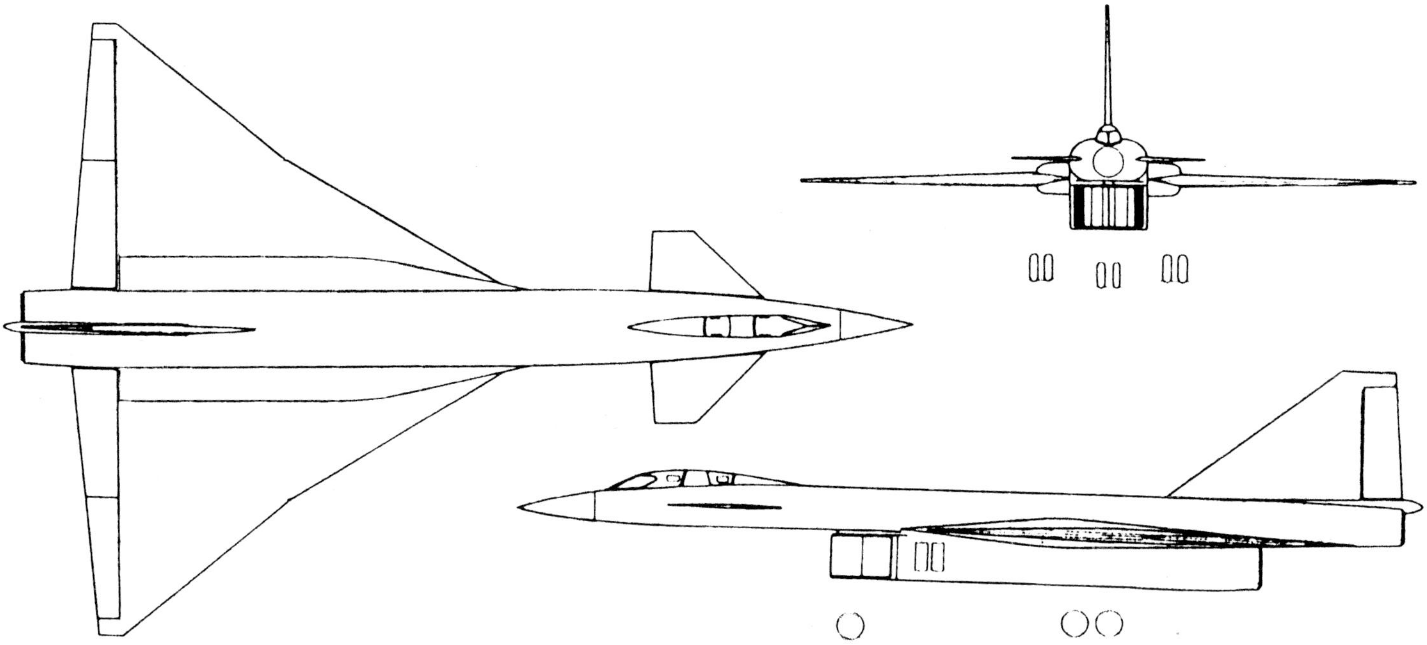

Model of the Sukhoi K-2.

Sukhoi's T-4 bomber is seen just before touch-down after one of its precious few test flights.

able for a full-scale mock-up to be built and for the issue of production drawings.

In late 1963 a special decree was passed by the Central Committee of the Communist Party and the USSR Council of Ministers that cleared the aircraft for full-scale development and this document set a tentative first flight date of 1968. A factory in Tushino was ordered to build the prototypes and static test airframe but the plant's directors immediately rebelled against this and fought a bitter but unsuccessful battle to get the job moved elsewhere. Between 1963 and 1965 a massive wind tunnel effort was undertaken by TsAGI to assess configurations with variations in wing sweep angle, aspect ratio and planform, different fuselage lengths and with or without a protruding canopy and spine. Various shapes and areas of fin and canard foreplanes were also compared and the location of the canards themselves was chosen carefully; the position of the engines was also changed on several occasions. A new SovMin resolution followed in December 1965 which cleared the way for construction to begin.

In 1964 the first information was received about the American Lockheed SR-71 Mach 3 aircraft, which provided a big shock. A similar layout was assessed by Sukhoi but was found to be unsuited for the required task and the designers then realised that the SR-71 had not been produced as a bomber. During 1966 there were still two possible layouts for the T-4 – the earlier P-4 type with the engines separated in twin nacelles or an alternative with all four engines grouped under the fuselage. The second format cut down the level of drag and reduced surface area and gave a better integration between the engine package and the wings, which also increased the wing's aerodynamic efficiency. Therefore this 'all together' layout was selected as the definitive solution and approved by Sukhoi.

By 1965 both the R15BF-300 and RD17-15 had dropped out of the picture, the T-4 having outgrown them, and when the full-scale T-4 mock-up was demonstrated to the VVS in December 1966 (right on schedule) four

35,275 lb (156.8kN) RD36-41 afterburning turbojets had taken their place. This engine had been developed by Pyotr A Kolesov of the Rybinsk Engine OKB and was expected to give the T-4 a cruising speed of 1,989mph (3,200km/h) at 65,617ft (20,000m) and a maximum speed at sea level of 684mph (1,100km/h). Between 17th January and 2nd February 1967 the mock-up commission reviewed two versions of the T-4, for strike and reconnaissance, with different equipment packages fitted into the same airframe.

In the meantime several other aircraft had been adapted as test beds for the development of numerous items of equipment and in 1966 a converted Sukhoi Su-9 *Fishpot* interceptor fighter began to evaluate the T-4's wing planform. This aircraft became known as the '100L' – in other words the testbed for 'Project 100', L standing for *(letayuschchaya) laboratoriya* or testbed. Sections were added to the Su-9's standard delta to act as a scale model for the T-4 wing, but between 1966 and 1970 a selection of other alternative wing shapes was also assessed by this aircraft. This was just part of an enormous volume of research needed to support the T-4 programme.

In another part of the programme Sukhoi, who had traditionally dealt purely with aircraft and never missiles, designed the T-4's 9,921 lb (4,500kg) Kh-45 anti-ship missile, a semi-active radar homing weapon which had a 1,102 lb (500kg) warhead and an operational range of 373 miles (600km); in fact this weapon was so large that it proved difficult to carry it aboard the T-4 and this was one of the reasons why the bomber's layout had had to be revised. In addition kinetic heating of the airframe from air friction at high Mach numbers, to temperatures approaching 300°C, led to the development of some new heat-resistant titanium alloys (VT-20, VT-22) and steels (VNS-2), plus the new technologies that were needed to cast, shape, machine and weld these materials. (Note: VT stands for *vysokoprochnyy teetahn* or high-strength titanium, VNS means *vysokoprochnaya nerzhaveyuschchaya stahl* or high-strength stainless steel). The same problem saw the end of the protruding cockpit canopy; from now on the cockpit was completely enclosed in the fuselage and the nose forward of the crew position was to be lowered for take-off and landing, which gave the pilot an excellent field of view.

Within the OKB there was criticism of TsAGI, whose recommendations were sometimes taken with more than a little indifference, for example when it was suggested that the T-4's eight-wheel undercarriage should be replaced by an alternative fitted with 32 wheels. Oleg Samoilovich quotes Roberto Bartini who once declared that 'The TsAGI is the Temple of Science, but it is too marble!' The outbreak of the Arab-Israeli War in 1967 also, to a degree, helped the T-4's cause in that some of the USSR's military leaders now gave more attention to the aircraft. The Minister of Defence, A A Grechko, visited Sukhoi and expressed a wish to see the T-4 flying in 1970; Pavel Sukhoi agreed to this deadline but his design bureau was essentially too small to complete such a massive aircraft so soon and the date was missed.

The first T-4 sections were completed in 1969 and the second airframe, designated 'Aircraft 100S' (for *stateecheskiye ispytaniya* or static testing), was completed in 1972. The first flying prototype arrived in Zhukovsky on 30th December 1971 but the volume of preflight testing that it required, plus various other problems not associated with the aircraft, prevented the first flight from taking place until 22nd August 1972. In the event the aircraft made only ten flights (which seems an incredible waste, even from just a research point of view) but it did achieve Mach 1.28. Meanwhile, the construction of the next airframes got moving at Tushino and the second prototype was completed in 1973 and expected to fly towards the end of that year. The Air Force request for the 1970 to 1975 Five-Year Plan included no less than 250 T-4s, which gave the then Minister of Aircraft Industry, Pyotr Dementyev, quite a shock.

In the meantime Andrei Tupolev, sensing the threat to his position, kept urging everyone to 'leave well alone' and to go for improved versions of his Tu-22 (Chapter 4), arguing that modifying the *Blinder* would be much easier than trying to build the T-4. Concurrently the Air Force placed a big order for MiG-23 *Flogger* tactical fighters and Dementyev told Defence Minister Marshal Andrei Grechko that this could only be fulfilled if the T-4 programme was cancelled. That tipped the balance and, soon afterwards, a directive followed ordering Tupolev to go ahead with his Tu-22M ('modified'), a project which was to enter service as the *Backfire*.

This was the end for the T-4 which was cancelled in 1974 (without protest from Pavel Sukhoi). The tooling at the Kazan factory which had been assembled for the large scale manufacture of production aeroplanes was gradually dismantled, as were several unfinished airframes at Tushino. In 1982 the first prototype was donated to the VVS Museum in Monino near Moscow, and for a period parts of the second airframe were used as teaching aids at the Moscow Aviation Institute before eventually being scrapped. It was the Air Force's desire to have the Tu-22M that prevented any series production of the T-4 from taking place, but Tupolev's designation was totally misleading because the Tu-22M (initially called 'Aircraft 145') was actually an all-new design with little or nothing in common with the original Tu-22. During the late 1970s there were proposals to convert the T-4 into a long-range interceptor armed with air-to-air missiles for attacking American B-52s and B-1s loaded with cruise missiles, but these ideas were rejected as impractical.

Yakovlev Yak-33

At the time of writing most of the details for Yakovlev's submission are, somewhat surprisingly, still secret. The designation Yak-33 eventually covered two alternative designs, both of which used a relatively small delta wing and were fitted with additional vertical-mounted lift engines to improve their take-off performance. The first had two propulsive engines placed side-by-side in the rear fuselage and a battery of six lightweight lift jets in two rows ahead of them in between the box intakes; for take-off the main engine jet pipes were to be diverted downwards ahead of the afterburners to give extra lift. This design had a long slim-fuselage and a bicycle undercarriage, two crew seated in tandem and a large attack radar in the nose. The second project had wingtip engine nacelles, plus canard foreplanes for improved longitudinal control. Two of the lift jets were still placed (side-by-side) in the fuselage but each nacelle housed two more lift jets, a single propulsive unit and an outrigger for the bicycle undercarriage.

–

There is uncertainty concerning the involvement of the Tupolev '135' in the 'design competition'. Some sources state that the Government wished to see the '135' stopped but it could not close down every large aircraft project just like that, particularly taking in to account the personality of Andrei Tupolev. Pyotr Dementyev, as the Minister of Aircraft Industry and supporter or implementer of Government policy, was keen to find ways of opposing the Tupolev '135's development and he decided to invite the Myasishchev OKB, plus the Sukhoi and Yakovlev fighter specialist design bureaux, to provide project proposals for a new bomber. A side issue was to see how the two OKBs that were most familiar with building relatively small aeroplanes, with the emphasis on saving weight, would handle the development of a much larger type. In truth the Minister was playing a political game but Sukhoi, for one, took the programme seriously and thus created and submitted the T-4.

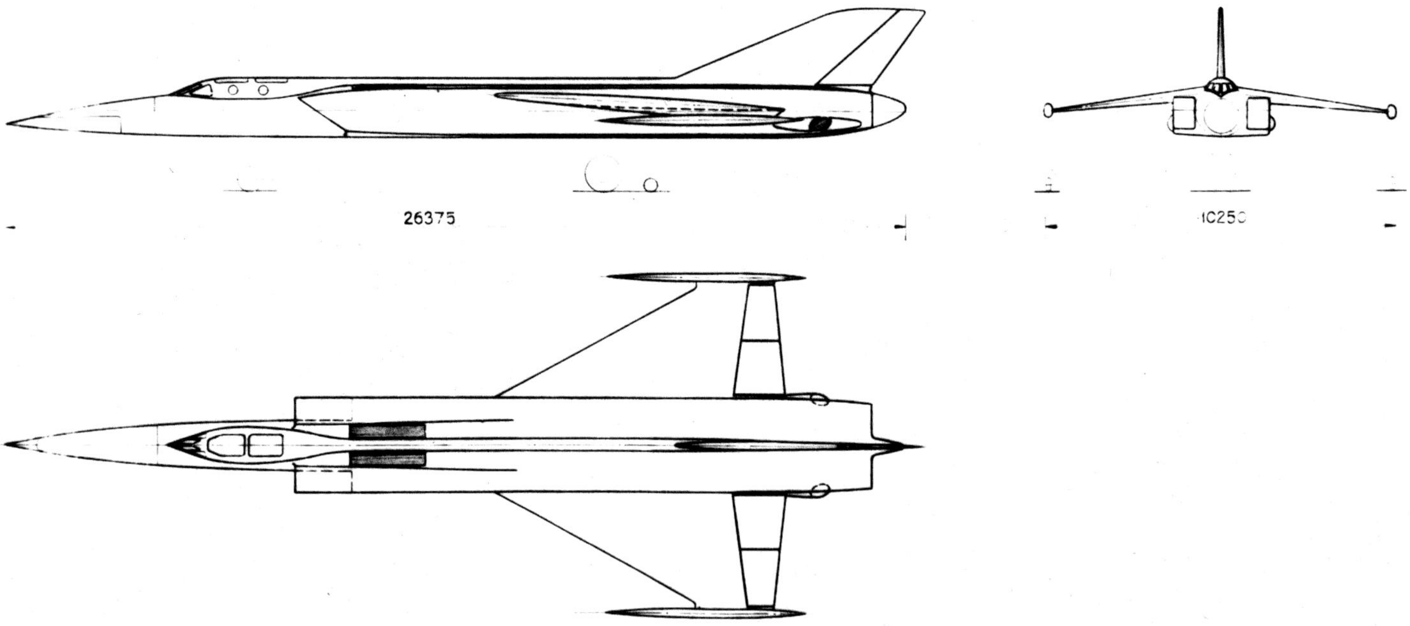

Above: **The Yakovlev Yak-33 design with fuselage-mounted main engines (c1962).**

Opposite page: **Yakovlev Yak-33 with engines in wingtip nacelles (c1962).**

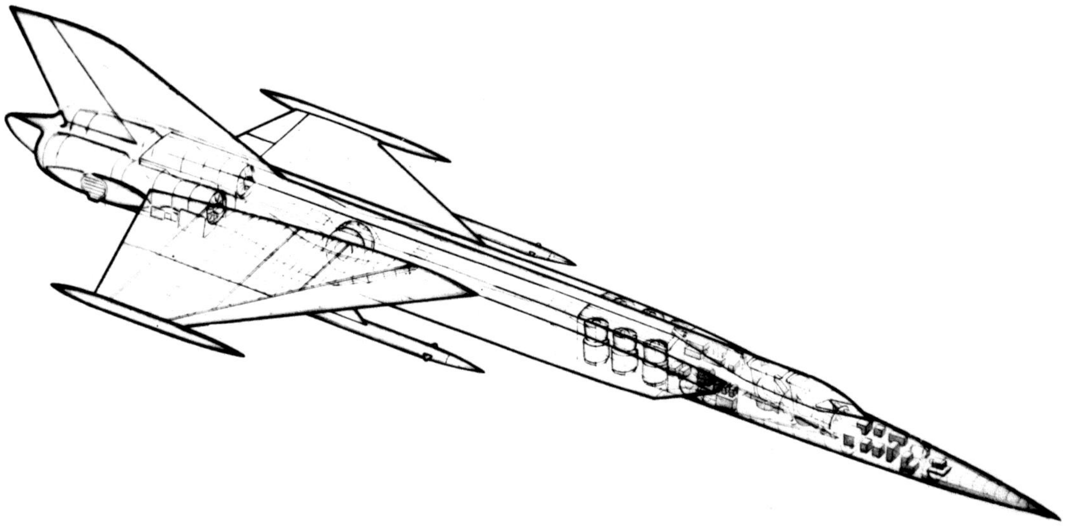

This sketch of the first version of the Yakovlev Yak-33 shows the main engines housed in the rear fuselage and two lines of lift jets behind the cockpit. Note the weapons hung on underwing pylons.

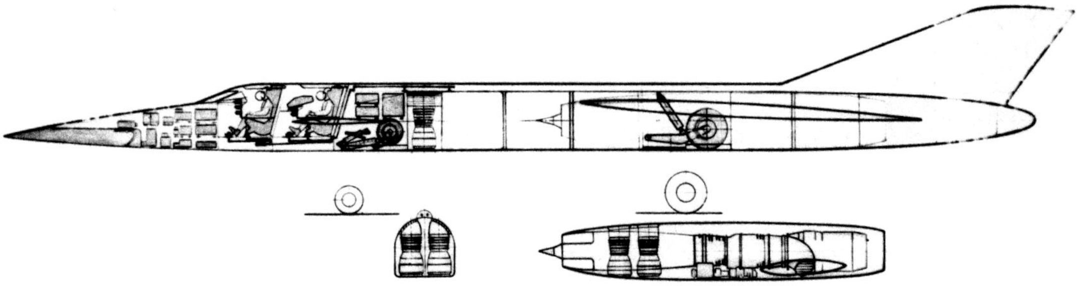

This cutaway of the second Yak-33 design reveals some internal detail for both the fuselage and nacelles.

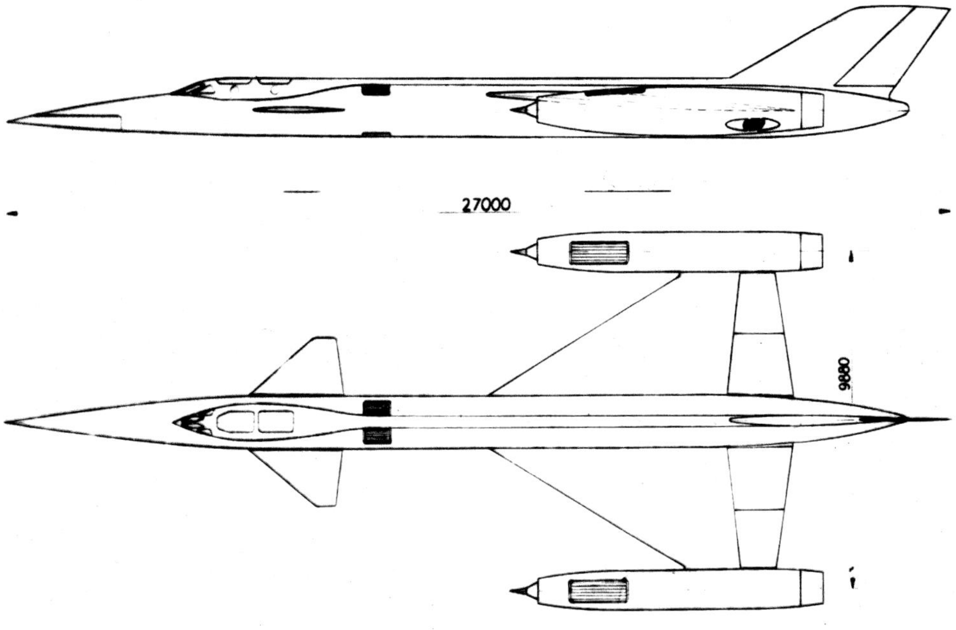

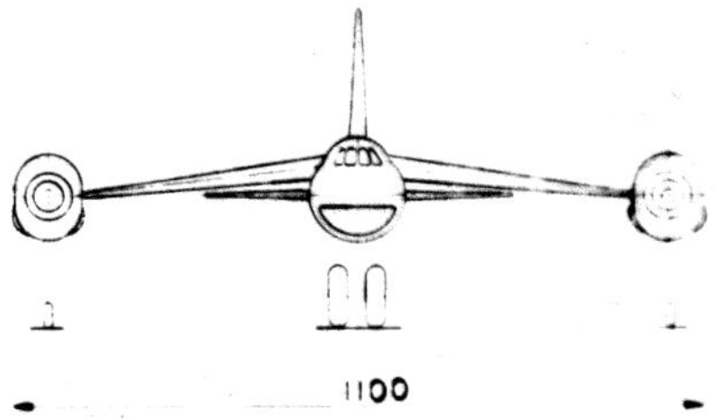

During one of the official evaluation meetings, held on 21st May 1963, the Tupolev OKB's Sergei Yeger argued that the competition had been arranged 'improperly' and he claimed that a speed of 1,865mph (3,000km/h) might, during an attack, give some gain in time, but it would also absorb a colossal amount of money to bring to fruition; in addition such an aircraft would weigh at least 440,917 lb (200,000kg). Both Pavel Sukhoi and Aleksandr Yakovlev presented their designs in person and by the next meeting, on 3rd June, Andrei Tupolev had acknowledged that the '135' would not be accepted and so offered his 'Aircraft 125' instead. Yakovlev's Yak-33 was soon rejected but the surviving Sukhoi and Tupolev projects were assessed for a long time. Nevertheless, it was Sukhoi's T-4 that won the day and support for Tupolev's designs was indeed withdrawn; yet, as already noted, there was never any real conviction for putting the T-4 itself into mass production.

Heavy Bombers – Data / Estimated Data

Project	*Span ft in (m)*	*Length ft in (m)*	*Gross Wing Area ft² (m²)*	*Max Weight lb (kg)*	*Powerplant Thrust lb (kN)*	*Max Speed / Height mph (km/h) / ft (m)*	*Armament*
Tupolev 'Aircraft 125' (1960s)	81 0.5 (24.7)	126 0 (38.4)	2,430 (226.0)	275,573 (125,000)	2 x NK-6B 54,675 (243.0)	Cruise up to 1,647 (2,650)	1 x Kh-22 (AS-4 *Kitchen*) cruise missile
Tupolev 'Aircraft 125' (1960s)	72 10 (22.2)	135 10 (41.4)	2,430 (226.0)	242,504 (110,000)	4 x R-15B-300 33,070 (147.0)	2,175 (3,500)	1 x Kh-22 (AS 4 *Kitchen*) cruise missile
Tupolev 'Aircraft 135' (Version A)	91 10.5 (28.0)	147 0 (44.8)	4,086 (380.0)	418,871 (190,000)	4 x NK-6 44,090 (196.0) reheat	1,554 (2,500)?	1 x Kh-22 (AS-4 *Kitchen*) cruise missile
Tupolev 'Aircraft 135' (Version B)	116 5.5 (35.5)	143 8.5 (43.8)	4,086 (380.0)	414,462 (188,000)	4 x NK-6 44,090 (196.0) reheat	1,554 (2,500)?	1 x Kh-22 (AS-4 *Kitchen*) cruise missile
Tupolev 'Aircraft 135' (Version C)	91 10.5 (28.0)	147 0 (44.8)	4,086 (380.0)	414,462 (188,000)	4 x NK-6 44,090 (196.0) reheat	1,554 (2,500)?	1 x Kh-22 (AS-4 *Kitchen*) cruise missile
Tupolev 'Aircraft 135' ('flying wing')	114 2 (34.8)	166 4 (50.7)	4,484 (417.0)	385,802 (175,000) to 451,940 (205,000)	4 x NK-6	1,865 (3,000)	1 x cruise missile or various other stores
Sukhoi T-4 (flown)	72 2 (22.0)	146 0 (44.5)	3,180 (295.7)	282,187 (128,000) or 299,824 (136,000) (sources differ)	4 x RD36-41 22,045 (98.0) dry, 35,275 (156.8) reheat	est 715 (1,150) at S/L, 1,989 (3,200) at 65,617 (20,000)+	1 x Kh-45 anti-ship missile; max warload 39,683lb (18,000kg)
Yakovlev Yak-33 (fuselage engines)	33 7.5 (10.25)	86 6.5 (26.375)	?	?	2 x main engines, 6 x lift jets	1,865 (3,000)	?
Yakovlev Yak-33 (canard/wing nacelles)	36 1 (11.0)	88 7 (27.0)	?	?	2 x main engines, 6 x lift jets	1,865 (3,000)	?

Chapter Eleven

Backfire and *Blackjack*

Prototype of the spectacular Tu-160 *Blackjack* bomber.

This chapter describes the background for the last examples of Soviet Union heavy bombers to enter service although some of the detail for these aeroplanes is currently still secret. The chapter also looks briefly at Russia's most recent studies into bomber design.

Tupolev Tu-22M *Backfire*

The previous chapter hinted at how Andrei Tupolev had pushed hard for a new and much simpler bomber design of his own, 'Aircraft 145', to be accepted by the Air Force rather than Sukhoi's T-4. The removal of Khrushchev in the autumn of 1964 allowed the Soviet aircraft OKBs to become far more active and Tupolev's work leading to the *Backfire* was just one example.

Tupolev 'Aircraft 145'

Despite completing a great deal of work in the search for a replacement for the Tu-22, in the form of 'Aircraft 106' (Chapter 4) and 'Aircraft 125' (Chapter 10), none of this was turned into hardware. This pair of designs represented entirely different aeroplanes, one a modification of the Tu-22 *Blinder* and the other an all-new type, but followed the same but now flawed concept in that a high-altitude bomber was, by the mid-1960s, a less important element in the Air Force than previously. The development of more advanced defensive weapons, particularly surface-to-air missiles, made the survival of such types less certain and some very complex solutions would be required for such an aircraft to get through to its targets. In addition a new-single-role bomber was not required; a multi-role type was considered to be more suitable and the Tupolev design bureau and the Soviet Air Force got together to prepare a concept for such a machine. This aircraft had to be capable of supersonic high-altitude flight, transonic low-level flight and long-range sorties at subsonic speed, plus offering an improved take-off and landing performance over earlier bombers; an aircraft fitted with a variable geometry swing wing was considered to be the best solution.

There would be a weight penalty for such a type, calculations suggesting about 3.5 to 4%, but this would be more than offset by the bomber's much greater flexibility. There were problems in retaining stability and controllability when changing the sweep angle, but research with TsAGI showed that adding fixed wing root extensions should cure this. The result was 'Aircraft 145' begun in 1965 and continued by Tupolev for a period as a private venture; in fact it was not until towards the end of 1967 that the Government issued a decree covering the development of a new aeroplane.

The final configuration of what was to become the Tu-22M in Soviet service did not

Model of the original Tupolev 'Aircraft 145' (autumn 1965).

come together at once and there were several intermediate projects which made use of the experience gained with the previous Tu-22 generation. In fact the first '145' design produced by Tupolev's Department of Preliminary Projects in autumn 1965 used the '106B' development of the Tu-22 (Chapter 4) as a basis and had a similar fuselage, engine position, weapon positions and defensive systems, together with a high-mounted VG wing complete with fixed centre section. The outer wing could be set at 20° sweepback for take-off and landing, 65° for maximum-range subsonic cruise and 72° for supersonic cruise, the angles corresponding to the optimal setting for the specific flight mode.

The powerplant was also new, namely two Kuznetsov NK-144 afterburning bypass turbofans rated at 48,500 lb (215.6kN) thrust in full afterburner, which were expected to give the '145' a top speed of 1,554mph to 1,678mph (2,500km/h to 2,700km/h) at 47,572ft (14,500m) and 684mph (1,100km/h) at 164ft to 328ft (50m to 100m). Estimated range was 3,729 to 4,972 miles (6,000km to 8,000km) in subsonic cruise and 2,486 miles (4,000km) in supersonic cruise. These figures represented a significant improvement over the original '106B' powered by NK-6 engines, the '145' at 231,481 lb (105,000kg) being about 7% heavier than its ancestor while the '106B' could 'only' do 1,367mph (2,200km/h) at 47,572ft (14,500m) and was incapable of flying for long periods at very low level; the take-off performance was also improved and 'Aircraft 145's service ceiling was 55,774ft (17,000m).

'Aircraft 145' was primarily developed as a missile strike aircraft but it was also capable of delivering free-fall bombs. If the enemy had strong air defences it was to operate at low altitude and engage the target with a 62 mile (100km) range air-launched cruise missile, but if the air defences were sparse (for instance, in a naval warfare scenario), the aircraft could carry a Kh-22 air-to-surface missile instead. At the initial design stage the '145' still shared a considerable degree of commonality with the Tu-22, a situation which Tupolev believed would facilitate its service entry, but it had many of the older aircraft's shortcomings removed; for example '145' had a crew of four, a co-pilot being added to reduce the crew workload during long missions. The crew section was totally redesigned – the pilot and co-pilot sat side by side in a much roomier and more comfortable cockpit, with the navigator/bomb aimer and gunner/radio operator sitting behind. Other proposed versions were the '145P' ECM aircraft with special equipment housed in the bomb bay, a dedicated anti-submarine warfare version developed in parallel with the basic bomber and equipped with sonobuoy launchers and depth charges (given the '145's estimated speed and endurance figures, it would quickly reach the area to be patrolled and have an adequate on-station time) and the '145R' long-range reconnaissance aircraft.

By the end of 1965 Tupolev's '145' preliminary project had been modified to eliminate some problems. The engine nacelles were moved a little higher above the fuselage to reduce the effects of the fuselage and wing airflow on the powerplant's operation at high speed, the fin and horizontal stabiliser were altered to improve their effectiveness, and the nose and radome were refined; also from now on it was decided to design the '145' as a cruise missile carrier only. Up to this point the Tupolev OKB had been working closely with TsAGI on the '145's design and had taken on board almost every recommendation made by this Research Institute. However, there was a good deal of sharp discussion and disagreement after TsAGI had declared that the engines should be moved to the rear of the

The first Tupolev Tu-22M prototype was known as the Tu-22M0 No 01.

Model of the Tupolev 'Aircraft 245' (c1980s).

fuselage, the bureau observing that the '145's relatively simple and light air intakes would have to be replaced by an altogether heavier arrangement, and this was after the VG wing had already increased the airframe weight.

TsAGI's opinion was swayed by experience with the Tu-22's powerplant, with engines mounted around the fin, where it was considered impossible to operate the engines to their maximum limits when positioned above the upper wing and fuselage. TsAGI's research indicated that, for a Tu-22-type nacelle position, at speeds between Mach 1.135 and 1.45 the pressure recovery ratio decreased sharply while the airflow irregularity at the intake inlets increased, even at comparatively low angles of attack. Tupolev eventually took these arguments on board and, as a result, 'Aircraft 145' in its original form was abandoned; by 1967 a new design had emerged with side intakes and the two engines mounted in the back of the fuselage.

This move required considerable changes to the overall layout and its systems and equipment and the wing was moved to a lower position; however, the original plan to install semi-circular intakes with half-cone centrebodies, rather like the Tu-28 *Fiddler* interceptor, was dropped. Analysis suggested that the operation of such intakes on a large bomber was uncertain and, after examining the flat intakes on the American F-4 Phantom fighter, this type was adopted with the final result being quite similar to the F-4. By now the bomber's design represented an all-new aircraft, including wing sweep settings re-set to angles between 20° and 60° plus a fin fitted with a leading edge root extension.

As such, the project finally gained official status through a Council of Ministers Decree issued on 28th November 1967, which tasked Tupolev with the development of the Tu-22M bomber and requested joint state testing to begin in the second quarter of 1969. Some Air Force officials wanted a rearward-facing defensive gun turret placed at the end of the fuselage, which was eventually fitted, the number of missiles to be carried rose to three and the maximum possible warload was increased from 24,250lb (11,000kg) to 52,910lb (24,000kg). After the mock-up examination was completed in November 1967, the decision was made to build a small production batch and the first Tu-22M (actually designated Tu-22M0) made its maiden flight on 30th August 1969.

However, some of the performance parameters exhibited by the new bomber fell short of both the VVS and OKB's requirements so, before flight testing had even begun, further modifications were under way. The weakest parts of the design were corrected, the airframe itself was modified which brought a reduction in empty weight of 6,614lb (3,000kg), and the aerodynamics were improved. Span was increased, other new systems were introduced and, as the Tu-22M1, the modified aircraft made its first flight on 28th June 1971. Production began during that year and the first deliveries were made in 1973 to start replacing Tu-16s. The Tu-22 was soon further upgraded as the Tu-22M2 and Tu-22M3, the latter flying for the first time on 20th June 1977 with more powerful NK-25 engines. In-house the bomber was known as the 'Project 45' and the West codenamed it *Backfire*; in error, the West initially designated the new bomber Tu-26.

Tupolev 'Aircraft 245'

This long-range multi-role bomber and missile carrier was a proposed further development of the Tu-22M made during the 1980s. In 1983 a requirement was issued for a replacement for the Tu-22M3 *Backfire-C* and the improved Tu-22M4 and M5 variants were prepared to it. However, 'Aircraft 245' itself actually represented a complete redesign with the engines moved out along a fixed wing into box nacelles, small outer wings were fitted outside the engines, horizontal stabilisers with pronounced anhedral were placed directly behind the outer wings and the aircraft had twin fins.

Tupolev Tu-160 *Blackjack*

The last heavy bomber to be built under Soviet colours was one of the biggest front line military aeroplanes to come out of the Cold War and the design competition and development programme that led to it once again brought out the old rivalry between

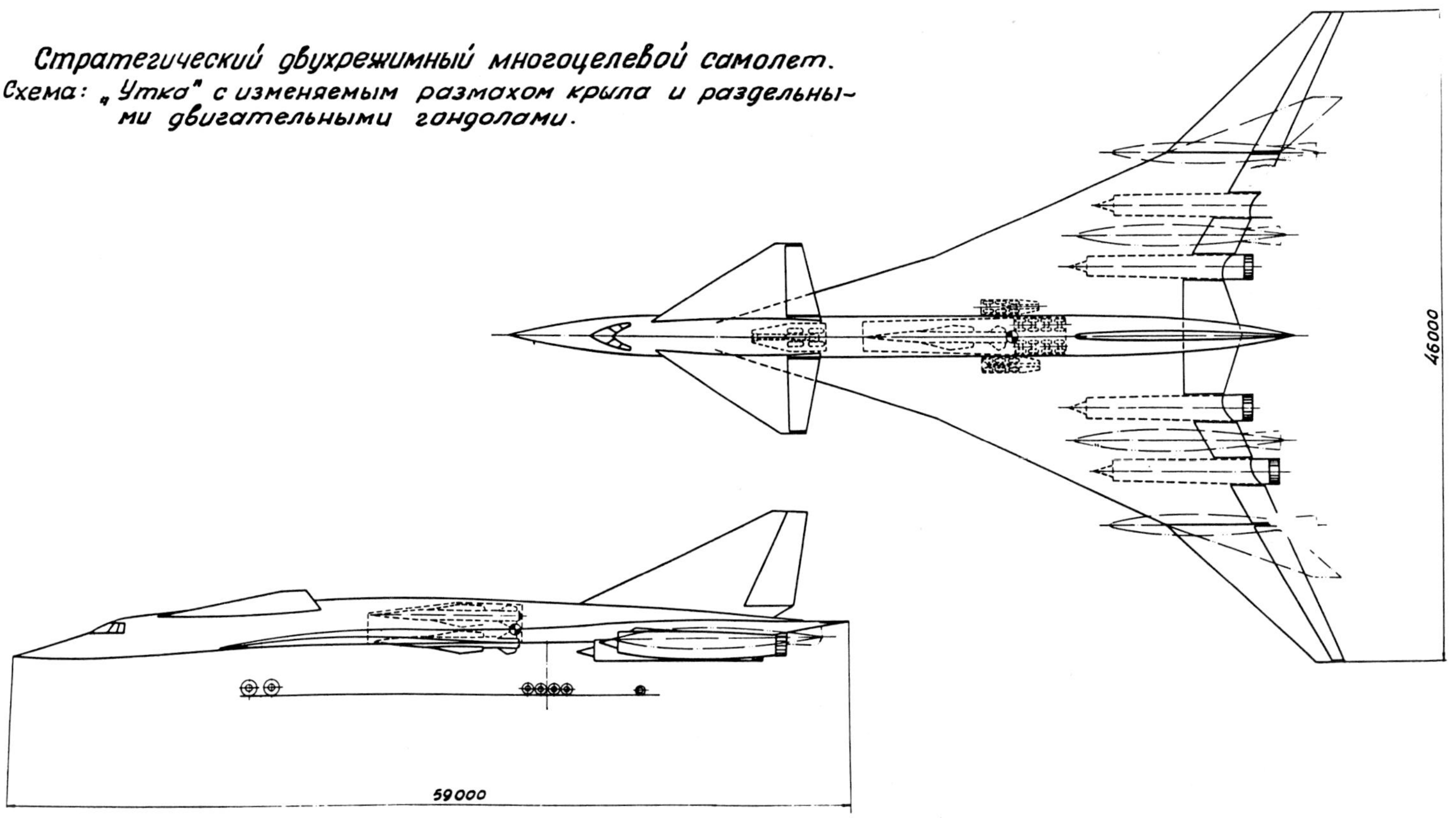

The fixed wing Myasishchev M-20-14 belonged to M-20 Group Two and could also droop its wingtips (late 1960s?).

Myasishchev and Tupolev. On 28th November 1967 a SovMin Decree was issued asking for competing designs to a new requirement for a multi-role strategic missile carrier. This had to have a cruise speed of at least 1,989mph (3,200km/h) at 59,055ft (18,000m), a range at this speed of over 6,837 miles (11,000km), plus a subsonic range of over 9,944 miles (16,000km) at height and over 6,837 miles (11,000km) at low level. Alternative warloads could include four Kh-45 Molniya (Lightning) cruise missiles, up to twenty-four 4,409 lb (2,000kg) bombs or a variety of other free-fall or guided weapons. The initial responses came from Myasishchev and Sukhoi.

Myasishchev M-20 and M-18

On 21st November 1966 a SovMin resolution was passed that authorised the establishment of a new Experimental Aircraft Works or EMZ (*Eksperimentahl'nyy Mashinostroitelnyy Zavod*), for which Vladimir Myasishchev was given the position of general designer. Besides maintaining support for the M-4 and 3M fleet, its first tasks were to create a new strategic bomber for the VVS and to design a high-altitude aircraft to shoot down balloons

Model of the M-20-14 which carried two Kh-45 cruise missiles in a centre fuselage bay, one in the upper fuselage and the other underneath protruding into the airflow. George Cox

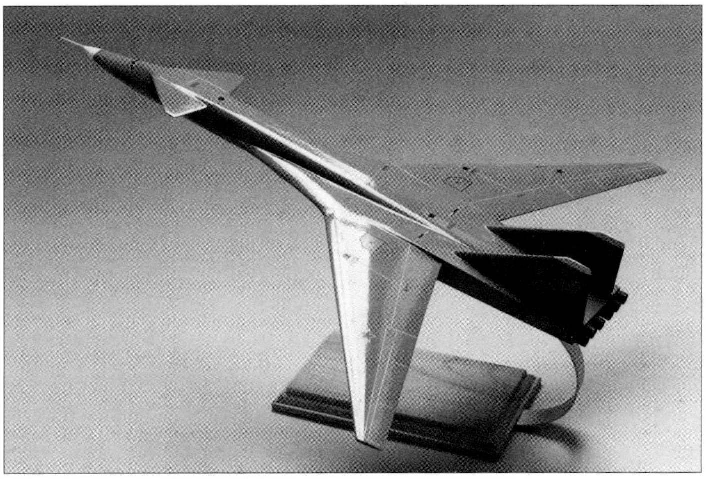

Model of the M-20-18. George Cox

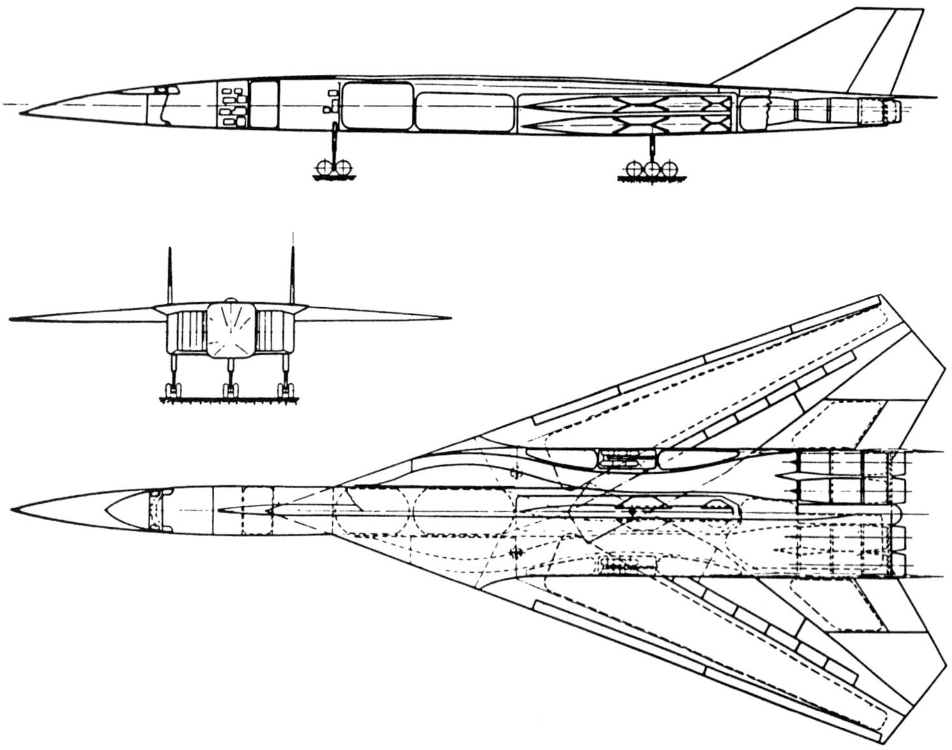

Left: **The Myasishchev M-20-2 project was an example of the designs produced in the M-20 Group One series, which had variable geometry wings in classic form with rear-mounted engines and side or chin intakes (1968).**

Below left: **Model of the M-20-2.** George Cox

Below right: **Model of the Myasishchev M-20-9 from M-20 Group Two (c1968/69). This particular project was fitted with a laminar airflow wing control system which blew air onto the upper wing surface. It carried two cruise missiles one behind the other in the fuselage plus further examples under the wings.** George Cox

Bottom left: **The fixed wing M-20-9 had four engines stacked in two pairs.** George Cox

Bottom right: **Model of the M-20-11. This project could droop its nose and wingtips while the fins were canted slightly inwards. There were four engines spread individually under the back of the wing – one under each fin and another just outboard of each side of the fuselage (late 1960s?).** George Cox

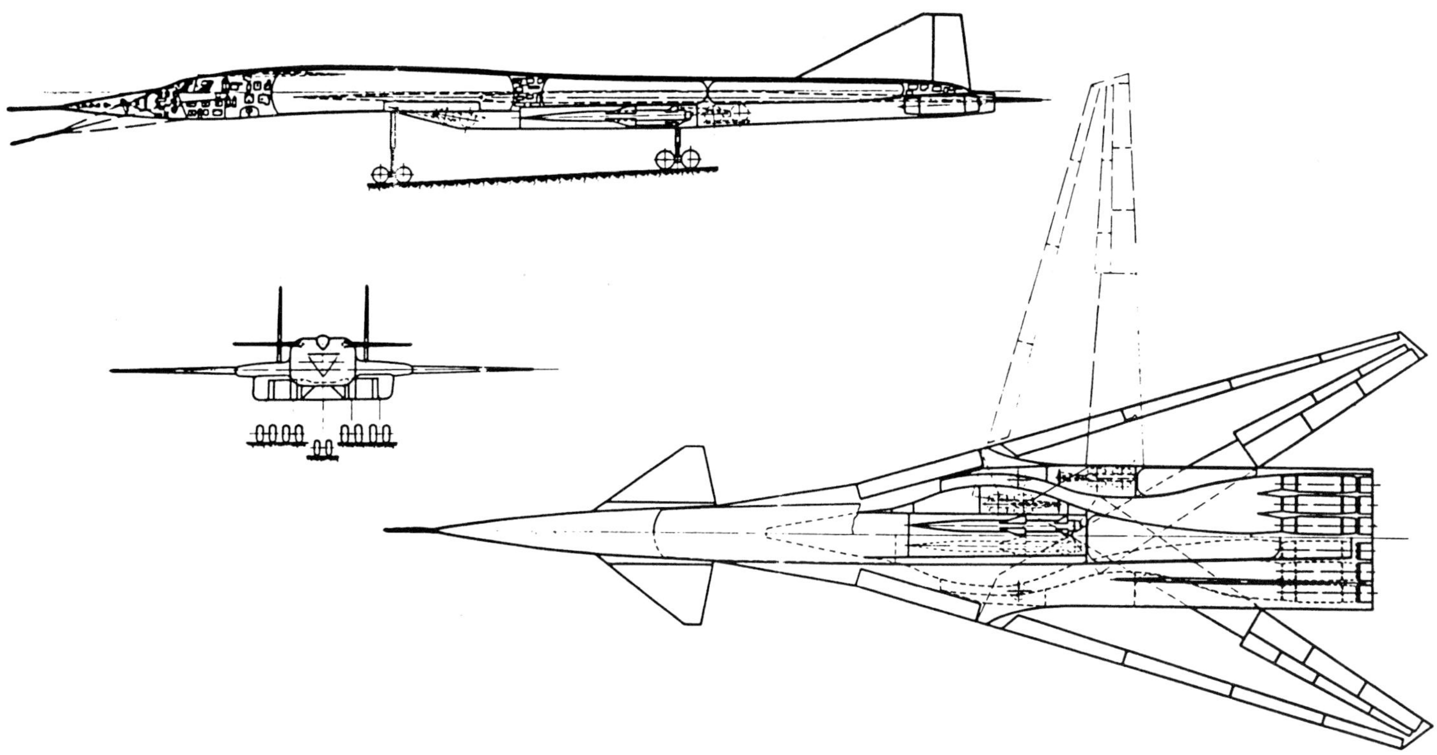

The Group Three Myasishchev M-20-18 had six engines, a large variable geometry wing, two fins placed quite close together and an undercarriage with no less than eight wheels on each main leg and four on the nose gear (late 1960s?). A pair of cruise missiles were carried side-by-side in a bay placed between the ducts that passed air through to the engines.

Three views of a model of the Group Three Myasishchev M-20-21 which had a variable geometry wing and four engines grouped together under the rear fuselage (late 1960s/early 1970s?). The views show the extremes of wing sweep and two cruises missiles were to be stored in a similar way to the M-20-18.

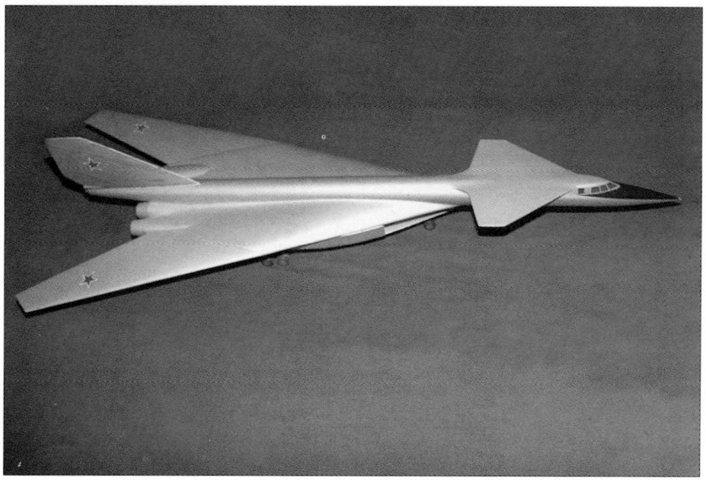

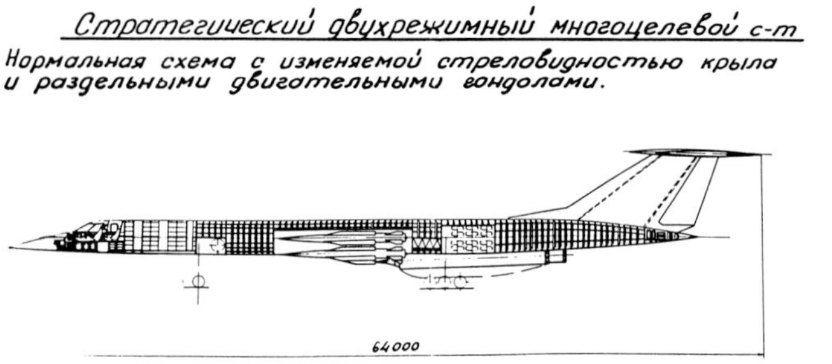

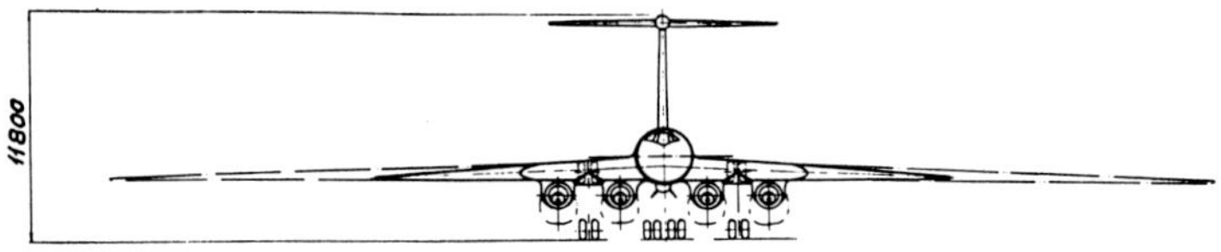

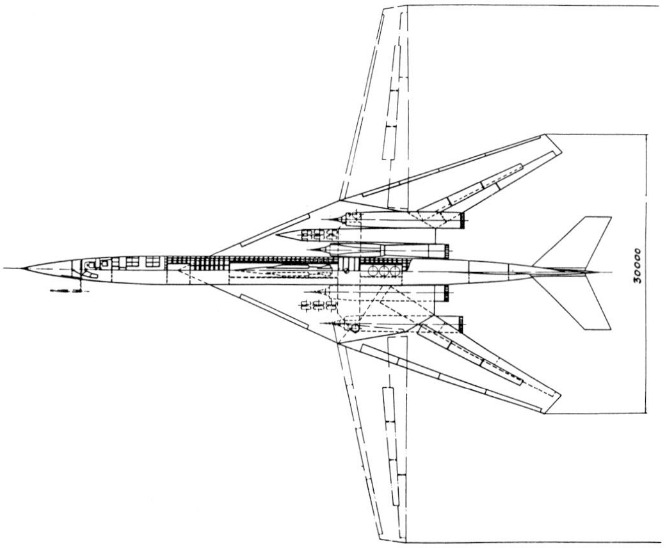

Основные данные

		Прототип 1975 г
1.	Вес с-та максимальный т	300
2.	Относительный вес топлива %	57,9
3.	Удельная нагрузка на крыло кг/м² при $\chi_{пк} = 13°$	608
4.	Тяговооруженность взлетная	0,3
5.	Дальность полета максимальная км при M=0,8; H=8-13.	14200
6.	Вес боевой нагрузки нормальной — максимальн. т	8,5-40
7.	Длина разбега по бетону — грунту м	1600-3000
8.	Двигатели: тип, число X, тяга в кг	ТРДДФ4×22000 ген. конст. Кузнецов Н.Д.
9.	Число членов экипажа	3-4

The T-tail Myasishchev M-20-23 belonged to the M-20 series Group Four (1975?). Two cruise missiles were carried one above the other and the aircraft used four Kuznetsov TRDDF-4 engines and had three or four crew.

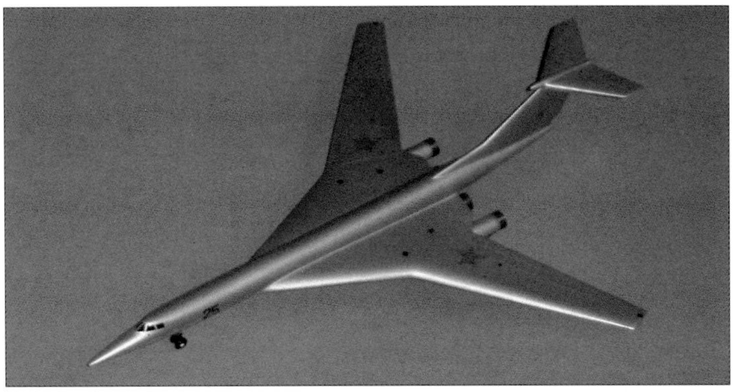

Model of the Myasishchev M-20-23. George Cox

Model of the M-20-24. George Cox

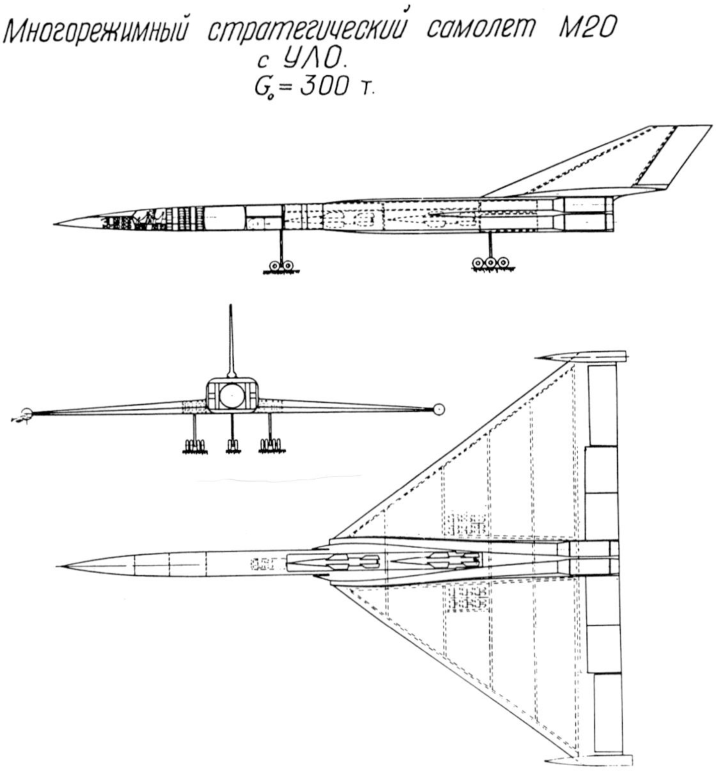

Myasishchev's M-20-24 was one of the last designs produced under the M-20 designation and belonged to Group Four (c1975?). It had a fixed delta wing and housed two Kh-45 cruise missiles one behind the other in the centre fuselage.

Model of the Myasishchev M-18 showing the wings in their minimum and maximum sweep positions (c1972).

sent across from Europe to spy on the Soviet Union's territory and facilities (the latter eventually flew as the M-17 Stratosfera high-altitude research and reconnaissance aircraft). The first preliminary research studies for the bomber, begun in 1967, were given the designation 'Subject 20' or M-20 and, after the preliminary design work on this had been completed, Myasishchev succeeded in getting his EMZ entered into the competition.

M-20 development was authorised by a further official directive of 28th February 1968 and there were to be three possible versions – an attack and reconnaissance bomber carrying missiles and bombs, an aircraft to attack enemy air transports and airborne early warning (AEW) aircraft, and a long-range anti-submarine aeroplane; they were to be powered by Kuznetsov bypass turbofan engines. A long series of designs were produced, split into four groups, and the drawings and models reproduced here indicate that a full range of configurations was explored. Group One comprised the M-20-1, -2, -5 and -6, Group Two was based on the M-20-7, -9 to -12, -14 and -15, Group Three the M-20-16 to -19 plus M-20-21, and Group Four the M-20-22 and -23. The estimated take-off weights for these designs showed a minimum of 330,688 lb (150,000kg) for Group One up to 716,490 lb (325,000kg) in the later Groups, Group Two designs featured wingtips that could deflect downwards at high speeds and there was the usual mix of canard/non-canard types, one or two fins plus variable geometry or fixed wings; most of the projects had three crew.

All of the M-20 concepts stayed on the drawing board but provided a strong basis for the follow-on 'Subject 18' or M-18 project. A new MAP directive authorised the development of the M-18 strategic missile launcher on 15th September 1969 and the basic project was derived from one of the M-20 Group Four series; further directives issued in 1970 agreed to the project development phase. The M-18 also had a crew of three and was quite similar to one of the last versions of the M-20 but, apparently, it was to be built in aluminium rather than employ a titanium structure, which limited the maximum speed to Mach 2.3.

On 15th February 1971 Myasishchev delivered a report to representatives of various

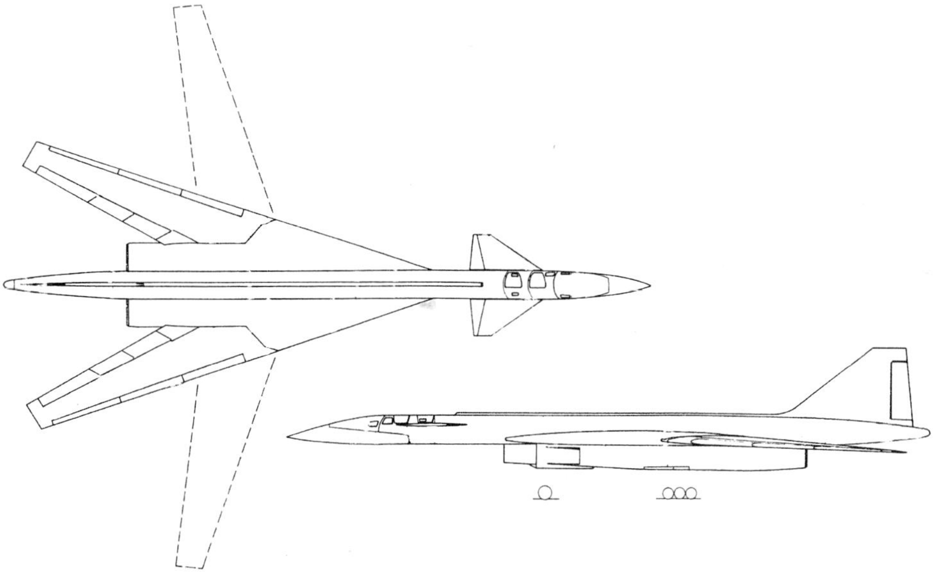

Sukhoi T-4M (1967). Russian Aviation Research Trust

Model of Sukhoi's T-4M-1 project (1967).

research establishments and OKBs that described the progress his EMZ had made on the programme together with TsAGI and several other research institutes. Myasishchev pointed out that the general operational requirement for the new bomber specified an increase in warload by a factor of 1.8 over aircraft currently in service with the Soviet Air Force, and this led to a higher all-up weight. The requirement also demanded the provision of special ECM equipment to assist the penetration of air defences, an improvement in thrust/weight ratio over existing aircraft by a factor of at least 1.5 to 1.7 (due to the need to be capable of operating from Class 1 unpaved airstrips), plus the cruising speed of 1,989mph (3,200km/h); according to Myasishchev's own calculations and those of his staff, these factors reduced the aircraft's range by 28-30%.

The OKB undertook a lot of research and development work for this strategic multi-mission strike aircraft, much of it performed jointly with TsAGI. An optimised size and weight was established for the type while the airframe strength and stiffness characteristics were determined for different aerodynamic layouts and structural materials. Finally Myasishchev established that its future aircraft should have variable geometry wings. The powerplant was selected according to the all-up-weight; a 150 tonne (330,688 lb) machine would be powered by four engines delivering 26,455 lb (117.6kN) of thrust each whereas the heavier 300 tonne (661,376 lb) type required engines in the 48,500 lb to 55,115 lb (215.6kN to 245.0kN) class. In general appearance the swing wing M-18 closely paralleled the American Rockwell International B-1 bomber and thus was considered to be, and promoted as, the more promising of the two Myasishchev projects.

Sukhoi T-4M

Sukhoi's first supersonic missile-carrying bomber and reconnaissance project was developed out of the T-4 described in the last chapter. SovMin's 28th November 1967 Decree ordered Sukhoi to design this new project, which had a variable geometry wing coupled with a canard. The design showed very long thin wings which could be swept between 15° and 72°, a fixed centreplane with leading edge root extensions and the T-4's powerplant of four engines in a pack beneath the fuselage with two flat intakes each feeding a pair of engines. The wing leading edges had extending slats along their entire length and the trailing edges had double-slotted extending flaps in two sections plus the ailerons. As a whole this design was really a T-4 with a new variable geometry wing and, in fact, the designation T-4M indicated a 'modification' of the earlier aircraft. The project utilised many of the T-4's systems and would also use similar construction materials; the droop nose was retained, fitted to a circular section fuselage, while the canard had extendable single-slotted flaps. Another feature of the research was new more reliable solid-state electronics to reduce weight.

Weapons were to have been carried on two hardpoints placed beneath the engine fairing and would include two Kh-45 long-range or eight Kh-15 short-range air-to-surface missiles, bombs or cluster bombs or mines; for supersonic flight the bombs would be loaded into standard 17,637 lb (8,000kg)

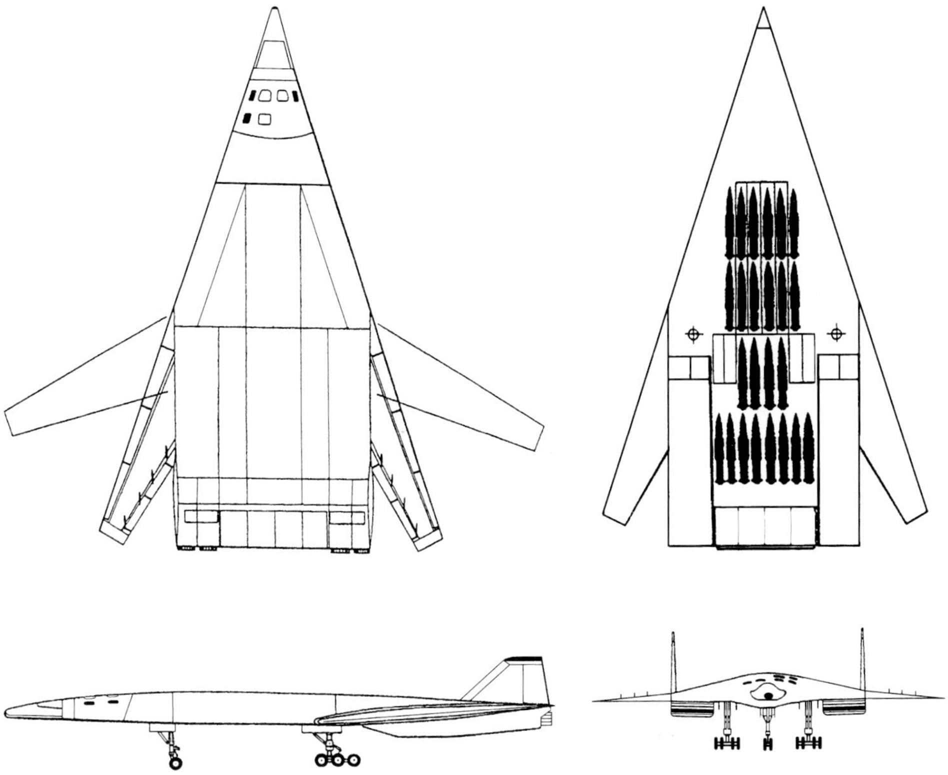

Sukhoi T-4MS drawing which indicates the positions for the external carriage of twenty-four Kh-15 missiles (1970/71).

Side view of a model of the T-4MS.

containers but, for the overload condition, they would also be carried on external racks. Total internal fuel, in ten fuel tanks, amounted to 149,912 lb (68,000kg) and three crew were carried. Cruise speed would be at least 1,864mph (3,000km/h) at an altitude between 65,617ft and 75,459ft (20,000m and 23,000m), subsonic range was 6,215 miles (10,000km) at height and 2,175 miles (3,500km) at sea level, and supersonic range 4,350 miles (7,000km); however, a subsonic maximum of 9,944 miles (16,000km) was possible with two in-flight refuellings.

Following the study of thirty preliminary designs and the completion of a preliminary project which was submitted to the Air Force and MAP at the end of 1969, it was realised that the basic T-4 could not be adapted to fit the other new requirements. In addition the engine arrangement precluded the inclusion of a weapons bay in an aircraft of this size, so all of the weaponry would have to be hung outside, while the narrow wings presented problems of rigidity. Consequently, after attempts to modify the design further, the T-4M was abandoned at the end of 1969 and Sukhoi's designers moved on to the T-4MS, which became the OKB's main proposal in the competition.

Sukhoi T-4MS

The Sukhoi OKB now started work on a strategic bomber designated T-4MS or 'Project 200' and the engineers paid special attention to ensuring maximum commonality with the earlier T-4 or 'Project 100'. The powerplant of four 35,275 lb (156.8kN) thrust Kolesov RD36-41 afterburning turbojets was retained, as were the T-4's systems and equipment, structural materials and various other features of its detail design. Several layouts were studied at the preliminary design stage and, at first, the engineers merely considered scaling up the earlier T-4M. However, it was quickly apparent that this was a bad idea because such an approach would lead to a dramatic increase in the bomber's overall dimensions and structure weight while still offering insufficient internal space to store weapons. The OKB had to find a different solution.

The future T-4MS's general arrangement had to meet the following main criteria: the internal volume had to be maximised while keeping the surface area (and hence drag) to a minimum, the weapon bays had to be big enough to accommodate the required armament and the aeroplane's structure had to be as stiff as possible to permit high-speed ultra-low-level operations. (Note: low-level flight increased the chances of penetrating the enemy's air defences but also placed high demands on an airframe's structural strength because of the high levels of turbulence that would be experienced and the need to make manoeuvres to avoid hilly terrain.) In addition the powerplant had to be located externally to facilitate a possible need to fit different engines – buried engines were out of the question because integrating new units might then require drastic structural changes.

As work progressed on the final versions of the T-4M project, which actually introduced an integral or blended wing/body layout where the fuselage contributed a large amount of lift, Sukhoi's engineers realised that such a 'flying wing' would meet the requirements. As described shortly, Tupolev also arrived at this conclusion but Sukhoi

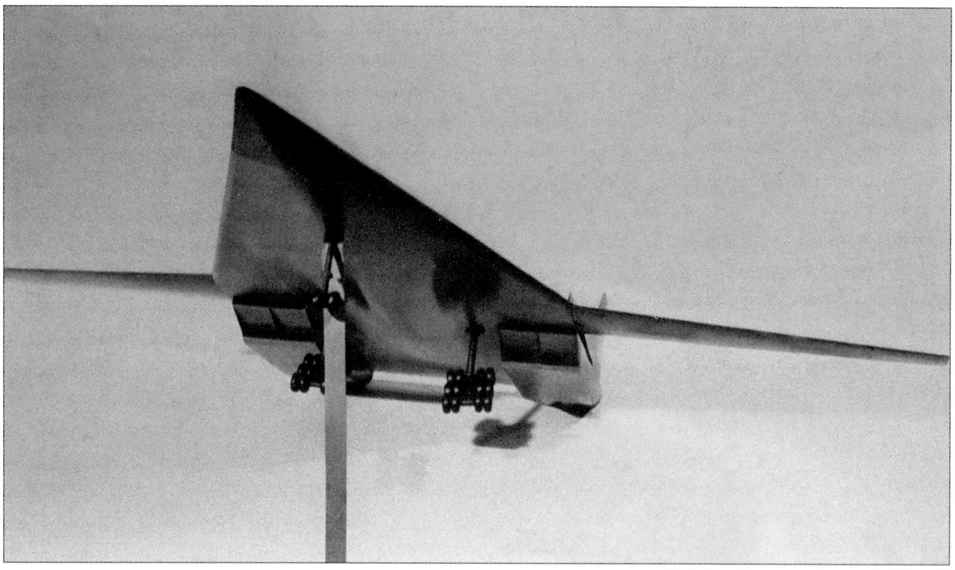

This angle illustrates the T-4MS's undercarriage with twelve wheels on each main gear.

Planview of a T-4MS model showing the wing sweep positions and the widely spaced engines.

added variable geometry wings with movable outer portions of relatively small area. The main body would house all of the fuel, weapons and equipment and the small VG wings would help the low-speed performance. Four large elevons positioned between the exhaust nozzle sections of the nacelles would provide control in pitch, roll and yaw and the aircraft was to be built in titanium, although high-temperature ceramic material would be used for the nose cone. Known in-house as 'Version 2B', the flying wing/swing wing layout was first developed in August 1970 by engineer L I Bondarenko and, in due course, was approved by the Project Development Department chief Oleg S Samoylovich, then by the T-4MS's chief project engineer N S Chernikov and finally by General Designer Pavel Sukhoi; this design would serve as the basis for an advanced development project.

Wind tunnel tests undertaken at TsAGI showed that the chosen layout offered a high lift/drag ratio in both subsonic and supersonic modes; in fact 'high' is a poor description because the results were truly fantastic – a lift/drag ratio of 17.5 at Mach 0.8 speed and 7.3 at Mach 3.0. The integral layout also took care of the aeroelasticity problem because the limited movable outer wing area, coupled with the inherently stiff structure created by the wing centre section/fuselage assembly, ensured that high-speed flight at low altitude would be possible. Work on defining and refining the advanced development project to a point where it could be submitted to the competition continued

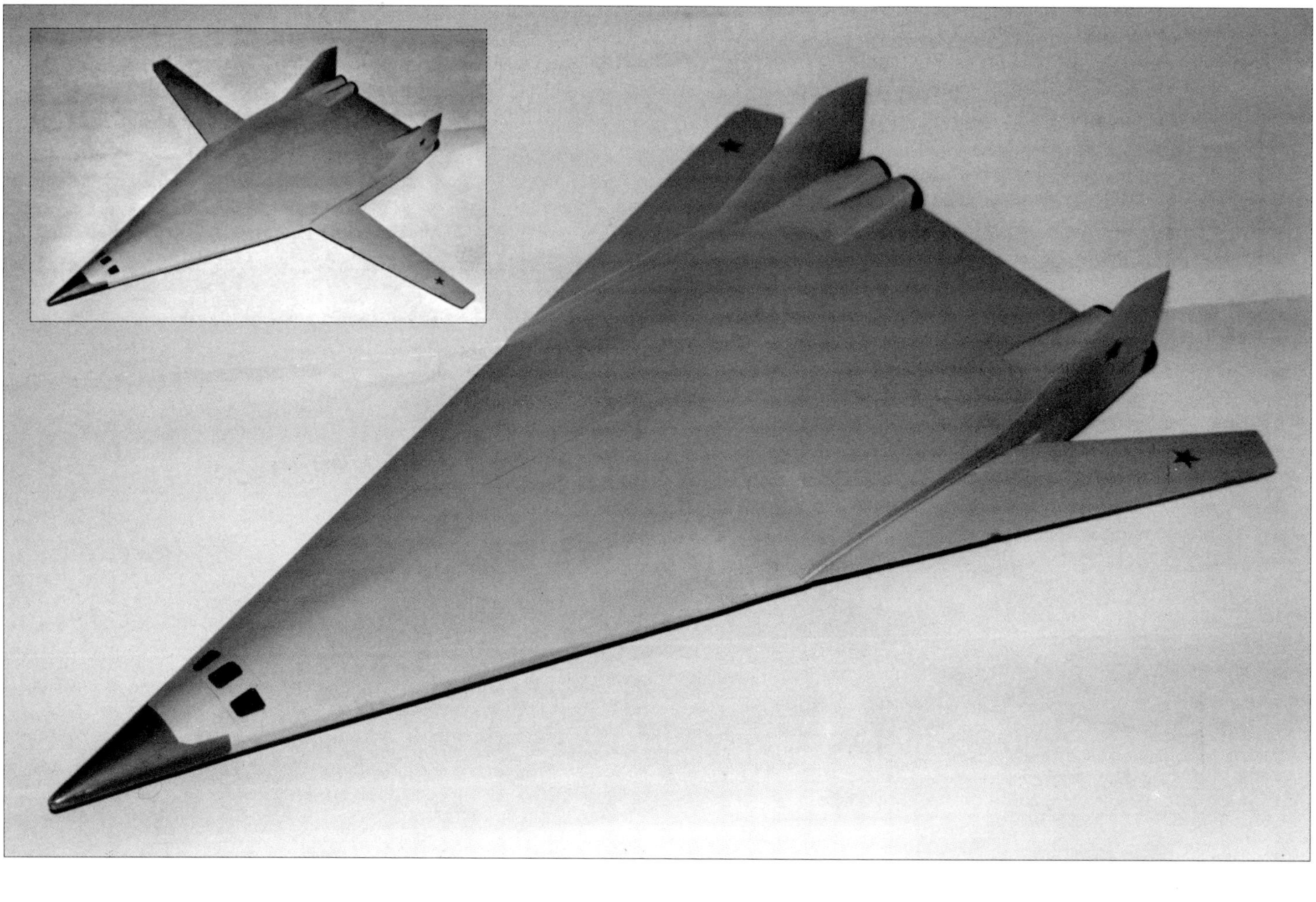

throughout 1971. However, wind tunnel testing also showed that the T-4MS was catastrophically unstable because the centre of gravity shifted too radically when the wing sweep angle was altered. As a result Chernikov decided to amend the design and versions now emerged featuring an extended nose and additional conventionally positioned horizontal tail surfaces; one of these, Version 8, also had a needle-sharp nosecone.

The final selection featured an extended forward fuselage with an extremely streamlined flightdeck canopy so that the upper fuselage contour was virtually unbroken but, apart from this, there were no changes from the original advanced development project. The Sukhoi T-4MS design was completed in September 1971 and the first examples were to be fitted with four Kolesov RD36-41s, but a later Stage B version would get four 44,090 lb (196.0kN) K-101 units. Minimum outer wing leading edge sweep was 30° and maximum 72°, internal fuel load 213,845 lb (97,000kg), normal internal offensive load 19,841 lb (9,000kg), cruising speed above 59,055ft (18,000m) 1,865mph to 1,989mph (3,000km/h to 3,200km/h) and at sea level

One of the earliest versions of Tupolev's 'Aircraft 160M' featured a cranked double-delta wing and twin fins. It also had the engines mounted in a single pack beneath the wing centre section which followed the practice used on the first Tu-144 supersonic airliner prototype flown in 1968 (1970/71).

This early '160M' layout had a single fin and resembled the Tu-144 even more closely; the airliner however, could not droop its wingtips (1970/71).

528mph (850km/h), and maximum range (with K-101s) at medium altitude and cruising speed with the normal warload and internal fuel only 8,701 miles (14,000km). Three crew were specified and the offensive load could include four Kh-45 air-to-surface missiles (two in the bomb bay with two more carried externally between the engine nacelles) or twenty-four Kh-15 missiles loaded externally.

Tupolev 'Aircraft 160'

In 1967 when the new bomber requirement was first raised, the Tupolev OKB was invited to compete along with Sukhoi and Myasishchev. However, although Tupolev was the Soviet Union's top design organisation for heavy bombers and possessed the greatest expertise in this field, and the new type clearly fitted into the OKB's traditional line of work, the bureau was at this time probably not in a position to join the contest because it had so many other ongoing civil and military aircraft programmes – the Tu-154 tri-jet medium-haul airliner, Tu-144 supersonic transport, Tu-22M *Backfire* bomber and Tu-142 *Bear-F* long-range anti-submarine aircraft all entered flight test in the mid to late 1960s. However, until 1970 Tupolev's top executives did attend all government meetings relating to the development and acquisition of new bombers, although their presence was strictly as observers.

However, after carefully analyzing the state and prospects for the programme, and assessing its own capabilities and those of its competitors, Tupolev did begin work on a new strategic bomber which used the 1967 specification as a guide. The work was performed by the design bureau's Section K under the co-ordination of Aleksey Andreyevich Tupolev, the General Designer's son and future successor. Later, the overall responsibility passed to Valeriy Ivanovich Bliznyuk, a designer who, as part of S M Yeger's department, had participated in the development of the unbuilt 'Aircraft 135' (Chapter 10) and the Tu-144 supersonic transport. Initially known as 'Aircraft 156', within a week the new bomber had been redesignated 'Aircraft 160'. It was also referred to as 'Article K' but, eventually, the '160' came to be known as 'Article 70' and this cipher is still in use today; the '156' designation was later applied to a proposed derivative of the Tu-154 airliner.

Initially Tupolev developed 'Article K' as a private venture and information about it was distributed only on a 'need to know' basis – a very small group of people working at the OKB and at MAP were the only ones to know of its existence. Therefore the designers were allowed to do pretty much what they pleased as far as the choice of aerodynamic layout and specific design features were concerned. They decided to rely heavily on the unique engineering experience gained from designing the Tu-144 and this served as the basis for the initial design, a layout that was markedly different from both Sukhoi's T-4MS and Myasishchev's M-18. The extremely high demands posed by the 1967 directive meant that the engineers faced an extremely difficult task, so at the start they decided to select the specified maximum speed and maximum-range cruise speed as the main performance targets and take it from there. Concurrent with this bomber project, Section

Even closer to the Tu-144 was this design fitted with a defensive tail gun barbette (1970/71).

Opposite: **During Tupolev's in-depth search into the most suitable position to put the engines on its Tu-160 bomber, one layout to be assessed was this version with the power units set in the wing roots. It also has large wedge-shaped intakes with variable-area ramps.**

K conducted research into future supersonic transports which later laid the foundation for the SPS-2 programme, the future Tu-244 airliner proposal. It was entirely logical therefore, that part of this work should go into the '160' and so, to begin with, the bomber shared the tailless delta layout used by the SPS-1 (the Tu-144) and SPS-2 (Tu-244).

Data accumulated under the next-generation supersonic transport programme showed that in theory it was possible to obtain a lift/drag ratio of 7 to 9 in supersonic cruise and as much as 15 in subsonic cruise. This, coupled with advanced fuel-efficient engines, made the specified range a realistic prospect, or at least a figure close to it. Given adequate engine power, the tailless delta seemed to ensure that the required speed and most other target performance figures could be met. The main problems were associated with the need to use new structural materials to contain the kinetic heat created by sustained cruise flight at high Mach numbers and so, in an effort to reduce the technical risk involved, the OKB decided to restrict the new bomber's cruise speed to Mach 2.3, which was rather lower than the competitor's figures.

Tupolev did assess variable geometry which offered certain advantages but imposed a weight penalty and also considerably complicated the design. One of the key requirements was long range over a complex mission profile involving the penetration of enemy defences at high altitude in supersonic mode (the so-called 'hi-hi-hi' mission profile) or at low altitude in subsonic mode (the 'hi-lo-hi' profile). Another less critical requirement was a good field performance that allowed the bomber to operate from relatively short runways and combining these virtues in a single aircraft was no simple task. An acceptable balance of subsonic and supersonic performance could only be obtained by utilising variable geometry wings plus unconventional compound engines which operated as turbojets in supersonic mode and as turbofans in subsonic mode. A comparison of a fixed swept wing and variable geometry versions revealed that the VG alternative had a 20 to 50% better lift/drag ratio at subsonic speeds, while in supersonic cruise with the wings at maximum sweep the L/D ratio was virtually equal to that of the fixed-sweep aircraft.

As noted, an inherent shortcoming of swing-wing aircraft is the increase in empty weight caused by the massive wing pivots and actuators. Calculations showed that if this assembly made up more than 4% of the empty weight, the weight penalty negated all the advantages conferred by using variable geometry wings. Assuming the powerplant was identical, a heavy VG aircraft offered around 30 to 35% longer range at medium altitude and 10% more at low altitude than a conventional fixed-sweep aircraft when flying at subsonic speeds. In supersonic cruise at high altitude the range was about equal but again a heavy swing-wing aircraft gained an advantage of approximately 15% at low altitude; variable geometry also offered the best field performance. Separate research also compared the range of a heavy swing-wing aeroplane at two different cruising speeds, Mach 2.2 and Mach 3.0. The results were unmistakable – reducing the cruise to Mach 2.2 enhanced the range considerably thanks to a lower specific fuel consumption and a better lift/drag ratio. In addition, an airframe designed to cruise at Mach 3 would include a large proportion of titanium alloys, which would increase unit costs because titanium is much more difficult to fabricate than aluminium.

After assessing the merits and shortcomings of the various layouts the design bureau eventually selected a tailless delta for its strategic bomber. Several versions were designed during 1970 to 1972 under the provisional designation 'Aircraft 160M' (or 'Article L'), and these were known as 'L-1', 'L-2' and so on. The preliminary development project for the '160M' was completed in 1972 and featured double-delta wings, that is, with a sharply kinked leading edge as used by the definitive production-standard Tu-144, and was to be powered by Rybinsk RD36-51 engines; this was submitted to the Soviet Air Force's Scientific & Technical Committee for appraisal.

–

In the autumn of 1972 MAP's Scientific and Technical Council convened to hear reports on the three OKB's competing designs and Tupolev's decision to use a Tu-144 style double delta proved fatal. Both MAP and the Air Force knew the Tu-144 airliner perfectly well and when the '160M' was unveiled the similarity was all too obvious; consequently it was rejected on the grounds that it 'did not meet the specification'. Commenting on the '160M', Colonel General Reshetnikov said that the Air Force was being offered 'a warmed-over airliner!' The fact that an excessively high lift/drag ratio had been unintentionally quoted in the project documents certainly did not help while the Tu-144 itself had fallen short of its performance target, was beset by reliability problems, was fuel-thirsty and difficult to operate. Consequently MAP would not consider the project any further because, even in revamped form, an aircraft designed originally for carrying passengers would still retain some inherent properties that were entirely unnecessary for a combat aircraft. In response Tupolev dumped the '160M' and began work on a new project.

Sukhoi's T-4MS drew a very favourable reaction and attracted a lot of attention but Myasishchev's project was less lucky. The latter was highly commended (the commission stated that it had been carefully designed and met the Air Force's specifications) but was nonetheless rejected because the recently reborn Myasishchev OKB lacked the necessary technological assets and manufacturing facilities to build a prototype; in fact, the OKB's new premises in Zhukovsky had little more than a flight test facility. As a result the T-4MS was declared the winner. However, to ensure that a T-4MS prototype could be built by No 22 Factory at Kazan (one of the Soviet Union's two factories that specialised in the production of heavy bombers) this major facility had to be assigned to the Sukhoi design bureau, a step that nobody within the industry wished to see happen except for Sukhoi itself. In addition, Sukhoi had its hands full developing the new T-10 advanced multi-role tactical fighter (which emerged as the Su-27 *Flanker*) and new versions of the Su-17M *Fitter-C* fighter-bomber and Su-24 *Fencer* tactical bomber, so any additional involvement with heavy bombers might actually jeopardize all of these other important programmes.

The final session of the tendering commission was summed up by Soviet Air Force C-in-C Air Marshal P S Kootakhov. 'Yes, the Sukhoi OKB's project is the best, we have given it due credit, but remember that the OKB is already heavily involved with the Su-27 fighter which we need badly. Therefore we will acknowledge that the Sukhoi OKB has won the tender and then order it to transfer all project materials to the Tupolev OKB so that the latter can proceed further with the project'. So the problem was solved and another recommendation followed that ordered the 'too small' Myasishchev Bureau to hand over all of its M-18 materials to the more powerful Tupolev design bureau as well.

It appears that this competition may have been held in two parts because some sources suggest that the Mach 3 1,989mph (3,200km/h) requirement would have made it extremely difficult for these OKBs to design a suitable aeroplane, and the resulting submissions suggested that this was true. The range and weapon limits ensured that the new type would be big and the specified top speed would make it essential to build the machine in high temperature materials, which would

then make the bomber excessively complex and expensive and require many new manufacturing facilities. This may be why Myasishchev (with the M-18) and Tupolev eventually centred on Mach 2.3 as their target speed since the new alternative requirements apparently specified this figure; the later specification also stated that the intercontinental range had only to be achieved at subsonic speed, which brought a reduction in the required fuel capacity.

Tupolev Tu-160 *Blackjack*

Although ordered to 'accept' Sukhoi's T-4MS project, Tupolev rejected the design, possibly because of the high level of technical risk involved, but it not reject Myasishchev's M-18. After the official decisions had been made Tupolev began to work on a design with a variable sweep wing and all of the bureau's projects with fixed wings were abandoned. Once the basic scheme had been chosen, the effort moved on to the aircraft's powerplant and equipment, the initial choice being the Kuznetsov NK-25 as used by the Tu-22M3. This supplied enough thrust but its fuel consumption needed to be improved before the required range could be achieved; fortunately Kuznetzov had a new three-shaft engine in the pipeline called the NK-32 which offered the same thrust but better rates of consumption.

Apart from the M-18's engine configuration, Tupolev looked at many other alternatives and finally settled on an underwing coupled unit installation with two-dimensional multi-mode air intakes and a vertical wedge. This VG aircraft ultimately became the Tu-160 and there were significant differences between it and the M-18, but the 'heritage' is still very apparent. On 26th June 1974 a Council of Ministers Directive was issued tasking Tupolev with the development of the Tu-160 multi-role strategic bomber and missile carrier powered by NK-32s and the construction of the first of three prototypes began in 1977. First flight was made on 18th December 1981 and there were plans to build a hundred examples; however, by 2000 only 35 had been completed, including the three prototypes, with three more unfinished. The Tu-160 still serves with the Russian Air Force.

Latest Developments

The end of the Cold War and a general lack of finance within Russia has brought the withdrawal and scrapping of a large proportion of the former Soviet Air Force's bomber fleet. In addition new programmes for replacement types have been drastically cut back and the aircraft industry itself is now much changed. Nevertheless, studies into future bombers have continued although relatively little information has, so far, been made available to the public regarding the industry's most 'recent' designs. The following presents a brief review of some of the work undertaken since the early 1980s but it is difficult to be sure if any of these programmes are still 'live'.

Model of the Mikoyan 301. John Hall

Mikoyan '301'

In the early 1990s Mikoyan began research into this hypersonic high-altitude reconnaissance aircraft which, it is thought, may also have been given a capacity for bombing. At speeds above Mach 3.5 special hybrid powerplants would operate as ramjets and, to cope with the build up of friction heat, the '301' would have been built in stainless steel. Its maximum take-off weight was estimated to be around 176,367 lb (80,000kg) and a variable geometry wing was employed. By the early years of the new century work on the design had been shelved or was progressing only very slowly.

Sukhoi T-60S

Few details have been made available for this supersonic stealth design. Reheat would not have been fitted, the aircraft being capable of supersonic cruise at high altitude on dry power, and weapons would have included cruise missiles, precision-guided conventional weapons and free-fall nuclear stores. Sources have suggested that by 1990 this aircraft had entered full-scale development and a prototype was actually being built, but the collapse of the Soviet Union at the end of 1991 prevented any flight testing and pretty well brought progress to a halt. In 1998 it was reported that the project was still ongoing as a possible replacement for Tu-16s and Tu-22Ms.

Tupolev 'Aircraft 260' and '360'

From about the early 1980s Tupolev began working on potential successors to the Tu-160 and the result was a pair of hypersonic projects. 'Aircraft 260' of 1983 onwards was intended to fly at Mach 4 at a height exceeding 82,021ft (25,000m) and was to achieve a range approaching 6,215 miles (10,000km). It was to be powered by four Soloviev D-80 jet engines mounted side-by-side beneath a double-delta wing and would have had a relatively fat fuselage. There was no tailplane and just a single tall fin and the maximum take-off weight was expected to be around 396,825 lb (180,000kg). A preliminary project was completed in 1985.

'Aircraft 360' had a similar layout but was rather larger and was to be capable of Mach 6 performance and a range of 9,323 miles (15,000km) with 22,046 lb (10,000kg) of ordnance aboard, although studies suggested that a constant Mach 6 might only give 6,215 miles (10,000km) range. To achieve such speeds would require engines that used cryogenic fuels and, as a result, six hydrogen-powered units were to be fitted of 'variable cycle' form that could operate in both turbojet and ramjet mode. There would be two crew and the weapons were to be housed in two wing root bomb bays. The development programme also envisaged the flight testing of a scale model aircraft weighing around 176,367 lb (80,000kg) but the project was halted in 1992, after some fuselage and fuel system parts had been manufactured, again due to a lack of money.

Tupolev B-90

It is believed that Tupolev began working on a subsonic flying wing bomber during the mid-1980s called 'Aircraft 202' and, with research still apparently ongoing during the late 1990s, it was hoped that a version of this design might actually reach hardware status. B-90 stood for 'Bomber for the 1990s' and the project was seen as an intercontinental strike aircraft replacement for the Tu-95MS *Bear-H*. Financing such designs is still a problem.

Heavy Bombers – Data / Estimated Data

Project	*Span ft in (m)*	*Length ft in (m)*	*Gross Wing Area ft² (m²)*	*Max Weight lb (kg)*	*Powerplant Thrust lb (kN)*	*Max Speed / Height mph (km/h) / ft (m)*	*Armament*
Tupolev 'Aircraft 145' (at 1965)	120 5 (36.7) (min sweep) 77 7.5 (23.66) (max sweep)	134 6 (41.0)	?	231,481 (105,000)	2 x NK-144 48,500 (215.6) reheat	1,367 (2,200)	1 or 2 x Kh-22 cruise missile
Tupolev Tu-22M0	103 8 (31.6) (min sweep) 74 7.5 (22.75) (max sweep)	136 0 (41.46)	1,974 (183.58) (min sweep) 1,890 (175.8) (max sweep)	266,755 (121,000)	2 x NK-144-22 44,090 (196.0)	951 (1,530)	3 x Kh-22 cruise missiles 2 x 23mm cannon fitted to some aircraft.
Tupolev Tu-22M3	112 5.5 (34.28) (min sweep) 76 5 (23.3) (max sweep)	139 0.5 (42.46)	1,974 (183.58) (min sweep) 1,890 (175.8) (max sweep)	273,369 (124,000)	2 x NK-25 55,125 (245.0)	1,429 (2,300)	2 x 23mm cannon. 3 x Kh-22 or 6 x Kh-15 cruise missiles. Max load 52,910lb (24,000kg) includes free-fall bombs or anti-ship mines
Myasishchev M-20-14	150 11 (46.0)	193 7 (59.0)	?	661,376 (300,000)	4 jet engines 55,115 (245.0)	Mach 3	2 x Kh-45 cruise missiles. Max warload 88,183lb (40,000kg)
Myasishchev M-20-21	180 5.5 (55.0) (min sweep) 87 11 (26.8) (max sweep)	189 3.5 (57.7)	?	661,376 (300,000)	4 jet engines 48,500 (215.6)	Mach 3?	2 x Kh-45 cruise missiles.
Myasishchev M-20-23	190 9 (58.14) (min sweep) 98 5 (30.0) (max sweep)	210 0 (64.0)	?	661,376 (300,000)+	4 x TRDDF4 48,500 (215.6)	Mach 3?	2 x Kh-45 cruise missiles
Myasishchev M-18	137 9.5 (42.0) (min sweep)	144 4 (44.0)	?	330,688 (150,000)	4 x TRDDF4 (?) 48,500 (215.6)	Mach 2.3	Up to 33,069lb (15,000kg) of stores
Sukhoi T-4M	142 4.5 (43.4) (min sweep) 73 10 (22.5) (max sweep)	164 0.5 (50)	1,898 (176.5) (min sweep) 2,591 (241) (max sweep)	328,483 (149,000)	4 x RD36-41 22,045 (98.0) dry, 35,275 (156.8) reheat	1,989 (3,200)	1 x 8,818lb (4,000kg) Kh-45 cruise missile or maximum of 39,683lb (18,000kg) of bombs and other stores
Sukhoi T-4MS	133 10 (40.8) (min sweep) 82 0 (25.0) (max sweep)	135 2 (41.2)	5,449 (506.8) (min sweep) 5,186 (482.3) (max sweep)	374,780 (170,000)	4 x RD36-41 22,045 (98.0) dry, 35,275 (156.8) reheat	684 (1,100) at S/L, 1,989 (3,200) at height	Up to max of 99,206lb (45,000kg) of offensive stores in weapon bay and on external hardpoints
Tupolev '160' fixed wing	No data available					4 x RD36-51	
Tupolev Tu-160 (flown)	189 3.5 (57.7) (min sweep) 116 9.5 (35.6) (max sweep)	177 6 (54.1)	3,152 (293.15) (min sweep) 2,497 (232) (max sweep)	606,261 (275,000)	4 x NK-32 55,115 (245.0)	1,367 (2,200)	Max 99,206lb (45,000kg) of stores Includes 12 Kh-55 cruise missiles or 24 Kh-15 missiles or 88,183lb (40,000kg) free-fall ordnance
Tupolev 'Aircraft 360' (approx)	133 6.5 (40.7)	328 1 (100)	13,441 (1,250) (lifting surface area)	771,605 (350,000)	6 x variable cycle engines	Mach 6	22,046lb (10,000kg) of stores

Soviet Secret Bomber Colour Chronology

Model of the Myasishchev RB-17. George Cox

Model of the Sukhoi P-4, a pre-project study that led to the T-4 prototype.

Side view of the Sukhoi T-4 bomber.

Opposite page:

Top left: **Nose view of the Sukhoi T-6 strike aircraft prototype.**

Top right: **Sukhoi Su-24M.**

Bottom: **Side view of the Sukhoi T-6 strike aircraft prototype.**

This page:

Top: **Tupolev Tu-160.**

Centre: **Sukhoi Su-7B.**

Below: **Beriev Be-10 jet flying boat.**

Top: **The Tupolev Tu-22KP was a version of the jet bomber used for air defence suppression attacking enemy ground radars, and the like.**

Centre: **Tupolev Tu-22M3 displaying nose and cockpit detail.**

Bottom left: **Model of the Myasishchev M-28/2M.**

Bottom right: **Model of the Baade EF 131.**
Both George Cox

Opposite page:

Top left: **Model of the Myasishchev M-32.**

Top right: **Model of the Myasishchev M-52.**
Both John Hall

Centre left: **Beriev SD-MBR supersonic flying boat.**

Centre right: **Model of the Moskalyov GSB-1 supersonic flying boat.** Both George Cox

Bottom: **Model of the Myasishchev M-31.** John Hall

32

52

70

03

31

Top left: **Myasishchev M-20-9.**
George Cox

Top right: **Model of the Tupolev Tu-135.**
George Cox

Centre: **Model of the Sukhoi K-2.**
John Hall.

Below left: **Myasishchev M-20-11.**
George Cox

Below right: **Myasishchev M-20-14.**
George Cox

Glossary

AAM Air-to-air missiles.
Anhedral Downward slope of wing from root to tip.
AoA Angle of attack, the angle at which the wing is inclined relative to the airflow.
Angle of Incidence Angle between the chord line of the wing and the fore and aft datum line of the fuselage.
Area Rule Principal law for keeping transonic drag to a minimum. States that cross-section areas of aeroplane plotted from nose to tail on a graph should form a smooth curve.
Aspect Ratio Ratio of wingspan to mean chord, calculated by dividing the square of the span by the wing area.
Chord Distance between centres of curvature of wing leading and trailing edges when measured parallel to the longitudinal axis.
CofG Centre of gravity.
Council of Ministers A Committee formed from those Ministries directly involved with the Soviet Aviation Industry – Ministry of Aircraft Industry, Defence, Civil Aviation, Higher and Specialised Education and Foreign Trade.
Critical Mach Number Mach number at which an aircraft's controllability is first affected by compressibility, that is, the point at which shock waves first appear.
Dihedral Upward slope of wing from root to tip.
Fly-by-Wire Flight control system using electronic links between the pilot's controls and the control surface actuators.
DVB *Dahl'niy Vyzotny Bombardirovschchik* (Long-range high-altitude bomber).
ECM Electronic Countermeasures.
GAZ *Gosudarstvenny Aviatsionny Zavod* (State Aircraft Factory).
GKAT *Gosudarstvenny Komitet Aviatsionny Tekniki* (State Committee for Aviation Equipment).
GKO *Gosudarstvenny Komitet Oborony* (State Committee for Defence).
GosNIIAS State Research Institute of Aviation Systems.
ICBM Intercontinental Ballistic Missile.
IFR In-flight refuelling.
Laminar Flow Wing Specifically designed to ensure a smooth flow of air over its surfaces with uniform separation between the layers of air.
LERX Leading edge root extensions.
Mach Number Ratio of aeroplane's speed to that of sound in the surrounding medium – expressed as a decimal.
MAI *Moskovskii Aviatsionii Institut* (Moscow Aviation Institute).
MAP *Ministerstvo Aviatsionnoy Promysh Lennosti* (Ministry of Aircraft Industry).
NATO North Atlantic Treaty Organisation.
NII VVS *Nauchno Issledovatelyskii Institut* (Soviet Air Force Scientific and Research Institute based at Akhtoobinsk).
NKAP *Narodny Komissariat Aviatsionnoi Promyshlennosti* (People's Commissariat for Heavy Industry).
NTS Scientific and Technical Committee.
OKB *Opytno Konstruktorskoye Byuro* (Experimental Construction/Design Bureau).
Politburo The Central Committee of the Communist Party. This was responsible for policy and worked with the Council of Ministers.
PVO *Protivovozdushnaya Oborona* (Air Defence Forces).
RAE Royal Aircraft Establishment.
RATOG Rocket-Assisted Take-Off Gear.
SibNIA Siberian Scientific Institute of Aerodynamics
SovMin Soviet Ministry.
STOL Short Take-Off and Landing.
t/c Thickness/chord ratio.
Transonic Flight The speed range either side of Mach 1.0 where an aircraft has both subsonic and supersonic airflow passing over it at the same time.
TsAGI *Tsentrahl'nyy Aero-i Ghidrodinameecheskiy Institoot* (Central State Aerodynamic and Hydrodynamic Institute), Zhukovsky.
TsIAM Central Institute of Aviation Motors.
VG Variable Geometry
VIAM *Vsesoyuzny Institut Aviatsionnykh Materialov* (All-Union Institute for Aviation Materials).
V/STOL Vertical/Short Take-Off and Landing.
VTOL Vertical Take-Off and Landing.
VVS *Voenno-Vozdushniye Sily* (Air Force of the USSR).

Appendix One

Soviet and Russian Bomber Project Summary

This list embraces all known Soviet and Russian post-war heavy, medium and light bomber projects, ground-attack aircraft (*Shturmovik*) and any research aircraft specifically proposed to advance the art of the bomber designer. Important reconnaissance developments are also included for completeness. Care must be taken with some designations because the re-use of project numbers was a common and infuriating habit of Soviet and Russian OKBs.

At the end of the Second World War the Lavochkin, Mikoyan, Sukhoi and Yakovlev OKBs (Experimental Construction Bureaux) specialised in the design of fighter aircraft and Ilyushin and Tupolev worked on multi-engined bombers; ground-attack aircraft were also the domain of Ilyushin. This situation changed relatively little until after the end of the Cold War in 1990. It appears that the Lavochkin OKB never got involved with bomber design before its move into spacecraft and missiles in 1960 but the following lists show that the other three fighter specialists did undertake a little bomber work, although generally with ground attack or strike aircraft in mind. The main addition to the post-war list was Myasishchev which devoted most of its energies towards heavy bombers during its relatively brief existence in the 1950s and 1960s.

Some of the bureaux enjoyed an uninterrupted existence throughout the period covered by this book, at least until the 1990s when the need arose for them to merge or to work together so that they could compete in the international marketplace. However, both Alekseyev and Baade soon disappeared. Myasishchev ended the war running OKB-482 but in February 1946 this was amalgamated with the Ilyushin OKB; Myasishchev's OKB was reopened in 1951. Then in late 1959 the Mya and Lavochkin OKBs were transferred to the control of the new, and quite small, Chelomyey OKB to serve as branch offices for new work on ballistic missiles, but the Mya OKB was resurrected as an independent concern in November 1966. Nyezval was closed in 1946, Sukhoi was closed in 1949 but restored in 1953 and Tsybin was closed in October 1959, much of the latter's work being absorbed by Myasishchev.

From 1954 a body called the Air Standards Co-ordinating Committee (ASCC), made up of representatives from America, the United Kingdom, Canada, Australia and New Zealand, introduced a series of 'reporting names' for Soviet aircraft. The idea was to assign a supposed unambiguous name that could be easily identified over a poor voice radio link with bombers represented by B-names, fighters by F, air-to-surface missiles by K, and so on. In the case of bombers a propeller-driven aircraft used a single syllable word (the Tu-4 was codenamed *Bull*) while jets had two-syllable codenames (the Tu-14 was *Bosun*) and this system was used to cover current and new types. Variants of the same aircraft received additional letters; Myasishchev's M-4 for example was *Bison-A* and the follow-on 3M was called *Bison-B*.

ALEKSEYEV (OKB-21, Gorkii)

I-218 *Shturmovik* project for which mock-up at least part built but prototype (I-218-I) never completed, 1948. Metal cutting may not have actually begun.

I-219 Revised I-218 (also known as I-218-Ib), 1948. Did not leave drawing board.

I-220 (I-218-III). I-221 variant with piston pusher engine, 1948. Project only.

I-221 (I-218-II). Larger design with straight wing, normal tail and jet pusher engine, 1948. Project only.

BAADE
(OKB-1, Dessau, Podberez'ye and later Kimry)

This design team was formed with German engineers who had been taken from the Junkers company to Russia at the end of the war. The team moved to Dresden in 1953 and then formed the VEB company.

EF 131 Bomber prototype flight tested in 1947.

EF 132 Strategic bomber project, 1945. Mock-up built and some parts made, but cancelled late 1948.

Type 140 Private venture experimental development of EF 131 with forward-swept wings and new engines first flown 20.9.48. Originally called EF 140.

RB-2 Short-range bomber project, 1948. Received official approval but, in due course, heavily redesigned to become Type 150.

Type 150 Bomber prototype first flown 14.5.51. One prototype only.

BARTINI

Roberto Bartini was a Soviet designer with an Italian background who from 1952 worked at the Siberian Scientific Institute of Aerodynamics (SibNIA). He had studied the concept of a supersonic strategic bomber for some time and consequently produced several such projects.

A-55 Strategic supersonic flying boat bomber project, 1955. R-55 reconnaissance aircraft also proposed.

A-57 Water-based nuclear bomber, 1957. Cancelled 1960. Other proposed versions included R-57 reconnaissance type.

E-57 Scaled-down A-57 for operations in European and Asian theatres, 1957/58.

A-58 Development of A-57, 1958.

BERIEV (OKB-49, Taganrog)

LL-143 Long-range maritime patrol and anti-ship flying boat first flown 6.9.45.

Be-6 Development of LL-143 fitted with more powerful engines. First flown 2.6.48. NATO recognition codename *Madge*.

Be-10 Patrol flying boat development of Be-6TR, 1948. Not built.

Flying Boat Patrol aircraft project with six VK-2 turboprop engines, late 1940s. Take-off weight 242,505 lb (110,000kg), not built.

R-1 Jet-powered flying boat first flown 30.5.52. Prototype only, not adopted for service.

R-2 Proposed improved R-1, early 1950s. Project only.

Be-10 Jet-powered maritime reconnaissance and torpedo-carrying flying boat first flown 20.6.56. Entered service as only all-jet flying boat to become operational. Codename *Mallow*. Be-10N missile carrier proposal not built.

SD-MBR Supersonic long-range maritime bomber-reconnaissance aircraft with amphibious ski-wing landing gear, 1957.

P-10B Submarine launched supersonic flying boat project, late 1950s. OKB ordered to develop P-10 guided 'missile plane' for deployment on submarines in 1957 and very advanced P.10B concept developed from it.

Be-12 Turboprop-powered flying boat first flown 18.10.60. Entered service as Tchaika, Western codename *Mail*.

Flying Boat Project for heavy amphibian intended to serve as cargo and anti-submarine warfare aircraft, 1961. Military cargo version to have substantial payload, ASW version standard anti-submarine torpedoes and depth charges, and the like, and capable of refuelling when afloat from surface ships or submarines. Four 6,000hp (4,474kW) Ivchenko AI-30 turboprop engines mounted above wing, span 163ft 4½in (49.8m), length 118ft 1in (36.0m), wing area 2,419ft² (225m²), maximum take-off weight 198,413 lb (90,000 lb), warload 9,921 lb (4,500kg), cruise speed 373mph (600km/h), ceiling 29,528ft (9,000m).

The large Beriev anti-submarine and military cargo flying boat project of 1961.

Beriev P-42 carrier-based anti-submarine attack aircraft (1972).

Flying Boat Project for long-range heavy anti-submarine aircraft with four NK-12 turboprop engines, 1962.

Be-26 Oceanic anti-submarine amphibian, 1963. Not built.

A-150 Large jet-powered flying boat design, 1965.

Light Attack Aircraft Counter-Insurgency (COIN) aircraft, c1968.

P-42 Carrier-based anti-submarine twin-jet attack aircraft, 1972. Appearance not unlike American Grumman S-2 Tracker but with longer more streamlined fuselage, slightly swept wings and one 14,220 lb (63.2kN) D-36 turbofan engine in nacelle under each wing. Span 83ft 0in (25.3m), length 65ft 7in (20.0m), wing area 882ft² (82m²), maximum take-off weight 63,933 lb (29,000kg), top speed 528mph (850km/h), service ceiling 42,651 ft (13,000m), range at least 2,486 miles (4,000km). Intended to be flown from 'Project 1160' aircraft carrier.

Be-42/A-40 Jet-powered anti-submarine flying boat first flown 12.86. Named Albatros, largest jet-propelled amphibian in the world. Eventually proposed in various military and civil versions. Western codename *Mermaid*.

A-42 Proposed maritime patrol aircraft development of A-40 fitted with fuel-efficient D-27A propfan engines, c2002.

CHETVYERIKOV (OKB-256, Podberez'ye)

'Arado Ar 234' OKB given task of restoring incomplete German Ar 234 jet bombers for flight test. Programme cancelled 1946.

RK-1 Medium bomber flying wing project, 1947. Go-ahead officially approved but aircraft never completed.

ILYUSHIN (OKB-240, Moscow)

At the time of writing little access had been given to the Ilyushin archive with the result that few details have emerged in regard to the OKB's unbuilt designs. Only some of the OKB's project designations have been traced but it seems likely that much of the OKB's post-Il-28 work was centred on civil design.

Il-14 High-speed bomber that formed possible competitor to Myasishchev and Tupolev piston-engined heavy bomber projects, 1944. Prototype begun in 1945 but work halted 1946.

Il-20 Piston-engined *Shturmovik* first flown early December 1948. One prototype only.

Il-22 Medium-range jet bomber prototype first flown 24.7.47. One prototype only.

Il-24 Development of Il-22 with two AM-TKRD-01 engines in single underwing nacelles and later four Derwents in twin nacelles, 1946/47. Prototype not completed.

Il-26 Piston-engined long-range heavy bomber project, 1947.

Il-28 Medium-range jet bomber first flown 8.7.48. NATO recognition codename *Beagle*. Designed and built in competition with Baade Type 140, Sukhoi Su-10 and Myasishchev RB-17. Swept-winged and re-engined Il-28S proposed 1949/1950 but offered no advantages and dropped.

Il-30 Swept wing development of Il-28. Prototype completed by 8.49 but never flown.

Il-38 Smaller development of Il-30, late 1940s. Not built.

Il-38 Designation re-used for land-based anti-submarine aircraft first flown 28.9.61. NATO codename *May*.

Il-40 Jet-powered *Shturmovik* first flown 7.3.53. Modified Il-40P first flown 14.2.55. Production started but project abandoned due to changes in military policy. Codename *Brawny*.

Il-42 Reworking of Il-38 project to take TR-3A (AL-5) engines, 1950. Not built. The Il-42 designation was later re-used for an attack aircraft developed out of the Il-40.

Il-46 Scaled-up Il-28 first flown 3.3.52. Lost out to Tupolev Tu-16 and did not enter production. Swept wing Il-46S of 1951 cancelled.

Il-54 Supersonic light bomber first flown 3.4.55. Early layout had engines in wing roots and T-tail. No production but received codename *Blowlamp*. Il-54T torpedo bomber not built.

Il-56 Supersonic tactical bomber project, 1955. Cancelled.

Il-42 Second use of designation. Ground-attack *Shturmovik* designed in competition with Sukhoi T-8, Mikoyan and Yakovlev designs, 1969. Project rejected.

Il-102 Jet-powered *Shturmovik* developed from Il-42 design first flown 25.9.82. Prototype only flown.

MIKOYAN (OKB-155, Moscow)

'7-23' Modified MiG-21PFM *Fishbed-F* fighter to match new requirements for jet *Shturmovik*, 1969.

MiG-27Sh Transonic attack aircraft (*Shturmovik*) hybrid with MiG-21 tail and MiG-23B nose, 1969. To same requirements as '7-23' but not built.

MiG-21LSh Hybrid of MiG-21I *Analog* research aircraft and MiG-23B, 1969. Also known as **'27II'** and produced to same *Shturmovik* requirement. Not built.

Ye-155K Proposed defence suppression variant of MiG-25 heavy interceptor, early 1970s.

MiG-25BM More advanced defence suppression MiG-25 variant first flown c1976. Built in small numbers.

Ye-155MF Strike variant of MiG-31 *Foxhound* heavy interceptor, late 1960s. Not built.

MiG-27 Attack aircraft based on MiG-23 fighter. First flown as MiG-23B 20.8.70. Design upgraded as MiG-23BN, then MiG-23BM flown 1974. Codenames *Flogger-F, H and D* respectively, sharing the original with MiG-23.

Type 301 Hypersonic high-altitude reconnaissance aircraft also thought to have bombing capability, early 1990s onwards.

MOSKALYOV (Research Institute)

During the 1950s A S Moskalyov worked in a special design department of the Leningrad Red Banner Engineering Academy of the VVS (the LKVVIA) where he designed several very advanced projects, the most important of which were as follows.

DSB-LK Long-range strategic flying wing bomber, 1957.

GSB-1 (GS-1) Long-range flying boat bomber, 1957-1960.

MYASISHCHEV
(OKB-23, Moscow [OKB-482 prior to 1946])

VM-22/DVB-202 Long-range piston-engined heavy bomber, 1944. Abandoned 1946 with closure of OKB.

VM-23/DVB-302 Long-range piston-engined heavy bomber related to DVB-202, 1944/45. Abandoned 1946 after closure of OKB.

Light Bomber Attempt to fit jet engines into Petlyakov Pe-2I wartime light piston-engined bomber, c1945. Pe-2I often called 'The Soviet Mosquito' after famous British multi-role aircraft.

Heavy Bomber Six-engine designs utilising either piston, turboprop or turbojets, 1946.

VM-24/DSB-17/RB-17 Fast day bomber project, 1945/46. Cancelled in 1946. Proposed DSB-17-D5 derivative had single large jet engine in each underwing nacelle.

VM-25 Began as SDB project for strategic jet bomber, 2.51. Many designs studied and chosen configuration first flown 20.1.53. Redesignated **M-4** after service acceptance, codename *Bison-A*. In Russia sometimes known as Molot (Hammer).

M-26 Long-range bomber (and transport) derivatives of VM-25, c1951.

M-28 DVB long-range high-altitude bomber project, 1952.

M-30 Medium bomber project, 1953. Designation re-used several years later (below).

M-31 Transonic heavy bomber proposal, 1952. Not built.

M-32 Supersonic heavy bomber proposal, 1952. Not built.

M-33 Single-seat delta wing research aircraft with one TRD-5 jet, 1951. Span 18ft 4½in (5.6m), length 24ft 7in (7.5m), wing area 320ft² (29.8m²), maximum speed 684mph (1,100km/h) at 32,808ft (10,000m).

M-34 Missile carrier with transonic cruise speed, 1953. Not built.

M-35 Designation covering in-flight refuelling development of M-4.

M-36 Improved version of M-4 first flown 27.3.56 and put into service as **M-6**, and later **3M** *Bison-B*. **3M-M** flying boat development not built.

M-50 Supersonic strategic bomber first flown 27.10.59. Prototype only. Codename *Bounder*. Originated from early studies for 'Composite Bomber', 1954.

M-51 Unmanned variant of M-50 capable of zero-length launch, 1956.

M-52 Development of M-50 designed to carry heavy stand-off missile, 1958. Mock-up built but prototype not flown and project abandoned.

M-54 Delta-wing supersonic bomber based on M-50, 1959. Different engine arrangements considered.

M-56 More advanced development of M-50/M-52 concept as supersonic cruise missile launcher, 1957 onwards. Later work embraced all new design. Work stopped when Myasishchev OKB became part of Chelomyey design team to concentrate on space/missile systems.

M-57 Supersonic cruise missile carrier version of M-56, 1959. Work later merged with M-56.

M-59 Advanced cruise missile carrier (M-59K) development of M-56, late 1959. Massive delta wing with combined turbojet and rocket powerplant including six VD-7-300 (initially) or four NK-11 jet engines mounted around lower rear fuselage and fed by single intake. Slim forward fuselage with canard foreplane level with cockpit. Twin fins well out on wing, single cruise missile carried beneath engines. Take-off weight approaching 529,101 lb (240,000kg), top speed 1,989mph (3,200km/h), ceiling 88,583ft (27,000m) and range up to 9,944 miles (16,000km); missile's speed after launch 2,486mph (4,000km/h). Brief study only.

M-60 Strategic missile launcher with nuclear propulsion unit, 1955. Many designs proposed for aircraft with unlimited practical range and work eventually became part of M-30 studies. **M-60M** project of 1956 onwards continued theme in series of flying boat designs.

M-62 Believed to be studies for strategic bomber with nuclear propulsion unit, late 1950s onwards. Practically unlimited range.

M-30 Supersonic nuclear-powered bomber, 1959 onwards.

M-70 Supersonic flying boat bomber, 1956.

M-20 Myasishchev OKB restored 11.66 and started M-20 strategic bomber in 1967. Project covered numerous alternative designs produced over long period.

M-18 Large supersonic bomber developed from variants of M-20, 1972. In competition beat Tupolev's 'Aircraft 160' but Tupolev given task of building M-18 as Tu-160.

NYEZVAL (OKB-22)

Heavy Bomber Design proposed in competition for four-engine bomber, 1944.

SUKHOI (OKB-51, Moscow)

Su-10 Tactical jet bomber, 1946. Prototype completed 1948 but project cancelled before flown.

'N' Attack Aircraft Armoured jet-powered attack aircraft, 1948. Not built.

S-1 First flown 7.9.55 as tactical fighter and entered service as **Su-7**. Appearance of types like North American F-100 as fighter-bomber prompted switch of Soviet type to fighter-bomber role as **Su-7B** (prototype designated **S-22-1** by OKB). Built in large numbers. NATO codename *Fitter-A*.

S-22I & S-32 Variable geometry wing development of Su-7 first flown 2.8.66. Entered service as **Su-17** fighter-bomber, codename *Fitter-C*. Further developed into **Su-20** (for export) and **Su-22**.

S-6 Tactical bomber/attack aircraft project, 1963. Two versions; fixed swept wing preferred to swing wing but type not built.

T-58VD Example of Su-15 fighter modified for research into lift engine technology. First flown 6.6.66.

T-58T Strike development of Su-15 with two AL-21 propulsive engines and four RD36-35 lift engines, 1963. Eventually redesignated T-6.

T-58M Tactical bomber/attack aircraft project developed from Su-15, mid-1960s. Design rejected as too expensive.

T-4 Also known as 'Project 100'. Mach 3 strategic bomber first flown 22.8.72. Prototype only flown and programme abandoned 1974. Early studies from 1962 to 1965 embraced many projects including I-2, K-2 and P4.

100L Modified Su-9 interceptor with wing sections added to standard delta shape to act as a scale model for T-4 bomber. Various alternative shapes assessed between 1966 and 1970.

T-4M Large variable geometry wing development of T-4 intended as missile carrier, 1967. Design showed weaknesses and not built.

T-4MS Large and very advanced bomber/missile carrier with variable geometry wings, 1969/70. Was favoured design in competition with Myasishchev and Tupolev but Sukhoi OKB too busy to undertake development. Requirement filled by Tu-160.

T-6 Fixed wing strike aircraft first flown 2.7.67 as T-6-1. Variable geometry wing T-6-2 development flown 17.1.70 and entered service as **Su-24**. NATO codename *Fencer*.

T-8 Ground-attack aircraft first flown 22.2.75. Winner of competition with Il-42, MiG-21LSh and Yak-LSh. Entered service as **Su-25**, codename *Frogfoot*.

T-58Sh Jet *Shturmovik* development of Su-15 (T-58) jet fighter, c1971. Not built.

Su-24BM Large-scale development of Su-24M with VG wings, 1979. Subsequently fitted with fixed wings but later abandoned.

Su-24MM Advanced development of Su-24M, 1985. Preliminary studies only.

S-54N & S-56 Proposed fighter-bomber development of Su-17 with switch to 45° sweep fixed wing, early 1980s. AL-31F engine and new equipment added but not built because Su-17 itself soon to be retired.

Su-27IB Fighter-bomber development of Su-27 *Flanker* fighter first flown 13.4.90. In service to be known as **Su-32**; has also been designated **Su-34**. Nicknamed *Platypus*.

Su-30 Multi-role version of Su-27 with equally effective air-to-air or air-to-ground capability, 1993.

S-37 Single or two-seat multi-role combat aircraft, c1990. Believed abandoned during 1990s but development contributed to design of S-37

Tupolev 'Aircraft 76' (1947). Based on Tupolev's 'Aircraft 74' which was very similar but had a slimmer more pointed nose, in service the '76' would have been called the Tu-24.

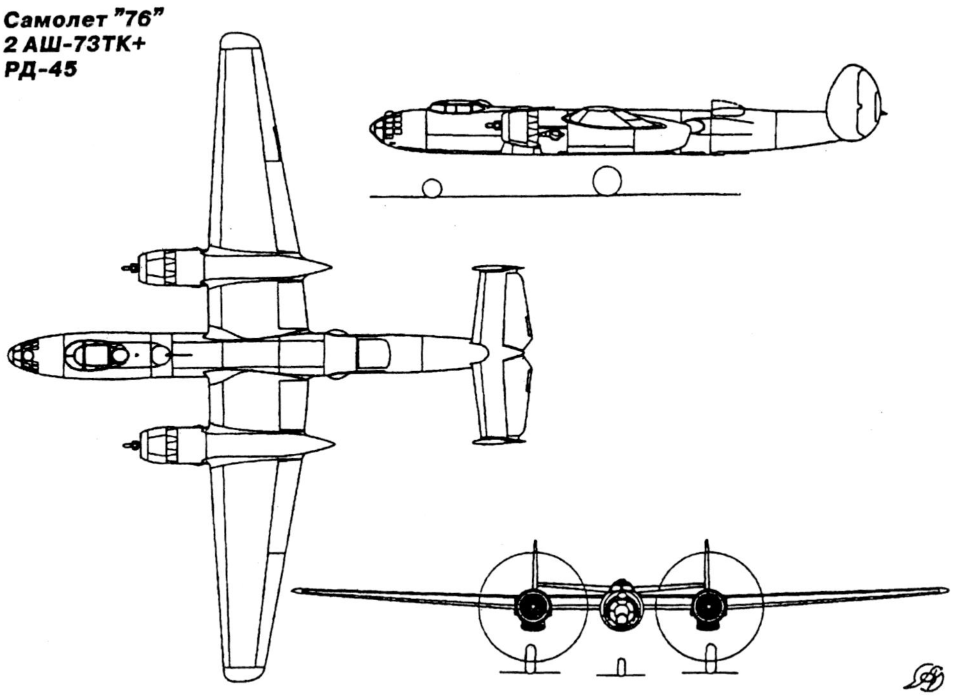

Berkut forward-swept wing technology demonstrator aircraft flown 25.9.97.

Su-39 Single-seat anti-tank development of Su-25, originally called Su-25T and first flown 17.8.84. Designation also allocated to much upgraded Su-25TM version flown 1996.

T-60S Future generation cruise missile carrier and bomber to replace Tu-22M3, 1983. Little information available. Possible construction of prototype started.

T-12 Project to Sh-90 (*Shturmovik* for 1990s) close air-support aircraft to replace Su-25, 1980s. Abandoned early 1990s before production commenced.

TSYBIN (OKB-256, Podberez'ye)

RS Strategic delivery vehicle, 1954 onwards. Abandoned for ballistic missiles but 2RS strategic reconnaissance variant of 1.56 continued. 2RS also abandoned in 1957.

RSR Modified version of RS as reconnaissance aircraft with conventional D-21 jet engines replacing ramjets, 1957. Construction begun but prototype modified before completion.

3RS Version of RSR with ability to be ground or air launched, mid-1950s.

NM-1 Scale model aircraft built to supply data for RSR first flown 7.4.59.

RGSR Water-based bomber derivative of 2RS/RSR, 1957.

RSR Mk.2 RSR modified during construction after receipt of NM-1 data, 1959. Version fitted with Tumansky engines known as RSR R-020. Uncertain if any airframes completed but programme cancelled 4.61 and aircraft scrapped.

TUPOLEV (OKB-156, Moscow)

Up until the period of Tupolev's imprisonment from 1936 to 1943 (for political reasons), his OKB prefixed its designs with ANT for Andrei Tupolev, but from then onwards new projects were described as 'Aircraft 71', and so on. Like most OKBs, prototypes which were put into or, at least, intended for production, received a separate service designation. However, 'Aircraft 95' does confuse the situation because it never received a new number, going straight into service as the Tu-95 *Bear*. In addition the Tu-4 never received its own Tupolev OKB number because it was a copy, not a new design.

'64' Long-range strategic bomber project, 1943 onwards. Abandoned when copy of American Boeing B-29 adopted. Service designation to have been **Tu-10**.

'65' High-altitude development of wartime Tu-2 light bomber with long span wing and advanced piston engines first flown 1.7.46. Single prototype only.

'67' Tu-2 variant with diesel engines first flown 12.2.46.

'68' High-speed front-line bomber and much improved development of Tu-2 first flown 19.5.45. Production aircraft officially termed **Tu-4**, and later **Tu-10**. Only ten production examples built.

'69' Redesign of Tu-2 as long-range bomber officially designated **Tu-8**. Prototype completed 5.47 but not adopted.

Tu-4 Service designation of Soviet copy of Boeing B-29 Stratofortress first flown 19.5.47. Initial designation B-4, codename *Bull*.

'71' Tu-2 short-range bomber development with new nose and engines, 1946. Not built. Official designation **Tu-14**.

'72' High-altitude medium bomber with two ASh-2TK piston engines, early 1946. Cancelled 1947.

'72' (second use of designation) Medium-range jet bomber development of the ANT-69, 1946. Prototype not completed. Official designation **Tu-18**.

'73' Medium-range jet bomber first flown 10.47. Official designation **Tu-20** and later **Tu-14**.

'74' High-altitude reconnaissance bomber development of Tu-2, 1946. Development passed through several stages including fitting auxiliary Nene jet in fuselage tail, but project abandoned 7.48 for alternative jet powered Tu-78. Official designation **Tu-22**.

'75' Short-range dive bomber project with two M-3 piston engines, 1947. Preliminary study only but official designation **Tu-26** allocated. Aircraft intended to be capable of 429mph (690km/h) at 19,685ft (6,000m), 31,168ft (9,500m) ceiling and possess range of 3,729 miles (6,000km) with 4,409 lb (2,000kg) bombs. Gun armament one forward-firing NS-23 cannon, two twin B-20 turrets firing rearward. State testing planned for July 1948 but no detail design undertaken. 'Aircraft 75' designation used again in 1947 for cargo aircraft.

'76' Medium bomber and torpedo carrier designs produced alongside 'Aircraft 75' dive bomber, 1946. Were bomber variants of 'Aircraft 74' reconnaissance type with two ASh-84TK piston and one RD-45 jet, but later pistons replaced with ASh-73TK from Tu-4 bomber. To be **Tu-28** and **Tu-24** respectively, prototype of latter planned but project abandoned 1948.

'77' Designation first allocated to medium bomber project powered by two Nenes with layout similar to North American B-45 Tornado bomber, 1946. Official designation **Tu-30**; not chosen for manufacture.

'77' Tactical jet bomber development of wartime piston-powered Tu-2 first flown 26.7.47. Small batch built as **Tu-12**.

'78' Derivative of 'Aircraft 73' for reconnaissance work first flown 7.5.48. Official designation **Tu-16** but Il-28 preferred choice. Previously **'Aircraft 78'** designation allocated to photo-reconnaissance version of 'Aircraft 69' piston aircraft, 1946/47. Official designation **Tu-28** allocated to this earlier project.

'79' Designation initially used for proposal to fit Tu-4 with new engines, 1947. In late 1948 **'79'** reused for further reconnaissance development of 'Aircraft 73' but work cancelled 5.49. Official designation **Tu-20**.

Heavy Bomber Projects for piston-engined bomber, late 1940s. **'Projects 471'**, **'473'**, **'474'** & **'485'**, developments of Tu-4, produced as part of this work. Provisional designation 'Tu-200' allocated by West but not used in USSR. Led to 'Aircraft 80'.

'80' Long-range piston-engined heavy bomber first flown 1.12.49.

'81' Medium-range jet bomber first flown 13.10.49. Official designation initially **Tu-18** but entered service as **Tu-14**. Codename *Bosun*. In competition with Il-28. Navalised 'Project 509' development not adopted.

'82' Swept wing modification of Tu-14. One prototype only flown 24.3.49. Official **Tu-22** designation used again.

'83' Front-line bomber version of 'Aircraft 82' with longer fuselage, 1948/49. Project

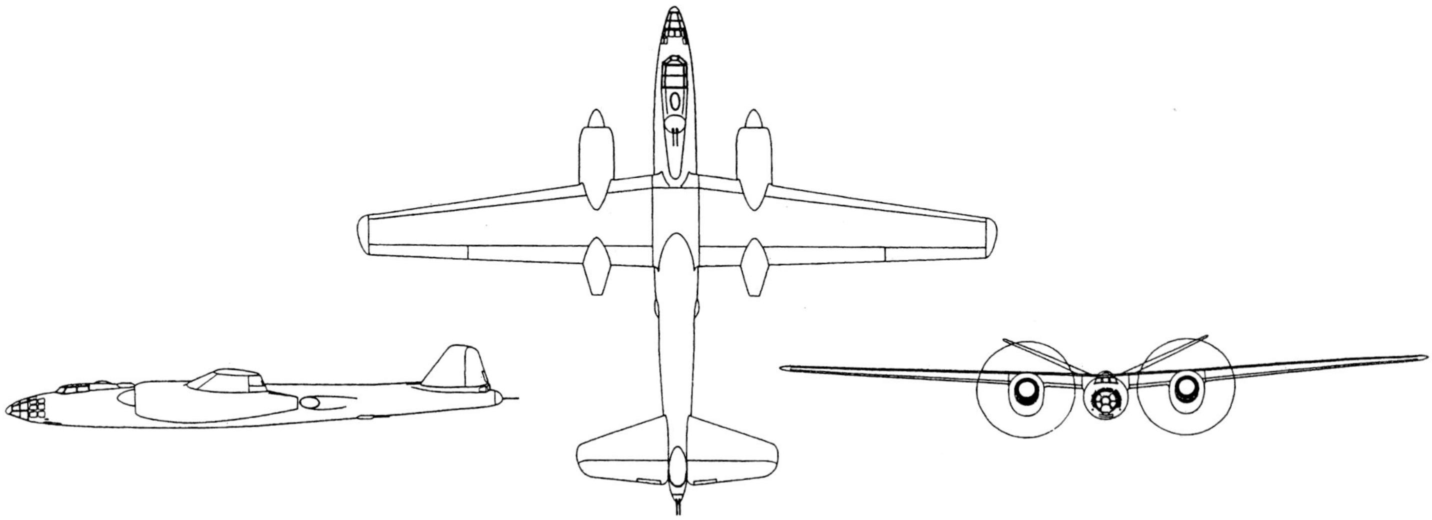

One version of Tupolev's 'Aircraft 84' (1948).

cancelled 1949, single prototype never completed. Also to be **Tu-22**.

'84' Photo-reconnaissance project, 1948/49. One of first native designs fitted with turboprop engines. V Ya Klimov OKB, designers of VK-2 turboprop, wanted Tupolev to help promote engine by producing new design to use it. Result was 'Aircraft 84' of which there were several variants. First was development of final 'Aircraft 74' design with turboprops replacing piston engines and VK-1 replacing tail-mounted RD-45 jet. Later completely redesigned with two VK-2s as high-altitude long-range reconnaissance aircraft with high wing, V-tail, two fixed forward-firing 23mm and dorsal, ventral and tail twin 23mm turrets, plus five crew. Span c111ft 6½in (34.0m), length c83ft 8in (25.5m), wing area c1,161ft² (108.0m²), estimated maximum speed 451mph (725km/h) at 19,685ft (6,000m), ceiling 42,979ft (13,100m) and range 3,480 to 3,605 miles (5,600km to 5,800km). Uncertainties with VK-2 engine meant Tu-84 not built.

'85' Long-range piston-engined heavy bomber first flown 9.1.51. Codename *Barge*. Resulted from several studies for long-range strategic bomber replacement for Tu-4 under project designations **'485'**, **'487'** and **'489'**, late 1940s. 'Aircraft 85' resulted from version of '487'; others never left drawing board. Proposed 'Aircraft 85A' had four VD-4K piston engines and two VK-1 turbojets.

'86' Twin jet medium bomber development of 'Aircraft 73', 1948 onwards. Initial work under **Project '486'**. 'Aircraft 86' abandoned for 'Aircraft 88' but **Project '491'** of 4.49 further development with 45° swept wing – results from this used for OKB's first transonic designs.

'87' Variant of 'Aircraft 86' with TR-3 engines, 1949.

'88' Strategic bomber first flown 27.4.52. Entered service as **Tu-16**. NATO codename *Badger*. Preliminary studies completed under designation **'494'** and **'495'**.

'89' Reconnaissance version of 'Aircraft 81' first flown 23.3.51. Development stopped mid-1951; for short period known as **Tu-16**.

'90' Long-range bomber variant of Tu-88 with four AL-5 engines, early 1950s. Later on fitting of Tu-16 production aircraft with turboprop engines also evaluated under 'Aircraft 90' designation.

Project '504' Long-range bomber/flying boat based on 'Aircraft 85', 1950.

'91' **'Project 507'** naval strike aircraft first flown 2.9.54. Selected for development ahead of navalised Tu-14T (**'Project 509'**) but not adopted for service. Codename *Boot*.

'92' Long-range reconnaissance variant of 'Aircraft 88' which entered service as Tu-16R.

'93' Development of Tu-14T with detail changes, 1951. Not built.

'94' Proposal to refit Tu-4 bombers with turboprop engines, 1950.

'95' Turboprop-powered strategic bomber first flown 12.11.52. Entered service as **Tu-95**. Codename *Bear*.

'96' High-altitude intercontinental development of 'Aircraft 95' first flown mid-1956. Development abandoned 3.56.

'97' Long-range transonic bomber based on 'Aircraft 88', early 1950s.

'98' Supersonic bomber first flown 7.9.56. One prototype, no production. Given codename *Backfin* and for some time identified in West as 'Yak-42'. Production examples of '98A' front-line bomber variant would have been **Tu-24**.

'99' Intercontinental bomber development of 'Aircraft 95' and 'Aircraft 96' with alternative turboprop or turbojet engines, 1953. Variations in airframe layout also studied.

'100' Nuclear bomb carrier designed to be suspended under larger aircraft as part of 'Composite' strategic strike system, 1953. Cancelled 1958.

'103' Long-range transonic bomber development of 'Aircraft 97', early/mid-1950s.

'105' Supersonic bomber first flown 7.9.59. Entered service as **Tu-22**, codename *Blinder*.

'106' Long-range supersonic bomber development of Tu-22, 1954. 'Aircraft 106A' more radical redesign of Tu-22 as long-range bomber and missile strike aircraft. Project not finally cancelled until 1965.

'108' Part of 'Composite' long-range intercontinental supersonic strategic carrier and strike system, 1952. Intended to be carrier for 'Aircraft 100'. Many variants proposed, cancelled 1958.

'109' Derivative of 'Aircraft 108', late 1950s. Drawing shows wing root engines, swept wing and tail and long slim fuselage with cruise missile beneath. Span 123ft 0½in (37.5m), wing area 3,763ft² (350m²).

'112' Front-line supersonic tactical bomber project, 1954/55.

'113' 'Suspended' air-to-ground unmanned cruise missile project, 1955. To be carried by Tu-95 or Tu-96.

'115' Designation appears allocated briefly to 'Aircraft 106A'.

'119' Experimental aircraft with nuclear powerplant, mid-1950s.

'120' Nuclear-powered family of supersonic bombers, mid-1950s.

'122' Frontline supersonic bomber development of 'Aircraft 98', 1957.

'125' Long-range bomber and missile strike aircraft intended to replace Tu-22, 1958. Developed into full-scale strategic system and design changed during 1960s. Not built.

'127' Frontline supersonic bomber and missile carrier, 1958.

'128' Long-range supersonic interceptor first flown 23.1.61, entered service as Tu-28 (NATO *Fiddler*). Front-line multi-role strike aircraft versions proposed 1963 by S M Yeger with S-5 or S-24 rockets, six-barrel 23mm AO-9 cannon, guided missiles including KSR-2 anti-ship weapon, or four 1,102 lb (500kg) or eight 551 lb (250kg) bombs. Proposals rejected but second strike version offered 1969 with 'Initsiativa-2' bombing and navigation radar and OPB-1b optical bomb sight, 3,307 lb (1,500kg) offensive stores carried internally plus 6,614 lb (3,000kg) externally. Range with internal load only 1,457 miles (2,345km) when flying at 1,100mph (1,770km/h) and 36,089ft (11,000m).

'129' Frontline supersonic bomber development of 'Aircraft 127', c1958. Little information available.

'132' Low-altitude bomber, 1958.

'135' (First use of designation) Combined aircraft and missile strategic system, 1958. Carrier aircraft based on Tu-95/Tu-96 and intended to carry pilotless cruise missile.

'135' (Second use of designation) Supersonic strategic bomber, 1960 onwards. Project abandoned by mid-1960s.

'136' Multi-role VTOL fighter with layout generally similar to British Hawker P.1127, 1963/64. (Note: designation 'Aircraft 136' also allocated to a hypersonic project of late 1950s and again for civil airliner during the 1970s.)

'137' Long-range supersonic strategic missile carrier developed from 'Aircraft 135', late 1950s. Preliminary studies only.

'142' Version of Tu-95 produced as dedicated long-range anti-submarine aircraft. First example, as Tu-142, flew 18.6.68.

'145' Major development of Tu-22 with variable geometry wing, 1965. Not proceeded with but evolved into '**Project 45**' and **Tu-22M** first flown 30.8.69. Entered service and codenamed *Backfire*.

'146' Long-range anti-submarine and patrol aircraft proposal, late 1980s.

'148' Long-range interceptor and multi-role strike project to replace Tu-128 fighter, 1965. Rejected by Air Force.

'156' Earliest designation for large strategic bomber, 1970. Quickly renumbered '160' so '156' reused for projected derivative of Tu-154 airliner.

'160' '160M' first proposed as strategic flying wing bomber 1972. Bomber competition won by Sukhoi but, under orders, its T-4MS design and Myasishchev M-18 taken over by Tupolev to become 'Aircraft 160'. M-18 formed basis for development and several versions studied with different engine configurations. Prototype first flown 19.12.81 and type entered service with same designation. Codename *Blackjack*.

'161' Intercontinental strategic multimode missile carrier, 1970s. Was development of Tu-160.

'170' Adaptation of Tu-160 to carry conventional weapons only, late 1980s. Parts of proposal incorporated into Tu-160 upgrade.

'245' Complete redesign of Tu-22M, c1980s/1990s. Not built. Designation possibly given to prototype Tu-22M4 upgrade completed 1990 but not adopted for production. Instead Tu-22Ms upgraded to Tu-22M5 standard.

'260' or **'Aircraft 230'** Proposed Mach 4 strategic bomber successor to Tu-160, 1983.

'342' Intercontinental strategic carrier of long-range cruise missiles proposed as further development of Tu-95MS, c1980s.

'360' Hypersonic Mach 6 strike aircraft project successor to Tu-160, 1980s.

B-90 Flying wing subsonic bomber project for 1990s, mid-1980s. Work possibly begun as **'Aircraft 202'**.

YAKOVLEV (OKB-115, Moscow)

Yak-123 Experimental tactical bomber developed from Yak-25 fighter first flown 1955. Pre-production batch completed as **Yak-26** but production plans replaced by Yak-28.

Yak-125B Prototype tactical nuclear strike aircraft based on Yak-25 flown 1955. Prototype only.

Yak-129 Tactical bomber first flown 5.3.58. Entered service as **Yak-28**, codename *Brewer*.

Yak-2VK-11 Two-seat high-altitude frontline bomber project development of Yak-26, 1956. Five pre-production aeroplanes requested but not built.

Yak-30/Yak-32/Yak-34 Proposed front-line reconnaissance aircraft with bomber developments to follow, 1957 onwards. Not built.

Yak-32Sh Yak-30 and Yak-32 designations also used for jet trainer and sports aircraft respectively and flown in 1960. Yak-32Sh proposed light attack aircraft based on Yak-32, 1961.

Yak-33 Supersonic bomber and missile launcher project proposed in two forms, c1962.

Low-level front-line fighter-bomber Proposed development of Yak-35MV interceptor project, 1959. To operate along similar lines to Sukhoi Su-7B but abandoned for Sukhoi's earlier aircraft.

Yak-36 VTOL research aircraft. First untethered hover 9.1.63 and first full transition to horizontal flight 16.9.63. Codename *Freehand*.

Yak-36M VTOL aircraft with primary role of attack but additional fleet air defence duties. First hover 22.9.70, first conventional flight 2.12.70, entered service as **Yak-38**. Codename *Forger*.

Yak-LSh Jet *Shturmovik* project in competition with Ilyushin, Mikoyan and Sukhoi designs, 1969. Believed to be highly modified version of the Yak-28 but not built. Sukhoi T-8 (Su-25) declared winner.

Yak-39 Proposed development of Yak-38 with bigger wing and more lift capability, 1983. Not built.

Yak-43 Proposed supersonic multi-role V/STOL aircraft for land and shipborne duties, 1986. Used large single engine with swivelling nozzles. Not built.

Model of Tupolev's 'Aircraft 230' project (1983). The engines are housed in underwing pods.

Select Bibliography

Aircraft of OKB Ilyushin: Yuri A Yegorov; Moscow Russavia, 2003.
Aircraft of OKB Tupolev: Vladimir G Rigmant; Moscow Russavia, 2001.
Aircraft of TANTK named after G M Beriev 1945-1968: G S Panatov; Russia, 2001.
Aircraft of TANTK named after G M Beriev 1968-2002: G S Panatov; Russia, 2002.
Beriev Be-10 'Mallow' – Russia's last flying boat: Aleksandr Zablotskiy and Andrey Salnikov; *International Air Power Review Volume 8*, 2003.
Early Russian Jet Engines: The Nene and Derwent in the Soviet Union and the Evolution of the VK-1: Vladimir Kotelnikov and Tony Buttler; Rolls-Royce Heritage Trust Historical Series No 33, 2003.
Early Soviet Jet Bombers: Yefim Gordon; Red Star Volume 17, Midland Publishing, 2004.
Illustrated Encyclopaedia of Aircraft from the V M Myasishchev OKB Volume 2 Part 1: A A Bruk, K G Udalov, S G Smirnov, B L Puntas; Aviko, 2001.
Illustrated Encyclopaedia of Aircraft from the V M Myasishchev OKB Volume 2 Part 2: A A Bruk, K G Udalov, S G Smirnov, A V Arkhipov, B L Puntas; Aviko, 2001.
Indigenous Bombers (1945-2000) Part 1: S M Ganin, A V Karpenko, V V Kolnogorov; Bastion, 2001.
Indigenous Bombers (1945-2000) Part 2: S M Ganin, A V Karpenko, V V Kolnogorov; Bastion, 2001.
Myasishchev M-4 and 3M – The First Soviet Strategic Jet Bomber: Yefim Gordon; Red Star Volume 11, Midland Publishing, 2003.
OKB Ilyushin: Y Gordon; Midland Publishing, 2004.
OKB Sukhoi: Vladimir Antonov, Yefim Gordon, Nikolai Gordyukov, Vladimir Yakovlev and Vyacheslav Zenkin; Midland Publishing, 1996.
Russian Aviation and Air Power in the Twentieth Century: Ed by Robin Higham, John T Greenwood and Von Hardesty; Frank Cass, 1998.
Russian X-Planes: Alan Dawes; Key Publishing 2001.
Soviet Seaplane Jet Bombers: Thomas Mueller with Jens Baganz; *Aerospace Projects Review*, July/August 2003.
Soviet X-Planes: Yefim Gordon and Bill Gunston; Midland Publishing, 2000.
Steps Towards 'Blackjack': Piotr Butowski; *Air Enthusiast No 73,* Jan/Feb 1998.
Supersonic Myasishchevs: Piotr Butowski; *Air Enthusiast No 82,* July/Aug 1999.
Tupolev Tu-160 Blackjack: Russia's Answer to the B-1: Yefim Gordon; Red Star Volume 9, Midland Publishing, 2003.
Tupolev Tu-160 Blackjack: Piotr Butowski; *International Air Power Review Volume 2*, 2001.
Tupolev Tu-22 'Blinder' Tu-22M 'Backfire': Yefim Gordon; Aerofax, Midland Publishing, 1998.
Tupolev Tu-4: Soviet Superfortress: Yefim Gordon; Red Star Volume 7, Midland Publishing, 2002.
Working with Sukhoi: Oleg Samoilovich; Published in Moscow, 1999.
Yakovlev Aircraft since 1924: Bill Gunston and Yefim Gordon; Putnam, 1996.
Yakovlev's VTOL Fighters: Anatoli Artemyev; *International Air Power Review Volume 10*, 2003.

The following articles, reproduced by the *Bulletin of the Russian Aviation Research Trust,* were treated as near primary source material, although some have previously been published in Russian language journals.

Aircraft by Myasishchev: N V Yakubovitch and V N Lavrov (*Russavia*).
From 'RS' to 'SR': Ivnamin Sultanov (*Aviatsiya i Kosmonavtika*).
Front-line Strike Aviation at the Threshold of the Missile Age: Alexandr Medved (*AviaMaster 4.98*).
MiG-27 Story: Vladimir Il'yin (*Aviatsiya i Kosmonavtika 20*).
Strange Contest: Piotr Butowski (*AeroHobby 4/1994*).
Su-10 – The First P O Sukhoi Bomber: Nikolay Gordyukov (*Samolyoty Mira*).
Sukhoi T-4 and T-4MS: Vladimir Antonov and Nikolai Gordukov (published as *Sotka: T-4M* in *Samolyoty Mira No 3 1996*).
T-4: A 'Russian Wonder' or Technological Adventure: B Yakovlev and G Grishayeva (*Aviatsiya i Kosmonavtika 9-10/1993*).

Finally a detailed and unpublished account of all Tupolev aircraft and project designs, compiled from the OKB archives by Vladimir Rigmant, proved to be a vital source of information for this famous design bureau.

Index

INDEX OF AIRCRAFT

Aerospatialé/BAC Concorde 62, 67
Alekseyev
  I-218 86, 87, 101
  I-219 87
  I-220 87, 101
  I-221 87, 101
Arado (Chetvyerikov) Ar 234 17, 19
Avro 730 74

Baade
  EF 131 33, 34, 164
  EF 132 33, 34, 51
  RB-2 35, 36, 51
  Type 140 20, 33-36, 51, 87
  Type 150 32, 35, 36, 51, 87
BAC TSR.2 108
Bartini
  A-55 125, 126
  A-57 74, 125-127, 131
  A-58 126, 127, 131
  E-57 126, 127, 131
  T-203 126
Beriev
  A-40/Be-42 123, 131
  A-42 123
  A-150 122, 123, 131
  Anti-Submarine Flying Boat 121, 131
  BB-288 117
  Be-6 117-119, 121, 130, 131
  Be-10 (1st) 118
  Be-10 (2nd) 53, 118, 131,163
  Be-10N 118, 120
  Be-12 120, 121, 123, 131
  Be-26 121, 122, 131
  LL-143 117-119, 131
  LL-600 121
  MDR-5 117
  MDR-7 117
  MDR-10 117
  R-1 17, 118, 119, 131
  R-2 118
  SD-MBR 128-131, 165
Boeing
  B-29 Superfortress 43, 48, 49, 141
  B-47 Stratojet 43, 45
  B-52 Stratofortress 43, 48, 49, 141
Bristol 188 71

Chetvyerikov
  MDR-6 117
  RK-1 83, 84
Convair
  B-36H (X-6) 79, 81
  B-58 Hustler 62, 66, 70, 80
  F-102 Delta Dagger 64

Douglas AD Skyraider 87

Fairchild-Republic A-10 Thunderbolt 91, 93

General Dynamics F-111 109, 112
Gloster Meteor 25

Hawker-Siddeley Harrier 99

Ilyushin
  Il-2 86, 87
  Il-10 87
  Il-14 7, 8, 15
  Il-18 20
  Il-20 86-89, 101
  Il-22 16, 17, 19, 20, 31
  Il-24 19, 20, 31
  Il-26 14, 15
  Il-28 14, 17, 20, 24-27, 29-31, 36-38, 46, 52, 53, 94, 106
  Il-28S 30
  Il-30 16, 29-31
  Il-38 (1st) 30, 31
  Il-38 (2nd) 123, 125, 131
  Il-40 91-94, 101, 114
  Il-42 (1st) 31
  Il-42 (2nd) 94, 95, 98
  Il-46 (& Il-46S) 31, 36-38, 41, 42, 51
  Il-54 52-54, 69, 106
  Il-56 54
  Il-102 95, 98, 99, 101

Junkers Ju 287 33

Lavochkin La-190 37
Lockheed
  P3V-1 Orion 125
  SR-71 Blackbird 74, 140

Martin XP6M Seamaster 118
McDonnell
  F-101 Voodoo 64
  F-4 Phantom 111, 146
McDonnell Douglas DC-10 61
Messerschmitt Me 262 18, 87
Mikoyan
  301 159
  I-350 37
  LSSh Studies (7-23, MiG-27Sh, MiG-21LSh) 94, 95
  MiG-9 18, 21
  MiG-15 13, 17, 89 100
  MiG-17 17, 100
  MiG-19 93, 114
  MiG-21 64, 73, 74, 94
  MiG-21I *Analog* 94
  MiG-21PD/23-31 110
  MiG-23 93, 94, 110, 114, 141
  MiG-23PD/23-01 110
  MiG-25 74, 109, 113, 114, 137
  MiG-25BM 113, 114
  MiG-27 94, 109, 114-116
  MiG-29 98
  MiG-31 113, 114
  Ye-8 110
  Ye-150 105
  Ye-155MF 113
Moskalyov
  DSB-LK 78, 84, 85, 127
  GSB-1 (GS-1) 127, 128, 131, 165
Myasishchev
  3M (M-36) 49-51, 62, 64, 65, 147
  3M-M Flying Boat 51
  DVB-202 (VM-22) 8, 9, 11, 15
  DVB-302 (VM-23) 8, 9, 11, 15
  Long-Range Bomber 14, 15
  M-4 (M & VM-25) 43-48, 50, 51, 62, 65, 147
  M-17 151
  M-18 147, 151, 152, 156, 158-160
  M-20 Series 147-151, 160, 166
  M-26 47, 48
  M-28 (2M) 48-51, 164
  M-30 (1st) 51
  M-30 (2nd) 82, 83, 85
  M-31 54, 55, 69, 165
  M-32 55, 56, 69, 165
  M-34 56, 57
  M-50 52, 62-66, 68-70, 77, 81-83, 129, 135
  M-51 64
  M-52 64-66, 69, 77, 130, 165
  M-54 66
  M-56 66-69, 83, 106, 130, 134, 136
  M-57 68, 69
  M-60 81, 82, 85
  M-60M 82, 85
  M-62 82
  M-70 64, 82, 129-131
  RB-17 (DSB-17 & VM-24) 17-19, 31, 161

North American
  B-45 Tornado 27
  OV-10A Bronco 93
  XB-70 Valkyrie 132, 135, 136
Nyezval Heavy Bomber 8, 11, 15

Petlyakov
  Pe-2 9, 18
  Pe-8 7, 9-11

Rockwell B-1 76, 136, 141, 152
Republic F-105 Thunderchief 108

SAAB 105 93
SEPECAT Jaguar 110
Short SC.1 99
Sukhoi
  100L 141
  I-2 137, 138
  K-2 139, 140, 166
  LVSSh 96, 97
  N 91, 101
  P-4 138-140, 161
  S-6 109, 110
  S-37 115
  Su-7 57, 73, 105-107, 110, 114, 116, 163
  Su-9 (1st) 21, 24
  Su-9 (2nd) 73, 141
  Su-10 17, 21-24, 31
  Su-11 16
  Su-12 24
  Su-15 95, 110
  Su-17 93, 106, 107, 109, 110, 116
  Su-24 104, 108, 110-116, 158, 162
  Su-24BM 112
  Su-24MM 112, 113
  Su-25 86, 91, 92, 97-99, 101, 110
  Su-27 113, 116, 158
  Su-27IB/Su-30/Su-32 115, 116
  Su-39 97
  T-4 112, 114, 132, 134, 136-141, 143, 144, 152, 153, 161
  T-4M 151-153, 160
  T-4MS 112, 153-156, 158-160
  T-6 102, 110, 111, 116, 162
  T-8 93-97, 101, 114
  T-12 97, 98
  T-37 105
  T-58M 110, 116
  T-58VD 110
  T-58Sh 95, 96
  T-60S 112, 159

Tsybin
  LL Series 70
  NM-1 70, 73, 74, 77
  RGSR 128, 129
  RS 70-72, 77
  2RS 72
  RSR (3RS) 72, 73, 77
  RSR 'Mk.2' (R-020) 73, 74, 77
  RSS 74, 125-127

INDEX continued

Tupolev
Aircraft 64 8-11, 15
Aircraft 66 11
Aircraft 72 20, 25, 31
Aircraft 73 20, 25-29, 31, 38
Aircraft 74 27
Aircraft 76 27
Aircraft 77 (Tu-30) 27, 31
Aircraft 77 (Tu-12) 16, 18, 20, 21, 31
Aircraft 78 26-28
Aircraft 79 26, 27
Aircraft 80 (Tu-80) 12, 15
Aircraft 81 (Tu-14) 17, 25, 26, 28, 31, 38, 89
Aircraft 82 28, 29, 31, 38
Aircraft 83 29, 30
Aircraft 85 (Tu-85) 11-13, 15, 123
Aircraft 86 38-40, 51
Aircraft 87 40, 51
Aircraft 88 (Tu-16) 32, 36-42, 47, 51, 57, 58, 106, 127, 159
Aircraft 89 28
Aircraft 90 42
Aircraft 91 (Tu-91) 16, 89-91, 101
Aircraft 93 28
Aircraft 94 13-15
Aircraft 95 (Tu-95) 16, 32, 43, 44, 46, 49, 51, 72, 74, 75, 77, 80, 124, 160
Aircraft 96 44, 45, 51, 74, 77
Aircraft 97 42, 43, 57
Aircraft 98 (Tu-98) 53, 57, 69, 102, 106, 107
Aircraft 99 45
Aircraft 100 45, 74, 75, 77
Aircraft 103 42, 43, 57
Aircraft 105 57-60, 69, 102, 106, 108
Aircraft 105A (Tu-22) 58-62, 69, 80, 102, 112, 127, 132, 141, 144-146, 156, 164
Aircraft 106 Series 59-62, 69, 144
Aircraft 108 45, 74-77
Aircraft 109 76, 77, 80
Aircraft 112 102, 103, 116
Aircraft 113 74, 76, 77
Aircraft 119 80
Aircraft 120 80, 81, 85
Aircraft 121 76
Aircraft 122 57, 106
Aircraft 123 76
Aircraft 125 132-136, 143, 144
Aircraft 127 107, 108, 116
Aircraft 128 (Tu-28) 29, 57, 106, 114, 146
Aircraft 129 107, 108
Aircraft 132 80, 108, 109, 116
Aircraft 135 (1st) 77
Aircraft 135 (2nd) 66, 77, 80, 81, 132, 134-138, 141, 143, 156, 166
Aircraft 137 136
Aircraft 142 (Tu-142M) 44, 124, 131, 156
Aircraft 144 (Tu-144) 67, 76, 94, 135, 156, 158
Aircraft 145 (Tu-22M) 62, 134, 136, 141, 144-146, 159, 160 ,164
Aircraft 146 124
Aircraft 148 114
Aircraft 160M 155-160
Aircraft 244 158
Aircraft 245 146
Aircraft 260 160
Aircraft 360 160
B-90 160
Heavy Bomber ('Tu-200') 14
Project 471 12, 14
Project 473 12, 14
Project 474 12, 14
Project 485 12, 15
Project 486 38, 39
Project 487 13, 15
Project 489 12, 13, 15
Project 491 39, 40
Project 494 41, 42
Project 495 42
Project 504 123, 124, 131
Project 507 89
Project 509 89
Tu-2 11, 18, 20, 21, 27
Tu-4 (B-29 copy) 7, 11, 12, 14, 15, 25, 40, 49, 66
Tu-95LAL 79, 80
Tu-160 60, 76, 136, 144, 146, 159, 160, 163

Yakovlev
BB-22 105
Yak-2 105
Yak-2VK-11 104, 105
Yak-4 105
Yak-15 18
Yak-23 21
Yak-25 (Yak-120) 53, 103
Yak-26 (Yak-123) 53, 92, 103, 104, 106, 116
Yak-27 (Yak-121) 104
Yak-28 (Yak-128) 95, 103-106, 116
Yak-30 (1st)/32/34 105, 106, 116
Yak-30 (2nd) 99
Yak-32Sh 98, 99
Yak-33 138, 141-143
Yak-35MV 106
Yak-36 99-101
Yak-38 99-101
Yak-39 101
Yak-41 101
Yak-43 101
Yak-122 103
Yak-125B 103
Yak-1000 37
Yak-LSh 95

INDEX OF PEOPLE

Alekseyev, Semyon 20, 35
Alexandrov, A P 78, 82

Baade, Brunolf 33, 35, 36
Babin, I B 42
Bartini, Robert L 126, 141
Baslavsky, I Ye 22, 93
Batukhov, A K 55
Bazenkov, N I 44
Beriev, Georgy M 17, 117, 118
Bliznyuk, Valeriy I 156
Bondarenko, L I 154
Bulganin, Nikolay A 24, 25

Cheryomukhin, A M 13, 44
Cheryomukhin, G A 42
Chetvyerikov, I V 17, 83
Chizhevsky, V A 89

Dementyev, P V 49, 64, 93, 97, 100, 141
Dobrynin 49

Fainshtein, A S 79

Grechko, Andrei 93, 141
Gunston, Bill 62

Ilyushin, Sergey 24, 29, 37, 52, 67, 95, 98
Ivanov, Yevgeny 95, 112, 138
Ivashechkin, Yuri 93

Judin, A A 42

Keldysh, M V 67
Khrunichev, M V 11, 18, 35, 118
Khrushchev, Nikita 68, 69, 91, 106, 134, 136, 144
Klimov, V Ya 16, 17
Kolesov, Pyotr A 141
Kolpakov, N G 105
Kondorsky, B M 13, 41, 42, 75
Konstantinov, A K 123
Kootakhov, P S 158
Kurchatov, I V 78
Kutakhov, Pavel S 99
Kuznetsov, N D 43, 45, 78
Kuznetsov, N G 118

Lyulka, A M 16, 78

Makarevsky, A I 67
Markov, D S 9, 26, 57, 58
Mazinovsky, R Ya 99, 106
Metskhvarishvily, N 110
Mikoyan, Artyom 73, 74
Mikulin, Alexander 16, 17
Mochalin, Ye D 57
Moskalyov, Aleksandr 84, 127
Myasishchev, Vladimir 18, 46, 66, 69, 72-74, 78, 134, 151

Nazarov, G N 45
Novozhilov, Ghenrikh V 98, 99
Nyezval, I F 79

Obrubov, P N 35

Preobrazhensky, Ye N 65, 118

Reshetnikov (Col) 158
Rutskoy, Aleksandr 99
Samoilovitch, Oleg 93, 109, 111, 137, 141, 154
Savchenko, Ivan V 93
Selyakov, Leonid L 46, 67, 135
Shakhurin, Aleksey 9, 11, 18
Shavrov, Vadim 126
Shekhtyer, L M 99
Shishkin 23
Shvetsov, A D 45
Silayev, Ivan S 99, 112
Simonov, Mikhail 96, 112
Sokolovsky, V D 106
Stalin, Josef 11, 27, 43, 69
Sterlin, V A 42
Sturninsky, V V 67
Sudets, V A 64
Sukhoi, Pavel 22, 23, 67, 89, 93, 95, 96, 105, 110-112, 137, 138, 141, 143, 154
Sultanov, Ivnamin 71, 74
Svishchev, G P 67

Tolstykh, I P 62
Tsvelev, V K 99
Tsybin, Pavel 70-74, 77, 126
Tumansky, S K 73
Tupolev, A A 42, 156
Tupolev, Andrei N 9, 26, 27, 29, 42-44, 46, 69, 71-74, 78, 134, 135, 137, 138, 141, 143, 144
Twining, N F 54

Ustinov, Dmitri F 99

Vershinin, Konstantin A 105, 106

Wocke, Hans 33, 35

Yakovlev, Aleksandr 53, 69, 99, 143
Yeger, Sergei M 42, 57, 58, 60, 76, 107, 108, 132, 143, 156
Yeltsin, Boris 112
Yoffe, Zelinka 96

Zalesski, A I 57
Zhigarev, P F 64
Zubets, P F 62
Zyrin, N 106